未来作物品种设计

中国科学院院士 李振声 题词

学术引领系列

国家科学思想库

中国学科发展战略

未来作物品种设计

中国科学院

科学出版社

北 京

内容简介

本书围绕未来农业对作物的需求,对相关领域基础研究国内外进展、发展趋势及关键突破口进行了全方位的论述,并对2035年和2050年可能实现的目标进行了展望。

本书可为相关领域科研工作者及社会大众了解未来农业发展提供全新的视角,也可为相关部门的政策制定、学科规划布局等提供参考。

图书在版编目(CIP)数据

未来作物品种设计 / 中国科学院编. —北京:科学出版社,2021.1
(中国学科发展战略)
ISBN 978-7-03-066245-3

Ⅰ. ①未… Ⅱ. ①中… Ⅲ. ①作物–品种–研究–世界 Ⅳ. ①S32

中国版本图书馆CIP数据核字(2020)第182470号

责任编辑:王 静 王 好 / 责任校对:严 娜
责任印制:赵 博 / 封面设计:刘新新

科学出版社出版
北京东黄城根北街16号
邮政编码:100717
http://www.sciencep.com
北京中科印刷有限公司印刷
科学出版社发行 各地新华书店经销
*

2021年1月第 一 版 开本:720×1000 1/16
2025年1月第二次印刷 印张:17 1/2 插页:1
字数:350 000
定价:168.00元
(如有印装质量问题,我社负责调换)

中国学科发展战略

指导组

组　长：侯建国

副组长：高鸿钧　秦大河

成　员：王恩哥　朱道本　傅伯杰

　　　　陈宜瑜　李树深　杨　卫

工作组

组　长：王笃金

副组长：苏荣辉

成　员：钱莹洁　赵剑峰　薛　淮

　　　　王　勇　冯　霞　陈　光

　　　　李鹏飞　马新勇

中国学科发展战略·未来作物品种设计

编辑委员会

主　编：李家洋

编　委：（按姓氏汉语拼音排序）

曹晓风	陈彩艳	陈晓亚	程祝宽	种　康
储成才	邓兴旺	丁　勇	丁兆军	傅向东
高彩霞	龚继明	郭　岩	郭庆华	韩　斌
韩方普	何光存	何祖华	黄三文	黄学辉
蒋才富	焦雨铃	景海春	康振声	赖锦盛
李　霞	李传友	李云海	廖　红	林荣呈
刘巧泉	刘耀光	罗　杰	漆小泉	孙其信
田志喜	万建民	王　台	王国栋	王汉中
王永红	吴殿星	吴孔明	谢　芳	熊立仲
徐国华	薛　淮	薛红卫	薛勇彪	严建兵
杨淑华	张启发	赵剑峰	钟上威	周俭民
周雪平	朱新广	朱玉贤		

编辑组成员：储成才　田志喜　王永红　于　昕

———— 总　　序 ————

九层之台，起于累土[①]

白春礼

近代科学诞生以来，科学的光辉引领和促进了人类文明的进步，在人类不断深化对自然和社会认识的过程中，形成了以学科为重要标志的、丰富的科学知识体系。学科不但是科学知识的基本单元，同时也是科学活动的基本单元：每一学科都有其特定的问题域、研究方法、学术传统乃至学术共同体，都有其独特的历史发展轨迹；学科内和学科间的思想互动，为科学创新提供了原动力。因此，发展科技，必须研究并把握学科内部运作及其与社会相互作用的机制及规律。

中国科学院学部作为我国自然科学的最高学术机构和国家在科学技术方面的最高咨询机构，历来十分重视研究学科发展战略。2009 年 4 月与国家自然科学基金委员会联合启动了"2011～2020 年我国学科发展战略研究"19 个专题咨询研究，并组建了总体报告研究组。在此工作基础上，为持续深入开展有关研究，学部于 2010 年底，在一些特定的领域和方向上重点部署了学科发展战略研究项目，研究成果现以"中国学科发展战略"丛书形式系列出版，供大家交流讨论，希望起到引导之效。

根据学科发展战略研究总体研究工作成果，我们特别注意到学科发展的以下几方面的特征和趋势。

一是学科发展已越出单一学科的范围，呈现出集群化发展的态势，呈现出多学科互动共同导致学科分化整合的机制。学科间交叉和融合、重点突破

[①] 题注：李耳《老子》第 64 章："合抱之木，生于毫末；九层之台，起于累土；千里之行，始于足下。"

和"整体统一",成为许多相关学科得以实现集群式发展的重要方式,一些学科的边界更加模糊。

二是学科发展体现了一定的周期性,一般要经历源头创新期、创新密集区、完善与扩散期,并在科学革命性突破的基础上螺旋上升式发展,进入新一轮发展周期。根据不同阶段的学科发展特点,实现学科均衡与协调发展成为了学科整体发展的必然要求。

三是学科发展的驱动因素、研究方式和表征方式发生了相应的变化。学科的发展以好奇心牵引下的问题驱动为主,逐渐向社会需求牵引下的问题驱动转变;计算成为了理论、实验之外的第三种研究方式;基于动态模拟和图像显示等信息技术,为各学科纯粹的抽象数学语言提供了更加生动、直观的辅助表征手段。

四是科学方法和工具的突破与学科发展互相促进作用更加显著。技术科学的进步为激发新现象并揭示物质多尺度、极端条件下的本质和规律提供了积极有效手段。同时,学科的进步也为技术科学的发展和催生战略新兴产业奠定了重要基础。

五是文化、制度成为了促进学科发展的重要前提。崇尚科学精神的文化环境、避免过多行政干预和利益博弈的制度建设、追求可持续发展的目标和思想,将不仅极大促进传统学科和当代新兴学科的快速发展,而且也为人才成长并进而促进学科创新提供了必要条件。

我国学科体系由西方移植而来,学科制度的跨文化移植及其在中国文化中的本土化进程,延续已达百年之久,至今仍未结束。

鸦片战争之后,代数学、微积分、三角学、概率论、解析几何、力学、声学、光学、电学、化学、生物学和工程科学等的近代科学知识被介绍到中国,其中有些知识成为一些学堂和书院的教学内容。1904年清政府颁布"癸卯学制",该学制将科学技术分为格致科(自然科学)、农业科、工艺科和医术科,各科又分为诸多学科。1905年清朝废除科举,此后中国传统学科体系逐步被来自西方的新学科体系取代。

民国时期现代教育发展较快,科学社团与科研机构纷纷创建,现代学科体系的框架基础成型,一些重要学科实现了制度化。大学引进欧美的通才教育模式,培育各学科的人才。1912年詹天佑发起成立中华工程师会,该会后来与类似团体合为中国工程师学会。1914年留学美国的学者创办中国科学社。1922年中国地质学会成立，此后，生理、地理、气象、天文、植物、动物、物理、化学、机械、水利、统计、航空、药学、医学、农学、数学等学科的学会相继创建。这些学会及其创办的《科学》《工程》等期刊加速了现代学科体系在中国的构建和本土化。1928年国民政府创建中央研究院，这标志着现代科学技术研究在中国的制度化。中央研究院主要开展数学、天文学与气象学、物理学、化学、地质与地理学、生物科学、人类学与考古学、社会科学、工程科学、农林学、医学等学科的研究，将现代学科在中国的建设提升到了研究层次。

中华人民共和国成立之后，学科建设进入了一个新阶段，逐步形成了比较完整的体系。1949年11月中华人民共和国组建了中国科学院，建设以学科为基础的各类研究所。1952年，教育部对全国高等学校进行院系调整，推行苏联式的专业教育模式，学科体系不断细化。1956年，国家制定出《十二年科学技术发展远景规划纲要》，该规划包括57项任务和12个重点项目。规划制定过程中形成的"以任务带学科"的理念主导了以后全国科技发展的模式。1978年召开全国科学大会之后，科学技术事业从国防动力向经济动力的转变，推进了科学技术转化为生产力的进程。

科技规划和"任务带学科"模式都加速了我国科研的尖端研究，有力带动了核技术、航天技术、电子学、半导体、计算技术、自动化等前沿学科建设与新方向的开辟，填补了学科和领域的空白，不断奠定工业化建设与国防建设的科学技术基础。不过，这种模式在某些时期或多或少地弱化了学科的基础建设、前瞻发展与创新活力。比如，发展尖端技术的任务直接带动了计算机技术的兴起与计算机的研制，但科研力量长期跟着任务走，而对学科建设着力不够，已成为制约我国计算机科学技术发展的"短板"。面对建设创新型国家的历史使命，我国亟待夯实学科基础，为科学技术的持续发展与创新能力的提升而开辟知识源泉。

反思现代科学学科制度在我国移植与本土化的进程，应该看到，20世纪上半叶，由于西方列强和日本入侵，再加上频繁的内战，科学与救亡结下了不解之缘，中华人民共和国成立以来，更是长期面临着经济建设和国家安全的紧迫任务。中国科学家、政治家、思想家乃至一般民众均不得不以实用的心态考虑科学及学科发展问题，我国科学体制缺乏应有的学科独立发展空间和学术自主意识。改革开放以来，中国取得了卓越的经济建设成就，今天我们可以也应该静下心来思考"任务"与学科的相互关系，重审学科发展战略。

现代科学不仅表现为其最终成果的科学知识，还包括这些知识背后的科学方法、科学思想和科学精神，以及让科学得以运行的科学体制、科学家的行为规范和科学价值观。相对于我国的传统文化，现代科学是一个"陌生的""移植的"东西。尽管西方科学传入我国已有一百多年的历史，但我们更多地还是关注器物层面，强调科学之实用价值，而较少触及科学的文化层面，未能有效而普遍地触及到整个科学文化的移植和本土化问题。中国传统文化以及当今的社会文化仍在深刻地影响着中国科学的灵魂。可以说，迄20世纪结束，我国移植了现代科学及其学科体制，却在很大程度上拒斥与之相关的科学文化及相应制度安排。

科学是一项探索真理的事业，学科发展也有其内在的目标，即探求真理的目标。在科技政策制定过程中，以外在的目标替代学科发展的内在目标，或是只看到外在目标而未能看到内在目标，均是不适当的。现代科学制度化进程的含义就在于：探索真理对于人类发展来说是必要的和有至上价值的，因而现代社会和国家须为探索真理的事业和人们提供制度性的支持和保护，须为之提供稳定的经费支持，更须为之提供基本的学术自由。

20世纪以来，科学与国家的目的不可分割地联系在一起，科学事业的发展不可避免地要接受来自政府的直接或间接的支持、监督或干预，但这并不意味着，从此便不再谈科学自主和自由。事实上，在现当代条件下，在制定国家科技政策时充分考虑"任务"和学科的平衡，不但是最大限度实现学术自由、提升科学创造活力的有效路径，同时也是让科学服务于国家和社会需要的最有效的做法。这里存在着这样一种辩证法：科学技术系统只有在具有高度创造活力的情形下，才能在创新型国家建设过程中发挥最大作用。

在全社会范围内创造一种允许失败、自由探讨的科研氛围；尊重学科发展的内在规律，让科研人员充分发挥自己的创造潜能；充分尊重科学家的个人自由，不以"任务"作为学科发展的目标，让科学共同体自主地来决定学科的发展方向。这样做的结果往往比事先规划要更加激动人心。比如，19世纪末德国化学学科的发展史就充分说明了这一点。从内部条件上讲，首先是由于洪堡兄弟所创办的新型大学模式，主张教与学的自由、教学与研究相结合，使得自由创新成为德国的主流学术生态。从外部环境来看，德国是一个后发国家，不像英、法等国拥有大量的海外殖民地，只有依赖技术创新弥补资源的稀缺。在强大爱国热情的感召下，德国化学家的创新激情迸发，与市场开发相结合，在染料工业、化学制药工业方面进步神速，十余年间便领先于世界。

中国科学院作为国家科技事业"火车头"，有责任提升我国原始创新能力，有责任解决关系国家全局和长远发展的基础性、前瞻性、战略性重大科技问题，有责任引领中国科学走自主创新之路。中国科学院学部汇聚了我国优秀科学家的代表，更要责无旁贷地承担起引领中国科技进步和创新的重任，系统、深入地对自然科学各学科进行前瞻性战略研究。这一研究工作，旨在系统梳理世界自然科学各学科的发展历程，总结各学科的发展规律和内在逻辑，前瞻各学科中长期发展趋势，从而提炼出学科前沿的重大科学问题，提出学科发展的新概念和新思路。开展学科发展战略研究，也要面向我国现代化建设的长远战略需求，系统分析科技创新对人类社会发展和我国现代化进程的影响，注重新技术、新方法和新手段研究，提炼出符合中国发展需求的新问题和重大战略方向。开展学科发展战略研究，还要从支撑学科发展的软、硬件环境和建设国家创新体系的整体要求出发，重点关注学科政策、重点领域、人才培养、经费投入、基础平台、管理体制等核心要素，为学科的均衡、持续、健康发展出谋划策。

2010年，在中国科学院各学部常委会的领导下，各学部依托国内高水平科研教育等单位，积极酝酿和组建了以院士为主体、众多专家参与的学科发展战略研究组。经过各研究组的深入调查和广泛研讨，形成了"中国学科发展战略"丛书，纳入"国家科学思想库—学术引领系列"陆续出版。学部诚挚感谢为学科发展战略研究付出心血的院士、专家们！

按照学部"十二五"工作规划部署，学科发展战略研究将持续开展，希望学科发展战略系列研究报告持续关注前沿，不断推陈出新，引导广大科学家与中国科学院学部一起，把握世界科学发展动态，夯实中国科学发展的基础，共同推动中国科学早日实现创新跨越！

序 一

国以民为本，民以食为天。粮食供给是决定人类社会发展的关键，自20万年前在非洲起源的那一刻开始，粮食安全一直就是人类追求和奋斗的目标。世界农业发展大体经过了原始农业、传统农业、现代农业三个阶段。经过长期的采集、渔猎生活，人类在公元前8000年左右开始驯养繁殖动物和种植谷物，人类进入了原始农业阶段。新石器时代后，农业工具和灌溉设施的利用，促进了由原始农业向传统农业的过渡。20世纪工业革命兴起，人类社会逐步进入现代农业阶段。以狩猎-采集方式的粮食供给最大限度只可支撑400万左右人口的生存，随着原始农业的产生和发展，新石器时代世界人口已经超过5000万，如今的现代农业使得全球人口已达70多亿。每一次新型农业文明的产生都是基于农业科技的发展，从某种意义讲，人类社会的文明史就是一部粮食生产的发展史。

纵观社会发展历史，世界粮食总量的增长一方面得益于科技的发展，如以"绿色革命"和杂交育种为代表的育种技术发展、以农药和化肥为代表的生物化学技术发展等；另一方面，粮食总量的增长还依赖于对耕地的开发利用。18世纪初，地球上95%的无冰土地还处于原生态或者半自然状态，到21世纪初，大约55%的无冰土地都已被用作耕地。伴随更多的森林和绿地开发利用，加之农药和化肥的大量施用，全球环境产生重大变化。例如，全球二氧化碳浓度一直保持在250 ppm或更低的水平，而自20世纪以来二氧化碳浓度持续上升，目前已经高达400 ppm左右。随之，全球气候变暖，极端天气频发，耕地肥力减弱。当今，世界人口还在逐渐增加，预计到2050年，全球人口将达到90亿。全球环境变化和世界人口持续增长对未来粮食安全提出新的挑战。

洪范八政，食为政首。自古至今，粮食安全都受到高度重视。早在《诗经》中就提到"国之大事，唯祀与戎"；汉代的《论贵粟疏》更是指出"粟者，王者大用，政之本务"。我国是一个拥有14亿多人口的发展中大国，面对世界百年未有之大变局，保障国家粮食安全极为重要，也是我们党治国理政长期以来一直高度关注的重大课题，十八大以来，更是受到高度重视。面对社会发展和环境变化，总结前期作物育种科技成果，积极研判未来作物重要发展方向及关键突破口，对保障我国乃至世界粮食安全具有重要意义。在此情形下，中国科学院生命科学和医学学部于 2018 年启动了由李家洋院士牵头的针对未来作物品种设计需求的相关战略性研究，经过国内农业基础研究相关专家历时两年的努力，完成了这份《未来作物品种设计》战略报告。该报告中每一个字句都凝集了编写者的心血和智慧，对于强化我国农业基础科技创新、驱动我国农业发展、保障粮食安全具有战略性指导意义。

中国科学院院长
中国科学院学部主席团主席
2020 年 10 月 13 日于北京

序 二

习近平总书记多次强调，中国人的饭碗任何时候都要牢牢端在自己手上，我们的饭碗应该主要装中国粮。粮食安全是经济和社会发展的重要前提，更是事关国家长治久安的重大战略问题。我国是一个有14亿人口的大国，这就决定了粮食供给不能也无法依赖其他国家。因此，保障粮食的绝对安全是永恒的课题。

种业是保障国家粮食安全和生态安全的根本。正如所有新技术、新工艺、新流程、新产品的出现在很大程度上都依赖于新的基础研究成果一样，种业发展也离不开农业基础科学研究的创新和进步。面对当前错综复杂的国际形势，我国种业乃至粮食安全面临的挑战前所未有。正因如此，加强未来作物相关基础研究、强化科技创新，对保障我国粮食安全，推动我国农业可持续发展具有重大战略意义。

如何适应未来农业对作物的需求？中国科学院生命科学和医学学部于2018年启动了由李家洋院士牵头的针对未来作物品种设计需求的相关战略性研究，经过国内农业基础研究相关专家历时两年的努力，完成了这份《未来作物品种设计》战略报告。

《未来作物品种设计》内容包括种子形成与萌发、植物形态建成、生育期及育性与杂种优势、光合作用、水资源高效利用、养分资源高效利用、生物固氮、病虫害与抗性、盐碱与极端温度适应、品质与营养、特殊功用作物改良、基因表达调控、高通量表型技术、功能基因高效解析、基因组编辑、驯化与多倍体育种16个大的研究领域，涵盖了作物形成的分子基础的方方面面。书中每章节都对相关研究领域国内外发展态势进行了系统的跟踪分析，为读者了解相关领域的发展水平和未来趋势提供了难得的窗口；同时对相关研究领域未来的发展方向及关键突破口也做出了前瞻性的研判，给出了2035年和2050年可能实现的目标，为从事植物学相关领域的科研人员指明了方向；该书在相关的政策保障和环境支持上提出了建设性的建议和方案，无疑也为相关管理部门及政策制定者提供了有益的参考和借鉴。读完全书，感觉到每字每句都凝集了编写者的心血和智慧。

  我很高兴看到《未来作物品种设计》能编写出版并为之作序。我相信该书不仅会成为科技决策者制定国家科技发展规划的重要基础，也会对社会大众了解未来农业发展具有重要参考价值。我国作物基础研究是最接近世界水平的学科领域之一，该书的出版也无疑会促进我国未来作物的基础研究工作百尺竿头更进一步。

李振声

中国科学院院士
国家最高科学技术奖获得者
2020年10月13日于北京

前　言

我国是人口大国，粮食安全是事关国家长治久安的重大战略问题。习近平总书记指出，中国人的饭碗任何时候都要牢牢端在自己手上，我们的饭碗应该主要装中国粮，要下决心把民族种业搞上去，抓紧培育具有自主知识产权的优良品种，从源头上保障国家粮食安全。

育种技术的发展为保证粮食产量和安全做出了巨大贡献，其发展得益于遗传学、分子生物学和基因组学的发展。早期通过驯化选育农家品种，进程慢，效率低。随着遗传学的发展，20世纪30年代通过遗传育种创制的杂交玉米开辟了农业革命；60年代起，在全世界范围内以矮化育种为标志的"绿色革命"使小麦、水稻等作物产量大幅度提高；80年代生物技术的发展促生了分子育种，使常规遗传育种有了一定的可跟踪性。但上述育种技术仍然不能满足日益增长的粮食需求，更加高效和精准的育种技术——"设计育种"出现，即通过品种设计进行多基因的复杂性状的定向改良与聚合，从而达到粮食高产优质的目标。设计育种技术的突破将依赖于遗传学、分子生物学和基因组学等学科的发展，尤其要依赖于对高产优质等复杂性状形成的分子机制的阐明。我国农业基础研究历经几十年发展，在农业生物功能基因组学等基础研究领域取得了长足进步，相继完成了水稻、小麦、玉米、大豆、油菜、棉花等重要农作物全基因组测序，在主要农业生物重要性状形成的遗传解析与分子机制研究方面取得了重要进展，水稻功能基因研究处于国际领跑地位。

中国科学院战略性先导科技专项"分子模块设计育种创新体系"实施以来，针对我国粮食安全和战略性新兴产业发展的重大需求，以水稻为主、小麦等为辅，初步建立了从"分子模块"到"设计型品种"的现代生物技术育种创新体系。在中国科学院学部学科发展战略研究项目"未来作物设计的分子生物学基础"的资助下，我们邀请从事植物基础研究和应用基础研究领域的一线科学家进行战略研究，他们中间80%以上的成员参与国家重大科学研究计划、中国科学院战略性先导科技专项，具有深厚的研究基础和对相关领域的前瞻性把握。

本书首次提出未来作物概念，针对未来作物品种设计的需求，围绕植物基础科学与现代农业、现代农业与环境、现代农业与人类健康研究领域，重点对种子生物学、植物形态建成、光合和营养高效利用、植物环境适应等的分子基础解析，植物代谢调控机制、多倍体形成的分子机制、复杂多倍体作物功能基因解析、基因组编辑与基因表达调控等的新技术新方法等进行了国内外进展综述，论述了上述领域未来发展趋势与关键突破口，并提出了2035年和2050年阶段性未来作物战略目标，同时对研究政策保障和环境支持建议进行了战略研究。希望本书的出版对强化农业基础科技创新、驱动我国农业发展、保障粮食安全起到战略性指导作用。

在项目立项、调研、报告撰写和本书的组织出版过程中，战略规划研究组专家投入了大量的心力。借此机会，向所有参与《未来作物品种设计》撰写的专家和同仁表示衷心的感谢！

本书内容涉及领域广泛，相关研究领域发展迅速，遗漏和不妥之处在所难免，恳请读者指正。

中国科学院院士
2020 年 9 月 22 日于北京

目 录

未来作物精准设计 / 韩　斌　薛勇彪　朱玉贤　康振声　张启发　　　1

种子形成与萌发 / 邓兴旺　薛红卫　王　台　钟上威　李云海　　　7

植物形态建成 / 焦雨铃　丁兆军　傅向东　　　19

生育期、育性与杂种优势 / 程祝宽　刘耀光　　　34

光合作用 / 朱新广　林荣呈　　　52

水资源利用效率与抗旱 / 熊立仲　　　62

养分资源高效利用 / 徐国华　廖　红　陈彩艳　龚继明　　　78

生物固氮 / 谢　芳　李　霞　　　96

病虫害与抗性 / 何祖华　周俭民　周雪平　李传友　何光存　　　114

盐碱与极端温度适应 / 种　康　杨淑华　郭　岩　蒋才富　　　126

品质与营养 / 刘巧泉　吴殿星　　　140

特殊功用作物改良 / 罗　杰　王国栋　漆小泉　陈晓亚　　　153

基因表达调控 / 丁　勇　曹晓风　　　162

高通量表型技术 / 郭庆华　　　181

功能基因高效解析 / 赖锦盛　严建兵　黄学辉　　　194

基因组编辑 / 高彩霞　　　208

驯化与多倍体育种 / 黄三文　韩方普　王汉中　　　221

政策保障与环境支持 / 景海春　孙其信　吴孔明　万建民　　　250

彩图

未来作物精准设计

韩 斌　薛勇彪　朱玉贤　康振声　张启发

对一个 14 亿人口的大国来说，保障粮食充足供应始终是国家安全的头等大事，也是农业农村现代化的前提。保障粮食绝对安全对中国来说是永恒的课题，任何时候都不能放松，中国必须能够自主解决吃饭问题，不能也无法依赖其他国家。因此，习近平总书记在中央农村工作会议上指出，中国人的饭碗任何时候都要牢牢端在自己手上，我们的饭碗应该主要装中国粮。

1 农业现代化必须依靠全新未来农业

绿色优质是未来农业发展的必然趋势，也是未来农业的重要标志。推进未来绿色农业的发展才能确保资源高效利用、生态系统稳定、产地环境良好和产品质量安全，实践绿水青山就是金山银山的理念。目前我国农业用水占水资源消耗的 60%，水资源短缺日益成为农业可持续发展的障碍。化肥的大量施用是目前提高粮食产量的重要手段之一，但给土壤和水体环境带来了严重问题，同时，也给生态环境与食品安全带来了严重风险。加快转变农业发展方式，确保粮食等重要农产品有效供给，提升农产品质量，实现绿色发展和资源永续利用，是农业生产亟待解决的难题。

供给侧改革是未来农业发展的必由之路。随着社会的发展和人们生活水平的提高，我国农业生产的主要矛盾已由总量不足转变为结构性矛盾，突出表现为阶段性供过于求和供给不足并存，矛盾的主要方面在供给侧，对品种开发要求超出了标准化大面积推广的传统路径，人们对食物提出了更多的个性化需求。2015 年中央一号文件《关于加大改革创新力度加快农业现代化建设的若干意见》首次提出要把追求产量为主，转到数量、质量、效益并重上来。确保粮食安全和种粮农民收入持续稳定增长，是我国粮食政策的两个基本目标。

农业工业化是未来农业发展的重要方向。农业现代化的实现是其他现代化的融入过程。农业现代化是利用现代科学技术和现代工业装备农业，是用现代经济科学来管理农业，来创造一个高产、优质、低耗的农业生产体系和一个既合理利用资源又保护环境的高转化效率的农业生态系统。例如，通过机械化、精简化和智能化实现大规模、高效率的工业化生产型农业，以及通过设施高精度环境控制实现农作物周年生产的植物工厂等。农业的工业化生产是科学技术发展到一定阶段的必然产物，是现代生物技术、机械工程、建筑工程、环境控制、材料科学、计算机科学等多学科集成创新、知识与技术高度密集的农业生产方式，是未来农业的重要发展方向。

2 未来农业急需创造新型种子

农业现代化无不伴随着作物品种的更新换代。以第一次"绿色革命"为例，20世纪60年代，半矮秆基因的应用将小麦和水稻单产提高了20%～30%，实现了作物品种的整体升级换代。20世纪70年代，杂交水稻三系配套成功并大范围推广，将水稻单产又提高了20%～30%，为解决我国14亿人的温饱问题做出了巨大贡献。

农业农村现代化急需符合我国国情、农情和时代发展，具备增产提质、减投增效、减损促稳（"一增二减"）特点的新型种子，以实现对现有品种的跨越升级。科技进步推动社会生产方式和人类生活方式不断变革，人类的需求越来越多样化和个性化，新的需求不断涌现，育种目标必将随之更新。传统作物以提供必需的食物和营养为目的，将人类的生命活动在很大程度上束缚于农业生产。新型种子及农业生物将是能够在更大程度上解放人类生产力，扩大生活维度，提升生活质量的全新种子。未来作物的精准设计和创造将是实现作物智能应对全球气候变化、适应生产方式变革、满足人们个性化需求的必然选择。要实现这一目标，必须以科技创新为引领，以技术集成和智能设计为导向，精准设计品种，促进品种的升级换代，为实现农业农村现代化，决胜全面建成小康社会，全面建设社会主义现代化强国提供战略支撑。

增产提质是未来作物精准设计的首要目标。从粮食需求增长的几个关键驱动因素来看，中国的人口规模、消费结构、城镇化水平仍未到达顶峰。据预测，中国人口将在2030年左右达到14.5亿峰值，比现在增加约5000万（2019年为14亿）；2019年，我国常住人口城镇化率首次超过60%（2020年政府工作报告），距离城镇化率峰值还有大约15%的增长空间；目前我国人均GDP略高于10 000美元（2019年为70 892元人民币），从日本、韩国和我国台湾地区的经验看，人均GDP达到2万美元以后，居民食物消费结构升级才基本稳定下来。所以，未来20年左右，在全面建设社会主义现代化国家的进程中，我国仍处于食物消费结构持续转变升级的过程。这一基本国情对品种提出了一个基本要求，即以高产稳产保证粮食供给，以优质保证营养健康和个性化需求。

减投增效是未来作物精准设计的基本要求。我国用占世界8%的耕地支撑占世界19%的人口，人均耕地只占世界平均水平的43%，人多地少是我国基本国情。随着城市化的加速，据2016中国国土资源公报数据，我国耕地每年持续减少约99万亩[①]。过去几十年，高强度、粗放式生产方式导致农田生态系统结构失衡、功能退化，生态系统退化。2019年，我国农药用量为145.6万吨（商品量），单位面积使用量比世界平均高2.5倍；农用化肥施用折纯量为5603.59万吨，占世界化肥消费总量的33%。据测算，我国养猪业每年排放700万～1130万吨氮污染物和130万～140万吨磷污染物。目前，我国干旱和半干旱农业区约占全国总耕地面积的51%，盐碱地面积达7亿亩。有限的耕地面积、不断增长的人口和农田的功能性退化是我国农业现代化面临的一个巨大挑战，现有品种已经不能满足我国未来农业的需求。为了保障我国绿色农业和未来农业可持续发展，急需培育资源节约型品种，以应对耕地资源挑战，推进农业现代化进程。

减损促稳是未来作物精准设计的显著特征。全球气候变化是农业生产和育种必须面对的最主要挑战之一。研究表明，在全球气候变化背景下，我国近100年地表年平均升温幅度为0.5～0.8℃，年降水量减少幅度约为0.86 mm/10年，极端气候事件（如极端气温、极端降水和干旱）出现的频率增加。同时，大气CO_2浓度增加，日照时数锐减，全国年平均日照时数从1956年至2000年减少了5%（130小时）左右。气候变化不仅对自然和人类

---

① 1亩≈666.7 m^2。

社会各方面产生了巨大影响，同时也影响到作物种植区域与种植制度，以及加重农作物病虫草害及加剧农业气象灾害等。除此以外，我国还必须面对环境污染严重、生态系统退化的严峻形势。为了应对气候变化带来的挑战，我国急需能适应气候变化的环境智能响应新型作物品种。

3 未来作物的精准设计依赖科技创新

习近平总书记指出，农业出路在现代化，农业现代化关键在科技进步。科技创新是现代化的发动机，是推动一个国家进步和发展最重要的因素之一。重大原始性科技创新及其引发的技术革命和进步也是产业革命的源头。现代科技正在朝高精尖、学科融合和技术集成方向迅猛发展。前沿和颠覆性技术也正在持续激活农业的潜在爆发力，并将带动农业产业格局重大调整和革命性突破。新一代测序技术、单分子技术、超高分辨率显微镜等技术，将人们对生命体系的观测精确到单分子水平，扩展到全系统层次；各种物理学即时定量检测技术的应用，将高通量大规模的"组学"研究，推向了动态和定量的高度，使得对生命活动的检测达到实时和原位；基因编辑、单细胞组学、多尺度成像和光遗传学等技术的崛起，使得对生命编码的精准改造和生命过程的精准操控成为可能；集成大数据分析、代谢工程及合成生物学等技术将从根本上改变农业生产和产业组织形式。

未来作物的精准设计离不开基础科学的创新和跨学科技术的融合和集成，基础科学研究的突破和颠覆性技术的应用，也将加速未来种子设计的广度和深度，而精准设计型品种的推出将为我国农业现代化提供升级换代的原动力。

4 国际在未来作物研究领域的发展布局

近年来，国内外政府、组织和机构相继发布咨询报告和研究计划，从不同层面对作物生物技术和产业化发展策略进行战略布局。

欧洲技术平台"未来植物"计划 欧洲技术平台（European Technology Platform，ETP）2007年发布的《"未来植物"战略研究议程2025》（"*Plants* for the Future" Strategic Research Agenda 2025）构建了一个用于指导和整合未来 20 年植物研究的路线图，以便充分发挥人力资本和该领域知识的全部潜力，提出了 5 个挑战和 18 个目标来支持以知识为基础的生物经济，维持欧洲经济竞争力，并提供以环境可持续方式确保未来燃料和粮食供应。

英国洛桑试验站科学战略　英国洛桑试验站（Rothamsted Research，UK）发布的最新科学战略（2017～2022）重点关注了英国和全世界可耕种与放牧 - 牲畜系统的可持续集约化，围绕"优质作物""确保生产力"和"未来的农业食品系统"三大主题 6 项挑战，提出了至 2022 年的 5 个战略研究计划：设计未来小麦、调整植物代谢、智能作物保护、土壤营养改良、实现可持续农业系统。

美国农业部 ARS 战略规划 (2018～2020)　美国农业部农业研究局（Agricultural Research Service，ARS）发布了 2018～2020 战略规划。在该战略规划中，分别从营养食品安全和质量、自然资源和可持续农业系统、作物生产与保护、动物生产与保护 4 个领域凝练了 15 个研究目标，并提出了系列国家计划。主要包括国家计划 211：水资源利用和流域管理（2016～2021）、国家计划 212：土壤和空气（2016～2020）、国家计划 215：草、饲料和牧场农业生态系统（2019～2023）、国家计划 216：可持续农业系统研究（2018～2022）、国家计划 301：植物遗传资源、基因组学和遗传改良（2018～2022）、国家计划 303：植物病害（2017～2021）、国家计划 304：作物保护与检疫（2015～2020）、国家计划 305：作物生产（2018～2023）等。

英国植物科学联合会《种植未来》　2019 年，英国皇家生物学会植物科学联合会发布了《种植未来》报告。该报告描述了植物科学在提升基础研究、改善饮食质量、提高作物产量、增强环境可持续性及创造新产品方面的潜力。在植物科学 4 个关键领域提出了战略机会和优先事项，这 4 个领域包括：改善作物和农业系统、植物健康和生物安全、植物生物技术、生物多样性与生态系统。报告特别强调通过基因组测序和编辑、计算生物学、代谢组学加速育种以及利用人工智能与精准农业等手段实现农业现代化，利用科技进步为未来农业发展创造机遇。

我国科学技术部"国家重点研发计划"　我国政府一直高度重视作物改良等研发，转基因生物新品种培育国家科技重大专项总投资达 240 亿元。

2016年，科技部发布国家重点研发计划"七大农作物育种重点专项"，按照"加强基础研究、突破前沿技术、创制重大品种、引领现代种业"的总体思路，以水稻、玉米、小麦、大豆、棉花、油菜、蔬菜等七大农作物为对象，重点部署优异种质资源鉴定与利用、主要农作物基因组学研究、育种技术与材料创新、重大品种选育、良种繁育与种子加工五大任务；围绕种质创新、育种新技术、新品种选育、良种繁育等科技创新链条开展研究；重点突破基因挖掘、品种设计和种子质量控制等核心技术，获得具育种利用价值和知识产权的重大新基因，创制优异新种质，形成高效育种技术体系，培育重大新品种并推广应用。

研究表明，到2050年世界人口将突破90亿。此外，环境恶化逐步加剧、极端气候日益频发。人口增长和环境变化的双重压力需要我们系统思考如何设计培育未来作物，以满足粮食需求。未来作物品种设计研究迫在眉睫。

致谢：本章在撰写过程中得到了牛敏杰、许璟的协助，特此致谢！

种子形成与萌发

邓兴旺　薛红卫　王　台　钟上威　李云海

1 国内外研究进展

作为植物的繁殖器官，种子发育在植物的生命周期中起到至关重要的作用。种子为胚胎提供营养，能够在胁迫环境下进行休眠使植物更好地适应环境。此外，种子是人类赖以生存的粮食的最主要来源，为人类提供了80%的主粮，是粮食产量和品质形成的基础。因此，对种子形成与萌发调控机制的研究将为农作物的产量和品质提高提供理论基础和技术支撑。

1.1 种子的形成

绝大多数种子由胚、胚乳和种皮3个主要部分组成，它们分别由合子（受精卵）、初生胚乳核（受精极核）和珠被发育而成。在被子植物的有性生殖过程中，来自花粉的两个精子分别与胚囊中的卵细胞和中央细胞融合形成合子和初生胚乳核。前者经过细胞分裂、分化、器官发生和休眠建立等过程形成成熟胚胎；后者经过游离核分裂、细胞化等过程形成胚乳。胚和胚乳均可作为植物的营养累积器官，在发育后期累积淀粉、脂肪酸和蛋白质等。

种子发育过程中涉及的细胞分裂、细胞分化、器官发生和胞间通信等过程的调控机制是生命科学领域的重大科学问题，也是发育生物学研究的核心命题。早期胚胎发育调控涉及的关键生物学过程包括合子激活、胚胎极性建立、胚胎模式形成、子叶形成等。近期的研究表明精细胞中特异表达的某些转录本受精后就出现在合子中，暗示合子胚胎发生的启动和胚胎的早期发育并非只有母本信息调控。在配子成熟时期建立了某种关键的母本机制（Wu et al., 2012），这些母本因子对于受精后胚胎发生的启动及早期胚乳发育起决定性调控作用，表明雌配子信息对受精与胚胎发生的重要意义。胚乳发育一般经历一个早期的核分裂时期和随后的细胞化过程，细胞化之后的胚乳开始进行营养累积。不同植物胚乳的后期发育差异很大，大部分双子叶植物的成熟种子中胚乳极少，有些单子叶植物（如多数禾本科植物）有大量的胚乳营养累积。这种营养累积策略的差异是由不同的细胞发育命运所决定的。研究表明，特异表达的基因、激素、调控基因转录的因子、表观修饰及 sRNA 等在胚乳早期发育中发挥重要作用。细胞分裂素信号途径相关基因在胚乳发育的多核体阶段有高水平的表达，对胚乳发育具有关键作用。植物生长素调控了胚乳的细胞化过程；胚乳中赤霉素的活性对种子的正常生长是必需的（Batista et al.，2019；Day et al.，2008）。MADS-box、Homeobox、B3 及 DOF 类转录因子家族成员在胚乳发育中发挥重要调节作用（Yin and Xue，2012；Zheng et al.，2019）。sRNA 参与了胚乳发育过程中基因组印记的形成。此外，利用水稻、玉米、大麦突变体开展的遗传学研究也鉴定出了一些参与胚乳早期发育的关键基因。最近的研究发现了胚乳和胚胎间信息交流的新机制。来源于胚胎的 TWS1 前体，通过胚乳产生的肽酶进行切割产生活性小肽，该小肽与胚胎中的受体结合，从而调控胚胎角质层的形成和种子发育（Doll et al.，2020）。

1.2 种子大小的调控

在种子的发育过程中，母体组织和合子组织的协同调控影响了种子的生长和最终大小。种子大小是作物产量的重要构成要素，也是作物遗传育种的主要改良目标之一。作物种子大小的研究是近年来快速发展的研究领域。近年来，科学家们克隆了多个影响种子大小的关键基因，发现 G 蛋白信号、泛素途径、MAPK 信号、转录调控因子以及激素和其他生长物质等都参与调控植物种子大小（Li and Li，2016；Li et al.，2019；Zuo and Li，2014）。在水稻中，张启发院士团队首先克隆了控制粒长的主效 QTL 位点 *GS3*，编码了一个 G 蛋白 γ 亚基（Fan et al.，2006），育种家们已利用 *GS3* 基因的优良等位变异改良水稻粒形和产量。最近的研究揭示了 G 蛋白亚基 GS3 和 DEP1 拮抗调控水稻粒长（Sun et al.，2018）。控制水稻粒宽的主效 QTL 基因 *GW2*，编码了一个 E3 泛素连接酶（Song et al.，2007）。其同源蛋白在小麦、玉米等植物中具有保守功能，负调控种子大小。在模式植物拟南芥中，也建立了以泛素受体 DA1 为中心的种子大小调控的分子遗传网络（Dong et al.，2017；Du et al.，2014；

Li et al., 2008；Xia et al., 2013)，并发现该网络成员在小麦、玉米、油菜和水稻等作物中具有保守功能，可以用来增加种子大小和产量（Wang et al., 2017；Xie et al., 2017)。最近，我国科学家在水稻中发现了 MAPK 级联信号途径调控了种子大小和穗粒数（Guo et al., 2018a；Xu et al., 2018a, 2018b)。水稻 GTP 酶 OsRAC1 影响了 MAPK6 的磷酸化水平，从而调控水稻种子大小（Zhang et al., 2019)。水稻 OsmiR396-OsGRF4-OsGIFs 信号途径在种子大小调控中起重要作用（Che et al., 2015；Duan et al., 2015)。SPL 家族的转录因子 OsSPL13、OsSPL14/IPA1 和 OsSPL16 参与调控水稻种子大小和株型等重要农艺性状（Jiao et al., 2010；Si et al., 2016；Wang et al., 2012)。OsSPL16 则通过直接调控 *GL7* 的表达调控水稻种子大小（Wang et al., 2015a, 2015b)。此外，染色质修饰和植物激素等也参与了种子大小的调控。育种学家正在利用这些基因的优良等位变异提高作物种子的产量和品质。

1.3 种子的萌发

种子成熟后即进入休眠状态，直到外界环境变得适宜植物生长时，种子才会进入萌发状态。种子萌发受到内源激素及多种外界环境因子的调控，如温度、光照和水分等（Jiang et al., 2016；Ma et al., 2019；Shi et al., 2015)。植物内源激素是调控种子萌发的内因，其中，脱落酸（ABA）和赤霉素（GA）起到主要作用。种子内 ABA 和 GA 的相对含量及种子对这两种激素的信号反应是决定种子萌发的关键因素（Bentsink and Koornneef, 2008；Guo et al., 2018b；Hu et al., 2019)。研究表明，ABA 可以诱导和维持种子的休眠，抑制种子萌发（Holdsworth et al., 2008；Kang et al., 2015；Yan and Chen, 2017)。GA 是一类类萜化合物，在种子萌发过程中起拮抗 ABA 的作用，促进种子萌发（Holdsworth et al., 2008；Liu et al., 2016)。除 ABA 和 GA 以外，乙烯、油菜内酯素（BR）和生长素在种子萌发中也具有调控作用（Finkelstein et al., 2008；Gazzarrini and Tsai, 2015)。乙烯可以通过促进胚轴的细胞伸长，增强种子的呼吸作用或者增加水势来促进种子萌发（Arc et al., 2013；Bentsink and Koornneef, 2008)。生长素在种子萌发中的作用还不明确，但转录组分析结果显示生长素的转运子 AUX1、PIN2 和 PIN7 的在萌发种子中的转录水平比在休眠种子中高（Carrera et al., 2008；Wang et al., 2016)，说明生长素可能参与种子萌发的调控。此外，多种环境因子是调控种子萌发的外因，其中光照、温度和水分起到主要调控作用。光诱导的种子萌发起始过程是由光受体 phytochrom B（phyB）所介导（Jiang et al.，2016；Shi et al.，2015）。外界光信号最终跟植物内源激素整合在一起，共同调控种子萌发。光照可以通过调控 ABA 和 GA 的合成、代谢及信号转导通路来调控种子萌发（Cho et al.，2012；Jiang et al.，2016；Vaistij et al.，2018）。温度也会影响种子萌发。低温处理会促进 GA 合成基因表达，有利于 GA 的积累及 GA 诱导基因表达的丰度，从而促进种子萌发（Topham et al.，2017；Yamauchi et al.，2004）。同时低温与光还能相互协调共同调控种子萌发（Penfield et al.，2005）；高温则会诱导种子的次级休眠（Chang et al.，2018）。因此，在播种前对种子进行激素和光温处理将直接影响农作物的种子萌发和出苗率。

2 未来发展趋势与关键突破口

尽管传统育种仍具有一定的增产潜力，但面对持续的粮食增产需求，有必要进一步发展育种技术和提高育种效率。未来作物设计对于种子产量、品质、活力的育种工作提出了更高的要求。对调控种子形成与萌发的分子机制的深入研究将有助于种子产量、品质、活力等育种工作，通过研究种子形成与萌发的调控机制，阐明作物种子产量和品质形成的分子基础，挖掘可用于粮食产量和品质提高的重要功能基因并应用于主要农作物的遗传改造，将为农作物的改良提供理论指导。

种子形成与萌发过程伴随着特定基因的时空表达，受体内生理信号和体外环境信号的多重调控，是一个多因素影响的复杂调控网络。最近的研究揭示了光照、温度、激素和多个信号途径对于种子形成与萌发的影响，鉴定了一系列种子形成与萌发的关键调控因子，但是种子形成与萌发的调控网络仍然不清楚，仍有很多关键问题亟待解决。例如，种子的形成和发育受到环境因素、非生物胁迫等的影响，但是这些外界因素是如何影响种子的形成和发育的？植物激素调控了大量生物学过程，但植物激素如何调控种子形成和发育等有待进一步完善。胚胎、胚乳和母体组织是如何进行信息交流，协同调控种子生长的？尽管种子的生长受到环境条件的影响，但是同一物种的种子大小相对恒定，而不同物种的种子大小差异很大，说明种子的生长上限是由植物的内源信号控制的，而植物种子生长过程中是如何感知生长信号并决定其最终大小的？种子的生长受细胞分裂和细胞扩展的协同调控，植物是如何协调这两个细胞过程，调控种子生长的？目前已经鉴定了一些调控种子形成和萌发的信号途径，这些途径之间是如何互作的？种子萌发受到多种内外界因素的综合调控，包括光照、水分、温度及内源植物激素等。综合考虑种子萌发过程中光、温、水等因子的耦合作用，环境因子与内源激素的协同变化，以及种子感知外源信号的分子传导途径等是将来的研究重点。

在作物种子的遗传改良中也有一系列问题需要解决。例如，提高粒重的同时经常伴随着种子数目的降低；有时提高了种子产量，而种子的品质也受到了影响；如何同时协同改良作物种子大小、数目和品质，培育既高产又优质的作物是育种的重要方向。作物种子发育过程中，在穗子不同位置的种子发育程度经常不一致。例如，水稻穗子基部种子的灌浆程度相对较差，如何提高水稻穗子基部的灌浆效率是生产中需要解决的实际问题。在生产中，人们通常根据种子在土壤中的萌发出苗率与整齐性，来决定品种选择与播种数量，并成为作物驯化过程中的关键指标。通常需要 2～3 倍种子来保证出苗率，且由于种子的过量播种，部分区域出土多颗幼苗，需要人工间苗，大幅增加了生产成本。如何能促进种子整齐萌芽出土是急需解决的关键问题。种子成熟后期遇到连续阴雨时会在植株上萌发（穗萌），这对种子与粮食生产造成巨大的经济损失。如何能促进种子整齐萌芽出土同时又能抑制穗萌？此外，先前的种子萌发研究主要在没有土壤覆盖的条件下进行，与农业生产的实际情况有较大差异，这个问题在将来的生产中更为突出。如何最大限度地模拟现代农业实际生产，深入研究真实土壤条件下的作物种子萌发出土的特性及分子调控机制，有效提高直播后种子萌发出土的比率，已经成为我国目前亟待解决的重要课题。

未来的关键突破将是整合近年来发展起来的转录组学、蛋白质组学、代谢组学、表型组学等高通量组学分析手段及信息生物学和系统生物学与传统研究方法相结合，从多层次研究种子形成与萌发的遗传调控网络；系统分析不同调控因子、不同遗传途径的互作关系，建立种子形成与萌发调控的数学模型；比较不同作物种子形成和萌发的相同和不同之处，从进化和人工选择的角度解析种子形成和萌发的规律；其结果不但会系统阐明种子形成与萌发的调控机制，也将为主要农作物的产量和品质育种提供重要线索，最终实现多基因控制性状的精准改良。

3 阶段性目标

种子的形成与萌发是一个包括多种细胞学过程和多个器官发生的复杂的生物学过程，受母体和合子的多种发育信号的调控，以及体内外环境条件的影响。揭示种子发育的调控机制和分子网络需要长期的研究积累和多层次、多方向研究结果的整合。未来的研究目标是进一步阐明种子形成和发育过程的分子机制和调控网络，揭示不同调控因子、不同信号之间的互作关系以及它们对种子发育和作物产量和品质的影响；从进化和人工选择的角度理解种子形成和萌发的规律，为农作物性状改良提供理论依据。为了实现这一目标，我们将分两个阶段进行。

3.1 2035 年目标

虽然之前的研究已经鉴定了一些影响种子形成与萌发的因子，但是相对于种子形成与萌发的复杂的调控网络，目前仍知之甚少。为了进一步阐明种子形成与萌发的调控机制，2035 年目标主要是进一步通过正、反向遗传学结合现代生物学技术系统地分离鉴定种子形成与萌发过程中的调控因子，解析环境、激素等对种子形成和萌发的影响，建立种子形成与萌发调控的分子遗传网络。利用大数据信息，比较不同作物或物种的基因组数据，发现种子形成和萌发的调控规律，挖掘种子形成和萌发的进化保守基因与人工驯化和选择的关键因子。利用基因编辑技术，对种子形成和萌发的关键基因进行遗传改良。

早期种子发育调控涉及合子激活、胚胎极性建立、胚胎模式形成、子叶形成等重要发育生物学问题。目前虽然鉴定了多个种子发育的基因和途径，并对种子发育过程有较为深入的理解，但是种子发育过程中所涉及的多因子或多途径之间是如何协同调控，从而决定种子形成的分子遗传网络有待进一步解析。另外，种子发育过程也受到植物激素和外界环境条件的极大影响，从而影响种子的产量和品质。因此需要加强研究外界环境（如温度、养分等）对种子发育的影响，阐明其调控的分子机制，培育出能够最佳适应外界环境变化的作物新品种。

虽然鉴定了多个调控种子大小的因子和途径，但这些因子和途径间的遗传和生化关系仍不清楚，在作物中还没建立种子大小的调控网络。进一步研究需要搞清楚调控种子大小因子或途径间的分子关系，建立调控种子大小的分子遗传网络。另外，提高种子大小的同时经常伴随着种子数目和品质的降低，因此需要解析种子大小、数目和品质相互调控的分子机制。应用常规育种结合现代技术，利用种子大小、品质等关键基因和优良等位变异，进行精准分子设计育种，培育高产优质的作物新品种。

种子萌发是一个同时受到多种环境因子与多个内源激素的综合调控过程。过去的研究中，虽然取得了许多重要研究成果，但由于主要针对某一种条件或者单一激素情况下对种子萌发的调控进行研究，使得对种子萌发分子调控网络的整体认识不够。接下来，我们将综合研究种子萌发中光照、温度等环境因子的耦合作用，并与植物激素的调控作用进行整合，鉴定连接不同环境因子与内源激素信号通路的关键整合蛋白，揭示这些蛋白的上下游调控机制，构建出植物种子整合环境因子与内源激素信号调控种子萌发的分子信号网络。同时，机械化直播是未来作物的发展方向。与传统育秧移栽相比，怎样提高直播后种子萌芽出苗率和水淹条件下水稻种子萌发率是两个需要重点解决的难题。需要通过真实土壤实验，对直播后种子的萌发进行深入研究，系统研究土壤机械压力，光温变化，以及低氧胁迫多重环境因子的调控机制，阐明土壤条件下种子萌发调控的分子机制，并选育出多个具有显著增强萌芽出土能力，适合机械化直播的作物品种。

3.2 2050年目标

在2035年目标的基础上，整合多方面的研究结果，建立种子形成与萌发过程中不同途径、不同网络和不同信号之间的互作关系，建立调控种子大小的数学模型。同时，利用这些理论基础，通过现代基因编辑等手段，对控制不同性状的基因优化组装设计，实现多基因和多网络控制性状的精准改良，培育出高产、优质、耐储藏、萌发整齐和适合机械化种植的作物新品种。

致谢：本章在撰写过程中得到了李娜教授和陈良碧教授的协助，特此致谢！

参考文献

Arc, E., Sechet, J., Corbineau, F., Rajjou, L., and Marion-Poll, A. (2013). ABA crosstalk with ethylene and nitric oxide in seed dormancy and germination. Front. Plant Sci. *4*, 63.

Batista, R.A., Figueiredo, D.D., Santos-Gonzalez, J., and Kohler, C. (2019). Auxin regulates endosperm cellularization in *Arabidopsis*. Genes Dev. *33*, 466-476.

Bentsink, L., and Koornneef, M. (2008). Seed dormancy and germination. *Arabidopsis* Book *6*, e0119.

Carrera, E., Holman, T., Medhurst, A., Dietrich, D., Footitt, S., Theodoulou, F.L., and Holdsworth, M.J. (2008). Seed after-ripening is a discrete developmental pathway associated with specific gene networks in *Arabidopsis*. Plant J. *53*, 214-224.

Chang, G.X., Wang, C.T., Kong, X.X., Chen, Q., Yang, Y.P., and Hu, X.Y. (2018). AFP2 as the novel regulator breaks high-temperature-induced seeds secondary dormancy through ABI5 and SOM in *Arabidopsis thaliana*. Biochem. Biophys. Res. Commun. *501*, 232-238.

Che, R., Tong, H., Shi, B., Liu, Y., Fang, S., Liu, D., Xiao, Y., Hu, B., Liu, L., Wang, H., et al. (2015). Control of grain size and rice yield by GL2-mediated brassinosteroid responses. Nat. Plants *2*, 15195.

Cho, J.N., Ryu, J.Y., Jeong, Y.M., Park, J., Song, J.J., Amasino, R.M., Noh, B., and Noh, Y.S. (2012). Control of seed germination by light-induced histone arginine demethylation activity. Dev. Cell *22*, 736-748.

Day, R.C., Herridge, R.P., Ambrose, B.A., and Macknight, R.C. (2008). Transcriptome analysis of proliferating *Arabidopsis* endosperm reveals biological implications for the control of syncytial division, cytokinin signaling, and gene expression regulation. Plant Physiol. *148*, 1964-1984.

Doll, N.M., Royek, S., Fujita, S., Okuda, S., Chamot, S., Stintzi, A., Widiez, T., Hothorn, M., Schaller, A., Geldner, N., et al. (2020). A two-way molecular dialogue between embryo and endosperm is required for seed development. Science *367*, 431-435.

Dong, H., Dumenil, J., Lu, F.H., Na, L., Vanhaeren, H., Naumann, C., Klecker, M., Prior, R., Smith, C., McKenzie, N., et al. (2017). Ubiquitylation activates a peptidase that promotes cleavage and destabilization of its activating E3 ligases and diverse growth regulatory proteins to limit cell proliferation in *Arabidopsis*. Genes Dev. *31*, 197-208.

Du, L., Li, N., Chen, L., Xu, Y., Li, Y., Zhang, Y., and Li, C. (2014). The ubiquitin receptor DA1 regulates seed and organ size by modulating the stability of the ubiquitin-specific protease UBP15/SOD2 in *Arabidopsis*. Plant Cell *26*, 665-677.

Duan, P., Ni, S., Wang, J., Zhang, B., Xu, R., Wang, Y., Chen, H., Zhu, X., and Li, Y. (2015). Regulation of OsGRF4 by OsmiR396 controls grain size and yield in rice. Nat. Plants, 15203.

Fan, C., Xing, Y., Mao, H., Lu, T., Han, B., Xu, C., Li, X., and Zhang, Q. (2006). GS3, a major QTL for grain length and weight and minor QTL for grain width and thickness in rice, encodes a putative transmembrane protein. Theor. Appl. Genet. *112*, 1164-1171.

Finkelstein, R., Reeves, W., Ariizumi, T., and Steber, C. (2008). Molecular aspects of seed dormancy. Annu. Rev. Plant Biol. *59*, 387-415.

Gazzarrini, S., and Tsai, A.Y. (2015). Hormone cross-talk during seed germination. Essays Biochem. *58*, 151-164.

Guo, G., Liu, X., Sun, F., Cao, J., Huo, N., Wuda, B., Xin, M., Hu, Z., Du, J., Xia, R., et al. (2018b). Wheat miR9678 affects seed germination by generating phased siRNAs and modulating abscisic acid/gibberellin signaling. Plant Cell *30*, 796-814.

Guo, T., Chen, K., Dong, N.Q., Shi, C.L., Ye, W.W., Gao, J.P., Shan, J.X., and Lin, H.X. (2018a). GRAIN SIZE AND NUMBER1 negatively regulates the OsMKKK10-OsMKK4-OsMPK6 cascade to coordinate the trade-off between grain number per panicle and grain size in rice. Plant Cell *30*, 871-888.

Holdsworth, M.J., Bentsink, L., and Soppe, W.J.J. (2008). Molecular networks regulating *Arabidopsis* seed maturation, after-ripening, dormancy and germination. New Phytol. *179*, 33-54.

Hu, Y., Han, X., Yang, M., Zhang, M., Pan, J., and Yu, D. (2019). The transcription factor INDUCER OF CBF EXPRESSION1 interacts with ABSCISIC ACID INSENSITIVE5 and DELLA proteins to fine-tune abscisic acid signaling during seed germination in *Arabidopsis*. Plant Cell *31*, 1520-1538.

Jiang, Z., Xu, G., Jing, Y., Tang, W., and Lin, R. (2016). Phytochrome B and REVEILLE1/2-mediated signalling controls seed dormancy and germination in *Arabidopsis*. Nat. Commun.

7, 12377.

Jiao, Y., Wang, Y., Xue, D., Wang, J., Yan, M., Liu, G., Dong, G., Zeng, D., Lu, Z., Zhu, X., et al. (2010). Regulation of OsSPL14 by OsmiR156 defines ideal plant architecture in rice. Nat. Genet. *42*, 541-544.

Kang, J., Yim, S., Choi, H., Kim, A., Lee, K.P., Lopez-Molina, L., Martinoia, E., and Lee, Y. (2015). Abscisic acid transporters cooperate to control seed germination. Nat. Commun. *6*, 8113.

Li, N., and Li, Y. (2016). Signaling pathways of seed size control in plants. Curr. Opin. Plant Biol. *33*, 23-32.

Li, N., Xu, R., and Li, Y. (2019). Molecular networks of seed size control in plants. Annu. Rev. Plant Biol. *70*, 435-463.

Li, Y., Zheng, L., Corke, F., Smith, C., and Bevan, M.W. (2008). Control of final seed and organ size by the *DA1* gene family in *Arabidopsis thaliana*. Genes Dev *22*, 1331-1336.

Liu, X., Hu, P., Huang, M., Tang, Y., Li, Y., Li, L., and Hou, X. (2016). The NF-YC-RGL2 module integrates GA and ABA signalling to regulate seed germination in *Arabidopsis*. Nat. Commun. *7*, 12768.

Ma, W., Guan, X., Li, J., Pan, R., Wang, L., Liu, F., Ma, H., Zhu, S., Hu, J., Ruan, Y.L., et al. (2019). Mitochondrial small heat shock protein mediates seed germination via thermal sensing. Proc. Natl. Acad. Sci. USA *116*, 4716-4721.

Penfield, S., Josse, E.M., Kannangara, R., Gilday, A.D., Halliday, K.J., and Graham, I.A. (2005). Cold and light control seed germination through the bHLH transcription factor SPATULA. Curr. Biol. *15*, 1998-2006.

Shi, H., Wang, X., Mo, X., Tang, C., Zhong, S., and Deng, X.W. (2015). *Arabidopsis* DET1 degrades HFR1 but stabilizes PIF1 to precisely regulate seed germination. Proc. Natl. Acad. Sci. USA *112*, 3817-3822.

Si, L., Chen, J., Huang, X., Gong, H., Luo, J., Hou, Q., Zhou, T., Lu, T., Zhu, J., Shangguan, Y., et al. (2016). OsSPL13 controls grain size in cultivated rice. Nat. Genet. *48*, 447-456.

Song, X.J., Huang, W., Shi, M., Zhu, M.Z., and Lin, H.X. (2007). A QTL for rice grain width and weight encodes a previously unknown RING-type E3 ubiquitin ligase. Nat. Genet. *39*, 623-630.

Sun, S., Wang, L., Mao, H., Shao, L., Li, X., Xiao, J., Ouyang, Y., and Zhang, Q. (2018). A G-protein pathway determines grain size in rice. Nat. Commun. *9*, 851.

Topham, A.T., Taylor, R.E., Yan, D., Nambara, E., Johnston, I.G., and Bassel, G.W. (2017). Temperature variability is integrated by a spatially embedded decision-making center to break dormancy in *Arabidopsis* seeds. Proc. Natl. Acad. Sci. USA *114*, 6629-6634.

Vaistij, F.E., Barros-Galvao, T., Cole, A.F., Gilday, A.D., He, Z., Li, Y., Harvey, D., Larson, T.R., and Graham, I.A. (2018). MOTHER-OF-FT-AND-TFL1 represses seed germination under far-red light by modulating phytohormone responses in *Arabidopsis thaliana*. Proc. Natl. Acad. Sci. USA *115*, 8442-8447.

Wang, J.L., Tang, M.Q., Chen, S., Zheng, X.F., Mo, H.X., Li, S.J., Wang, Z., Zhu, K.M., Ding, L.N., Liu, S.Y., et al. (2017). Down-regulation of BnDA1, whose gene locus is associated with the seeds weight, improves the seeds weight and organ size in *Brassica napus*. Plant Biotechnol. J. *15*, 1024-1033.

Wang, S., Li, S., Liu, Q., Wu, K., Zhang, J., Wang, S., Wang, Y., Chen, X., Zhang, Y., Gao, C., et al. (2015a). The OsSPL16-GW7 regulatory module determines grain shape and simultaneously improves rice yield and grain quality. Nat. Genet. *47*, 949-954.

Wang, S., Wu, K., Yuan, Q., Liu, X., Liu, Z., Lin, X., Zeng, R., Zhu, H., Dong, G., Qian, Q., et al. (2012). Control of grain size, shape and quality by OsSPL16 in rice. Nat. Genet. *44*, 950-954.

Wang, Y., Xiong, G., Hu, J., Jiang, L., Yu, H., Xu, J., Fang, Y., Zeng, L., Xu, E., Xu, J., et al. (2015b). Copy number variation at the *GL7* locus contributes to grain size diversity in rice. Nat. Genet. *47*, 944-948.

Wang, Z., Chen, F., Li, X., Cao, H., Ding, M., Zhang, C., Zuo, J., Xu, C., Xu, J., Deng, X., et al. (2016). *Arabidopsis* seed germination speed is controlled by SNL histone deacetylase-binding factor-mediated regulation of AUX1. Nat. Commun. *7*, 13412.

Wu, J.J., Peng, X.B., Li, W.W., He, R., Xin, H.P., and Sun, M.X. (2012). Mitochondrial GCD1 dysfunction reveals reciprocal cell-to-cell signaling during the maturation of *Arabidopsis* female gametes. Dev. Cell *23*, 1043-1058.

Xia, T., Li, N., Dumenil, J., Li, J., Kamenski, A., Bevan, M.W., Gao, F., and Li, Y. (2013). The ubiquitin receptor DA1 interacts with the E3 ubiquitin ligase DA2 to regulate seed and organ size in *Arabidopsis*. Plant Cell *25*, 3347-3359.

Xie, G., Li, Z., Ran, Q., Wang, H., and Zhang, J. (2017). Over-expression of mutated *ZmDA1* or *ZmDAR1* gene improves maize kernel yield by enhancing starch synthesis. Plant Biotechnol. J. *16*, 234-244.

Xu, R., Duan, P., Yu, H., Zhou, Z., Zhang, B., Wang, R., Li, J., Zhang, G., Zhuang, S., Lyu, J., et al. (2018a). Control of grain size and weight by the OsMKKK10-OsMKK4-OsMAPK6 signaling pathway in rice. Mol. Plant *11*, 860-873.

Xu, R., Yu, H., Wang, J., Duan, P., Zhang, B., Li, J., Li, Y., Xu, J., Lyu, J., Li, N., et al. (2018b). A mitogen-activated protein kinase phosphatase influences grain size and weight in rice. Plant J. *95*, 937-946.

Yamauchi, Y., Ogawa, M., Kuwahara, A., Hanada, A., Kamiya, Y., and Yamaguchi, S. (2004). Activation of gibberellin biosynthesis and response pathways by low temperature during imbibition of *Arabidopsis thaliana* seeds. Plant Cell *16*, 367-378.

Yan, A., and Chen, Z. (2017). The pivotal role of abscisic acid signaling during transition from seed maturation to germination. Plant Cell Rep. *36*, 689-703.

Yin, L.L., and Xue, H.W. (2012). The MADS29 transcription factor regulates the degradation of the nucellus and the nucellar projection during rice seed development. Plant Cell *24*, 1049-1065.

Zhang, Y., Xiong, Y., Liu, R., Xue, H.W., and Yang, Z. (2019). The Rho-family GTPase OsRac1 controls rice grain size and yield by regulating cell division. Proc. Natl. Acad. Sci. USA *116*, 16121-16126.

Zheng, X., Li, Q., Li, C., An, D., Xiao, Q., Wang, W., and Wu, Y. (2019). Intra-kernel reallocation of proteins in maize depends on VP1-mediated scutellum development and nutrient assimilation. Plant Cell *31*, 2613-2635.

Zuo, J., and Li, J. (2014). Molecular genetic dissection of quantitative trait loci regulating rice grain size. Annu. Rev. Genet. *48*, 99-118.

植物形态建成

焦雨铃　丁兆军　傅向东

1 国内外研究进展

在植物个体生长发育过程中,由于不同细胞向不同方向不断分裂与分化,从而形成了具有各种特殊构造和机能的细胞、组织和器官,这个过程称为植物形态建成(morphogenesis),该过程同时受到外源环境变化及内源多种因子的共同影响。植物株型是指植物个体的三维形态及其动态变化,包括根、茎、叶、穗和花的形态。株型的形成过程主要取决于器官形成的数目、大小、位置及其形态,是典型的发育生物学问题,也是植物发育生物学研究的难点和热点(Wang and Li, 2008)。此外,农作物的产量是受多基因控制和环境影响的复杂数量性状,构成农作物产量的三要素包括:单位面积有效穗数、每穗粒数和千粒重。植物的株型和穗型直接影响这三个要素,所以一直是作物遗传改良研究的重要对象。

1.1 植物株型

影响作物株型的常见要素包括:分枝数目与角度、叶片形态、穗形态与穗粒数等。这些要素分别对应于侧生分生组织的形成与侧芽外生、叶片发生与塑形、花序分枝等发育生物学过程。叶片是植物的主要光合作用器官,其形态、面积和夹角是影响植物个体和群体光合效率的重要因素。同时,叶片也是植物呼吸作用和蒸腾作用的主要场所。因此,有效合理调控叶片的位置、形态、面积和夹角是作物育种选择中的重要形态指标之一。叶片的面积取决于叶片的长度、宽度、复杂度等因素,其中复杂度指复叶中小叶的数目和单叶的分裂程度。叶片的展开取决于背-腹极性的建立,这一直是植物发育生物学的经典问题。利用正向遗传学手段,在模式植物拟南芥和水稻中发现并克隆了一批能够影响叶片背-腹极性的关键基因,特别是编码转录因子(如

AS2 和 HD-ZIPIII 等）(Iwakawa et al., 2002；Emery et al., 2003) 和小 RNA（如 miR165/166 和 tasiR-ARF 等）(Juarez et al., 2004；Nogueira et al., 2007) 的基因。近年来，研究还发现植物激素生长素和机械力信号在叶片展开与复杂性调控中扮演非常重要的作用 (Bar and Ori, 2015；Du et al., 2018；Lewis and Hake, 2016；Qi et al., 2017)。此外，叶片中特定的组织和细胞类型也对植物生长发育和环境适应具有重要意义，如气孔、表皮毛、维管束等的数目与形态。

分枝的形成首先依赖于侧生分生组织的形成。侧生分生组织具有和胚胎期顶端分生组织类似的发育潜能，使其不断生长，能够形成侧芽，进而成为侧枝。研究表明侧生分生组织的形成受到外界环境因素影响较少，而后续侧芽外生和侧枝形成受到外源及内源多种因子的共同影响 (Wu et al., 2020a, 2020b)。虽然分枝形成的早期发育过程一直被认为是非常重要的一个发育阶段，但由于侧生分生组织难于观察，同时也很难通过正向遗传学筛选到特异影响早期发育的因子，其早期发育过程及其调控机制尚不清楚 (Wang and Li, 2011)。近年来，通过基因组学、系统生物学和计算生物学等研究方法，科研工作者对侧生分生组织的早期发育和侧芽形成的遗传调控网络有了新的认识，发现了一批新的关键基因和激素调控作用机制 (Wang et al., 2018a；Wang and Jiao, 2018；Cao and Jiao, 2019)。侧芽形成后，或者处于休眠，或者外生成为侧枝。一方面，侧枝具有与主茎相同的发育模式，既能够形成叶片，也能形成花序并结实，因此增加侧枝数目是提高农作物产量的有效途径之一。另一方面，侧枝的数目绝非多多益善。例如，过多的侧枝带来不能结实的无效穗或无用的小果实与种子，在生产中需要通过人工打叉去除多余侧枝。近年来，侧枝外生调控机制研究领域取得了重要突破，特别是新激素独脚金内酯的发现及其信号转导机制研究 (Jiang et al., 2013；Yao et al., 2018；Zhou et al., 2013)、生长素介导的远距离抑制调控机制研究 (Domagalska and Leyser, 2011) 和赤霉素介导的表观遗传调控机制研究等 (Wu et al., 2020a)。此外，通过对重力反应的差异转录组分析，科研工作者发现了调控分枝夹角的新通路 (Zhang et al., 2018)。分枝夹角和叶夹角是决定植株紧凑程度的主要性状，直接影响农作物密植条件下个体和群体光能利用效率，因此，紧凑株型关键基因的克隆与育种应用对培育密植高产新品种具有重要意义（Yu et al.，2007；Tian et al.，2019）。

1.2 作物穗型

穗部形态结构包括穗轴长度、一级枝梗数目和长度、二次枝梗数目和长度、穗层整齐度、着粒密度等，主要受到茎尖分生组织（stem apex meristem，SAM）和叶腋分生组织（axillary meristem，AM）的细胞活力所控制。此外，穗型紧凑、小穗间距、芒等花器官形态、落粒性等性状也是作物驯化和人工选择的关键农艺性状（Li et al.，2006；Ishii et al.，2013；Gu et al.，2015）。穗型是受多基因和环境影响的复杂数量性状，直接影响农作物产量，一直以来是农作物遗传改良的重点对象。自 20 世纪 90 年代以来，随着植物分子遗传学的发展和多个作物基因组测序的完成，使得对控制农作物穗型的数量性状位点的基因克隆得以实现。

目前，许多调控水稻等作物穗形态、穗粒数和种子大小的关键基因已经被相继克隆（Ali et al.，2019；Li et al.，2019；Zhang and Yuan，2014；Zuo and Li，2014），但是在农业生产中具有重要育种应用价值的基因数目并不多。控制水稻穗粒数的基因 *Gn1a* 编码一个细胞分裂素氧化酶 OsCKX2，该基因表达量降低能够提高 SAM 活力，显著提高水稻产量（Ashikari et al.，2005）。水稻直立穗基因 *DEP1* 是从超级稻'沈农 265'中克隆的高产基因，其优异等位基因 *dep1* 能抑制 OsCKX2 基因表达，提高 SAM 活力，增加每穗穗粒数，进而显著提高水稻产量（Huang et al.，2009）。控制水稻理想株型基因 *IPA1* 编码转录因子 OsSPL14，突变后 *ipa1* 基因的 mRNA 不能被 OsmiR156 降解，导致基因表达量升高，使得水稻分蘖数减少、穗粒数和千粒重增加，进而提高水稻产量（Jiao et al.，2010；Miura et al.，2010）。水稻新株型基因 *NPT1* 编码一个去泛素化酶，它通过调控 OsSPL14 蛋白稳定性，从而实现对水稻株型和穗型的调控（Wang et al.，2017）。水稻穗粒数相关基因 *NOG1* 编码烯酰-CoA 水合酶/异构酶蛋白，该基因的过量表达可以在不影响其他农艺性状的前提下，大幅提高水稻穗粒数和产量（Huo et al.，2017）。2009 年，美国农业部建立了代表全球玉米种质多样性的巢式关联图谱群体，利用该群体，通过 QTL 定位和 GWAS 分析鉴定了一批与玉米产量性状关联的候选基因或位点。Bommert 等（2013）克隆了控制玉米穗行数基

因 *Fea2*，该基因突变引起穗行数的数量变异；Chuck 等（2014）克隆了控制玉米穗行数基因 *UB3*，其外显子区的一个 SNP 变异与穗行数变化显著关联；Jia 等（2020）克隆到一个新的控制玉米穗行数基因 *KNR6*，该基因编码丝氨酸/苏氨酸蛋白激酶，通过影响雌穗小花数目、穗长和行粒数，进而提高玉米产量。还有研究发现，玉米 G 蛋白复合体的 Gα 亚基（FEA2）和 Gβ 亚基在控制顶端分生组织大小、株（穗）型和产量等方面起重要作用（Bommert et al.，2013；Wu et al.，2020b）。

此外，生殖发育时期的穗分枝虽然与营养生长期的分枝产生过程在形态学上有一定相似性，但所涉及的基因调控网络有显著的差异。特别值得一提的是，在水稻中发现的穗型调控基因往往具有多效性，如 *DEP1* 不仅能增加水稻穗粒数，还可以提高水稻光能和氮肥利用效率（Sun et al.，2014），*IPA1* 不仅能显著增加水稻产量，还可以提高水稻对稻瘟病的抗性（Wang et al.，2018b）。因此，解析"一因多效"基因及其遗传调控网络，将对实现高产、稳产、高效等优异性状的协同改良具有重要意义。

1.3 根系构型

植物根系对水分和养分的吸收、植物固定、与根际微生物的相互作用在植物生长发育过程和农作物产量与品质性状形成中发挥着重要的作用。有 3 个重要的发育过程影响着根系构型，包括：① 根尖分生组织不断的细胞分裂与分化导致根系无限生长；② 侧根的形成和生长产生庞大的根系；③ 根毛的形成扩充根系的整个表面积。因此，根系构型的改变会明显影响植物水分和养分的吸收能力。由于植物根系生长在地下，无法直接观察其自然生长状况；加之许多植物（特别是重要农作物）的根系结构非常庞大和复杂，因而作物根系生长发育及其构型的研究在技术上仍存在较大难度。基于这些原因，过去人们对根尖干细胞稳态调控、侧根发生发育、根系构型决定机制研究大多数是以模式植物拟南芥进行的。根的向性生长是植物根系构型对土壤水肥供应的动态响应机制。研究表明，Ca^{2+}-CPK-NLP 信号通路在调节拟南芥根系发育适应硝态氮浓度变化过程中发挥重要作用（Liu et al.，2017）；生长素、脱落酸和细胞分裂素信号转导途径在植物根系向水性生长的调控过程中起着重要作用（Chang et al.，2019；Dietrich et al.，2017；Orosa-Puente et al., 2018)。近年来，随着植物根系构型研究方法的不断改进和表型组学的快速发展（Cahill et al., 2010），水稻和玉米等主要农作物的根系研究取得了较大进展，并成功克隆了一些控制作物根系构型形成相关的关键基因（Hochholdinger et al., 2018；Meng et al., 2019）。

根长是植物根系构型的重要形态指标之一，目前在水稻中已经克隆的关键控制基因包括 *RL6*、*RL7* 和 *RT9* 等（Li et al., 2015；Obara et al., 2010；Wang et al., 2013）。水稻中克隆的第一个调控不定根发育的基因是 *CRL1/ARL1*，该基因缺失突变后导致不定根形成缺陷，侧根数目减少（Inukai et al., 2005）。*CRL4/OsGNOM1* 编码鸟嘌呤核苷酸交换因子，缺失突变后不定根缺失，导致侧根数目减少和根向地性减弱（Liu et al., 2009）。CRL6/OsCHR4 编码类解旋酶/ATP酶的CHD家族蛋白，通过调控生长素途径来影响不定根的发育（Wang et al., 2016）。*OsWOX11* 基因缺失后植株表现不定根缺失或稀少，而过表达该基因则产生大量的不定根（Zhao et al., 2009）。此外，水稻 osiaa23 突变体表现为无不定根和无侧根（Ni et al., 2011），OsWOX和OsCKX4可能是不定根的调控网络中生长素和细胞分裂素信号传导途径的交叉点（Gao et al., 2014）。研究还发现控制水稻根生长角度的主效QTL基因 *DRO1* 通过改变根系构型提升水稻耐旱性（Uga et al., 2013）。蛋白激酶Pstol1的积累通过促进水稻根系的生长使水稻对低磷土壤具有耐受性（Gamuyao et al., 2012）。在玉米中，Majer等（2012）证明了 *RTCS* 编码一个LBD蛋白，并且ZmARF3结合在RTCS的启动子上激活下游基因的表达，从而调控种子根和茎生根的起始和维持。控制玉米侧根基因 *RUM1* 编码一个生长素信号蛋白IAA10，*rum1* 突变体主根上的侧根和种子根的起始方面有缺陷（Woll et al., 2005）。此外，编码MATE转运体蛋白的 *BIGE1* 基因突变后会导致玉米不定根增多（Suzuki et al., 2015）。

自20世纪60年代以来，以矮化育种为特征的"绿色革命"解决了高产与倒伏的矛盾，使得全世界水稻和小麦产量翻了一番。"绿色革命"通过株型改良实现了产量潜力提升的目标，但也伴随着氮肥利用效率下降的缺点。在目前农业生产上水肥资源匮乏、生态环境恶化、产量潜力徘徊不前的大背景下，深入研究农作物株型、穗型和根系构型发育及其环境可塑性的遗传和表观遗传调控规律，使作物能够最大限度地吸收利用光能、水分和养分等，培育出符合"少投入，多产出，保护环境"的绿色高产高效农作物新品种。因此，尽快开展植物株型、穗型和根系发育可塑性调控机制的研究是保障我国粮食安全和农业可持续发展的重大和迫切需求。

2 未来发展趋势与关键突破口

经过多年的研究，人们对模式植物形态建成的遗传调控网络的认识有了很大的提高。然而，对作物株型、穗型和根系构型等重要农艺性状形成的分子基础的了解还明显具有片面化、碎片化的特点。决定作物产量的各个重要农艺性状间的关系错综复杂，彼此关联，相互影响。因此，想要实现作物产量潜力的进一步提高，需要协同调控株型、穗型和根系构型三者之间的动态平衡。产量是受多基因与环境控制的数量性状，尤其是当前主栽品种间遗传多样性逐步减小，常规育种瓶颈效应愈来愈明显，通过现有常规育种技术已经很难育成产量突破性新品种，且育种周期长。随着基因组学、系统生物学和计算生物学等学科的兴起和大数据处理技术的不断进步，为复杂性状形成的遗传调控网络解析和分子设计育种理论突破带来了新的机遇，也为作物精准育种技术体系创新奠定了科学基础。针对未来作物品种设计的需求，对作物理想株型、穗型和根系构型性状形成的遗传调控网络的进一步了解，乃至实现人工智能设计，是未来研究的重点。

2.1 干细胞微环境的建立及维持机制研究

侧生分生组织的形成与发育、分枝数、穗粒数及根系形态的调控都受到不同时期侧生干细胞活性的影响。叶片形态取决于叶片分生组织的活性，与茎尖分生组织、侧生分生组织和花分生组织类似，叶片分生组织同样具有能够自我复制和分化的干细胞，但在叶片成熟过程中逐渐分化并消失。如何对各个组织中不同类型的干细胞活性进行精准调控是未来作物分子设计中的株型设计的关键。在今后的研究中，应该综合利用分子遗传学及新近发展起来的多维组学技术，特别是单细胞测序技术、高通量的蛋白组学技术、表观遗传学、3D基因组学等，来鉴定调控干细胞微环境的建立及维持的新基因，而且应该特别关注干细胞的形成、维持、活性调控以及终止的新机制。

2.2 适应和响应全球气候变化的可塑性机制研究

作物株型、穗型和根系构型是受多基因控制和环境影响的复杂数量性状，如何培育适应全球气候变化，并能够精准响应环境变化的未来作物，需要解析调控株型、穗型和根系构型各个内源因子与环境因素互作的遗传调控网络，并揭示植物适应环境的可塑性发育的分子机制。如何在未来全球气温上升、高 CO_2 浓度、干旱等环境下维持并提高产量，如何发展适应未来机械化密植、设施化栽培、工厂化生产的新株型作物，将是作物株型研究的新方向与趋势。

2.3 植物形态建成可视化技术研发与计算辅助设计

由于植物干细胞及侧生分生组织发育早期，无法直接观察其动态变化过程，尤其是在田间种植条件下的根系表型分析困难，成本高且耗时，并且通常需要在特定时间点对根系进行破坏性取样。近年来，强大的3D成像系统提供了高分辨率原位可视化的新机会，如通过MRI和多层螺旋CT技术实时监测根系发育和水分运输，实现作物根系原位形态构型观察。随着科学家对植形态建成的环境适应调控网络研究的不断深入，目前已经找到了一批影响株型形成的基因"零件"。理解这些"零件"如何组装、如何工作是进一步理解株型建立与调控的必经之路。在认识株型调控的基础上，将有望改造株型，特别是通过计算辅助设计实现对未来作物株型的精准设计，定向改良株型并耦合多个株型要素与其他性状要素，创造未来作物。对株型建立的组装、模拟与设计需要多学科交叉，特别是生命科学与应用数学、生物力学、计算机等学科的有机融合。

2.4 理想株型塑造与分子设计育种体系建立

基因组学、系统生物学、计算生物学和合成生物学等学科的兴起和大数据技术的不断进步，为作物理想株型塑造和绿色高产高效分子设计育种带来了新的机遇。首先，"理想株型"塑造由重视"个体产量"转向侧重"群体产量"。目前，我国的农业生产模式正处于大变革时期，由传统的农户种植逐步转变为大户种植，清洁栽培和人工智能种植的生产模式日趋成型，因此对育种提出了新的要求。譬如，"理想株型"塑造注重"增源-扩库-畅流"；"扩库"不仅是增加穗粒数和粒重，提高库容量，而且还是进一步提高收获指数；"增源"不仅需要提高根系对水分和养分的吸收和转运，还需要提高叶片光合效率和减少蒸腾作用，这些会涉及植物生长-代谢平衡的协同作用机制研究。目前，分子设计育种在水稻中已经得到了实践。期望在不久的将来，可以借助计算机和数学建模等交叉学科，以及基因合成等技术来实现包括水稻、小麦、玉米等主要农作物的分子设计育种。

3 阶段性目标

3.1 2035年目标

目前，我国植物科学家在水稻株型、穗型和根系构型等研究领域已经取得了重要的进展，在独脚金内酯信号转导途径、理想株型和穗型性状形成的分子基础等方向做出了国际领先的工作。在此基础上，进一步凝聚我国在植物形态建成的分子机制研究领域的优势研究力量，协同攻关，有望在2035年全面提升对于株型、穗型和根系构型等影响作物产量和品质性状形成的遗传调控网络的研究水平，尤其是在以下领域取得重要进展。

解析植物干细胞的时空特异调控和产量性状形成的分子基础。阐明干细胞微环境建立、干细胞维持与分化以及根尖干细胞可塑性调控的分子机制，取得一批具有重大科学意义的原创性成果，使我国植物干细胞研究获得理论性突破，产量等复杂性状解析研究处于国际前沿。

阐明作物株型、穗型和根系构型适应环境变化的分子基础。揭示植物整合外界环境信号和内源发育信号调控植物形成建成的可塑性发育的遗传调控网络，为基于株型、穗型和根系构型改良的节水育种、抗逆育种和营养高效育种提供理论和技术支撑；研发可视化研究技术新体系，并拥有自主知识产权并开发其应用价值，推动我国精准农业和智能农业的发展。

植物形态建成可视化技术研发与理想株型3D计算模拟。在解析植物形态建成的遗传调控网络的基础上，明确影响作物株型、穗型和根系构型的关键元件（或称为"分子模块"）。理解这些"分子模块"如何组装、如何耦合、如何集成是塑造理想作物株型、穗型和根系构型的必经之路。在此基础上，通过计算辅助设计实现对未来作物株型、穗型和根系构型株型的精准设计，定向改良作物。对作物株型、穗型和根系构型模拟与设计需要多学科交叉，特别与应用数学、计算生物学、生物力学、大数据技术等多学科的有机融合。

3.2 2050年目标

在2035年目标基础上，通过进一步凝练作物株型、穗型和根系构型研究领域的关键科学问题和核心技术，有望在2050年实现作物理想株型、穗型和根系构型性状的创造，以及新品种的工业化设计。

实现主要农作物理想株型、穗型和根系构型性状的创造。阐明植物尤其是作物株型、穗型和根系的生长发育的遗传调控规律，在主要农作物中通过精准操控干细胞和侧生分生组织形成和发育过程，从头设计并创造出包括叶型、分枝在内的理想株型、穗型和根系构型，精准培育"超级"农作物新品种。

适应新环境、新生产方式的跨界融合式株型、穗型和根系构型的创造。阐明植物形态建成与环境互作的遗传和表观遗传调控规律，基于株型、穗型和根系构型性状适应与响应环境变化分子调控机制，设计并培育出适应生产方式变革和精准响应环境变化的新株型、新穗型和根系新构型，培育出节水、耐盐、抗病虫害、养分高效等优异性状协同改良的"超级"作物新品种。

作物株型、穗型和根系构型的精确计算辅助设计与新品种创造。基于大数据智能分析，对光、温、水、土、肥等基本要素优化配置，发展可视化、自动化、自然语言处理和深度学习超级计算机平台，利用理想株型、穗型和根系构型的从头设计与创造，在优质高产、抗病耐逆、资源高效利用等方面开展计算辅助设计，实现主要农作物规模化与智能化品种设计与创造。

参考文献

Ali, A., Xu, P., Riaz, A., and Wu, X. (2019). Current advances in molecular mechanisms and physiological basis of panicle degeneration in rice. Int. J. Mol. Sci. *20*, E1613.

Ashikari, M., Sakakibara, H., Lin, S., Yamamoto, T., Takashi, T., Nishimura, A., Angeles, E.R., Qian, Q., Kitano, H., and Matsuoka, M. (2005). Cytokinin oxidase regulates rice grain production. Science *309*, 741-745.

Bar, M., and Ori, N. (2015). Compound leaf development in model plant species. Curr. Opin.

Plant Biol. *23*, 61-69.

Bommert, P., Nagasawa, N.S., and Jackson D. (2013). Quantitative variation in maize kernel row number is controlled by the *FASCIATED EAR2* locus. Nat. Genet. *45*, 334-347.

Cahill, J.F. Jr., McNickle, G.G., Haag, J.J., Lamb, E.G., Nyanumba, S.M. St, and Clair, C.C. (2010). Plants integrate information about nutrients and neighbors. Science *328*, 1657.

Cao, X., and Jiao Y. (2019). Control of cell fate during axillary meristem initiation. Cell Mol. Life Sci. *77*, 2343-2354.

Chang, J., Li, X., Fu, W., Wang, J., Yong, Y., Shi, H., Ding, Z., Kui, H., Gou, X., He, K., and Li, J. (2019). Asymmetric distribution of cytokinins determines root hydrotropism in *Arabidopsis thaliana*. Cell Res. *29*, 984-993.

Chuck, G.S., Brown, P.J., Meeley, R., and Hake, S. (2014). Maize *SBP*-box transcription factors *unbranched2* and *unbranched3* affect yield traits by regulating the rate of lateral primordia initiation. Proc. Natl. Acad. Sci. USA *111*, 18775-18780.

Dietrich, D., Pang, L., Kobayashi, A., Fozard J.A., Boudolf V., Bhosale R., Antoni R., Nguyen T., Hiratsuka S., Fujii N., et al. (2017). Root hydrotropism is controlled via a cortex-specific growth mechanism. Nat. Plants *3*, 17057.

Domagalska, M.A., and Leyser, O. (2011). Signal integration in the control of shoot branching. Nat. Rev. Mol. Cell Biol. *12*, 211-221.

Dong, G., Guo L., Zhu X., Gou Z., Wang W., Wu Y., Lin H., and Fu, X. (2014). Heterotrimeric G proteins regulate nitrogen-use efficiency in rice. Nat. Genet. *46*, 652-656.

Du, F., Guan, C., and Jiao, Y. (2018). Molecular mechanisms of leaf morphogenesis. Mol. Plant *11*, 1117-1134.

Emery, J.F., Floyd S.K., Alvarez J., Eshed Y., Hawker N.P., Izhaki A., Baum S.F., and Bowman J.L. (2003). Radial patterning of *Arabidopsis* shoots by class III *HD-ZIP* and *KANADI* genes. Curr. Biol. *13*, 1768-1774.

Gamuyao, R., Chin, J.H., Pariasca-Tanaka, J., Pesaresi, P., Catausan, S., Dalid, C., Slamet-Loedin, I., Tecson-Mendoza, E.M., Wissuwa, M., and Heuer, S. (2012). The protein kinase Pstol1 from traditional rice confers tolerance of phosphorus deficiency. Nature *488*, 535-539.

Gao, S., Fang, J., Xu, F., Wang, W., Sun, X., Chu, J., Cai, B., Feng, Y., and Chu, C. (2014). *CYTOKININ OXIDASE/DEHYDROGENASE4* integrates cytokinin and auxin signaling to control rice crown root formation. Plant Physiol. *165*, 1035-1046.

Gu, B., Zhou, T., Luo, J., Liu, H., Wang, Y., Shangguan, Y., Zhu, J., Li, Y., Sang, T., Wang, Z., and Han, B. (2015). *An-2* encodes a cytokinin synthesis enzyme that regulates awn length and grain production in rice. Mol. Plant *8*, 1635-1650.

Hetz, W., Hochholdinger, F., Schwall, M., and Feix, G. (1996). Isolation and characterization of *rtcs*, a maize mutant deficient in the formation of nodal roots. Plant J. *10*, 845-857.

Hochholdinger, F., Yu, P., and Marcon, C. (2018). Genetic control of root system development in maize. Trends Plant Sci. *23*, 79-88.

Huang, X., Qian, Q., Liu, Z., Sun, H., He, S., Luo, D., Xia, G., Chu, C., Li, J., and Fu, X. (2009). Natural variation at the *DEP1* locus enhances grain yield in rice. Nat. Genet. *41*, 494-497.

Huo, X., Wu, S., Zhu, Z., Liu, F., Fu, Y., Cai, H., Sun, X., Gu, P., Xie, D., Tan, L., and Sun, C. (2017). NOG1 increases grain production in rice. Nat. Commun. *8*, 1497.

Inukai, Y., Sakamoto, T., Ueguchi-Tanaka, M., Shibata, Y., Gomi, K., Umemura, I., Hasegawa, Y., Ashikari, M., Kitano, H., and Matsuoka, M. (2005). *Crown Rootless1*, which is essential for crown root formation in rice, is a target of an AUXIN RESPONSE FACTOR in auxin signaling. Plant Cell *17*, 1387-1396.

Ishii, T., Numaguchi, K., Miura, K., Yoshida, K., Thanh, P.T., Htun, T.M., Yamasaki, M., Komeda, N., Matsumoto, T., Terauchi, R., et al. (2013). *OsLG1* regulates a closed panicle trait in domesticated rice. Nat. Genet. *45*, 462-465.

Iwakawa, H., Ueno, Y., Semiarti, E., Onouchi, H., Kojima, S., Tsukaya, H., Hasebe, M., Soma, T., Ikezaki, M., Machida, C., and Machida, Y. (2002). The *ASYMMETRIC LEAVES2* gene of *Arabidopsis thaliana*, required for formation of a symmetric flat leaf lamina, encodes a member of a novel family of proteins characterized by cysteine repeats and a leucine zipper. Plant Cell Physiol. *43*, 467-478.

Jenkins, M.T. (1930). Heritable characters of maize. XXXIV-rootless. J. Hered. *21*, 79-80.

Jia, H., Li, M., Li, W., Liu, L., Jian, Y., Yang, Z., Shen, X., Ning, Q., Du, Y., Zhao, R., et al. (2020). A serine/threonine protein kinase encoding gene *KERNEL NUMBER PER ROW6* regulates maize grain yield. Nat. Commun. *11*, 988.

Jiang, L., Liu, X., Xiong, G., Liu, H., Chen, F., Wang, L., Meng, X., Liu, G., Yu, H., Yuan, Y., et al. (2013). DWARF 53 acts as a repressor of strigolactone signalling in rice. Nature *504*,

401-405.

Jiao, Y., Wang, Y., Xue, D., Wang, J., Yan, M., Liu, G., Dong, G., Zeng, D., Lu, Z., Zhu, X., et al. (2010). Regulation of *OsSPL14* by OsmiR156 defines ideal plant architecture in rice. Nat. Genet. *42*, 541-545.

Juarez, M.T., Kui, J.S., Thomas, J., Heller, B.A., and Timmermans, M.C. (2004). microRNA-mediated repression of *rolled leaf1* specifies maize leaf polarity. Nature *428*, 84-88.

Lewis, M.W., and Hake, S. (2016). Keep on growing: building and patterning leaves in the grasses. Curr. Opin. Plant Biol. *29*, 80-86.

Li, C., Zhou, A., and Sang, T. (2006) Rice domestication by reducing shattering. Science *311*, 1936-1939.

Li, J., Han, Y., Liu, L., Chen, Y., Du, Y., Zhang, J., Sun, H., and Zhao, Q. (2015). *qRT9*, a quantitative trait locus controlling root thickness and root length in upland rice. J. Exp. Bot. *66*, 2723-2732.

Li, N., Xu, R., and Li, Y. (2019). Molecular networks of seed size control in plants. Annu. Rev. Plant Biol. *70*, 435-463.

Li, S., Tian, Y., Wu, K., Ye, Y., Yu, J., Zhang, J., Liu, Q., Hu, M., Li, H., Tong, Y., et al. (2018). Modulating plant growth-metabolism coordination for sustainable agriculture. Nature *560*, 595-600.

Liu, K.H., Niu, Y., Konishi, M., Wu, Y., Du, H., Sun, C.H., Li, L., Boudsocq, M., McCormack, M., Maekawa, S., et al. (2017). Discovery of nitrate-CPK-NLP signalling in central nutrient-growth networks. Nature *545*, 311-316.

Liu, S., Wang, J., Wang, L., Wang, X., Xue, Y., Wu, P., and Shou, H. (2009). Adventitious root formation in rice requires OsGNOM1 and is mediated by the OsPINs family. Cell Res. *19*, 1110-1119.

Majer, C., Xu, C., Berendzen, K.W., and Hochholdinger, F. (2012). Molecular interactions of ROOTLESS CONCERNING CROWN AND SEMINAL ROOTS, a LOB domain protein regulating shoot-borne root initiation in maize (*Zea mays* L.). Philos. Trans. R. Soc. Lond. B Biol. Sci. *367*, 1542-1551.

Meng, F., Xiang, D., Zhu, J., Li, Y., and Mao, C. (2019). Molecular mechanisms of root development in rice. Rice. *12*, 1.

Miura, K., Ikeda, M., Matsubara, A., Song, X.J., Ito, M., Asano, K., Matsuoka, M., Kitano, H., and Ashikari, M. (2010). *OsSPL14* promotes panicle branching and higher grain productivity in rice. Nat. Genet. *42*, 545-550.

Ni, J., Wang, G.H., Zhu, I.X., Zhang, H.H., Wu, Y.R., and Wu, P. (2011). OsIAA23-mediated auxin signaling defines postembryonic maintenance of QC in rice. Plant J. *68*, 433-442.

Ni, J., Wang, G.H., Zhu, I.X., Zhang, H.H., Wu, Y.R., and Wu, P. (2011). OsIAA23-mediated auxin signaling defines postembryonic maintenance of QC in rice. Plant J. *68*, 433-442.

Nogueira, F.T., Madi, S., Chitwood, D.H., Juarez, M.T., and Timmermans, M.C. (2007). Two small regulatory RNAs establish opposing fates of a developmental axis. Genes Dev. *21*, 750-755.

Obara, M., Tamura, W., Ebitani, T., Yano, M., Sato, T., and Yamaya, T. (2010). Fine-mapping of *qRL6.1*, a major QTL for root length of rice seedlings grown under a wide range of NH_4^+ concentrations in hydroponic conditions. Theor. Appl. Genet. *121*, 535-547.

Orosa-Puente, B., Leftley, N., von Wangenheim, D., Banda, J., Srivastava, A.K., Hill, K., Truskina, J., Bhosale, R., Morris, E., Srivastava, M., et al. (2018). Root branching toward water involves posttranslational modification of transcription factor ARF7. Science *362*, 1407-1410.

Qi, J., Wu, B., Feng, S., Lv, S., Guan, C., Zhang, X., Qiu, D., Hu, Y., Zhou, Y., Li, C., et al. (2017). Mechanical regulation of organ asymmetry in leaves. Nat. Plants *3*, 724-733.

Sun, H., Qian, Q., Wu, K., Luo, J., Wang, S., Zhang, C., Ma, Y., Liu, Q., Huang, X., Yuan, Q., et al. (2014). Heterotrimeric G proteins regulate nitrogen-use efficiency in rice. Nat. Genet. *46*, 652-656.

Suzuki, M., Sato, Y., Wu, S., Kang, BH., and McCarty D.R. (2015). Conserved functions of the MATE transporter BIG EMBRYO1 in regulation of lateral organ size and initiation rate. Plant Cell *27*, 2288-2300.

Tian, J., Wang, C., Xia, J., Wu, L., Xu, G., Wu, W., Li, D., Qin, W., Han, X., Chen, Q., et al. (2019). Teosinte ligule allele narrows plant architecture and enhances high-density maize yields. Science *365*, 658-664.

Uga, Y., Sugimoto, K., Ogawa, S., Rane, J., Ishitani, M., Hara, N., Kitomi, Y., Inukai, Y., Ono, K., Kanno, N., et al. (2013). Control of root system architecture by DEEPER ROOTING 1 increases rice yield under drought conditions. Nat. Genet. *45*, 1097-1102.

Wang, B., Smith, S.M., and Li, J. (2018a). Genetic regulation of shoot architecture. Annu. Rev. Plant Biol. *69*, 437-468.

Wang, H., Xu, X., Zhan, X., Zhai, R., Wu, W., Shen, X., Dai, G., Cao, L., and Cheng, S. (2013). Identification of *qRL7*, a major quantitative trait locus associated with rice root length in hydroponic conditions. Breed. Sci. *63*, 267-274.

Wang, J., Zhou, L., Shi, H., Chern, M., Yu, H., Yi, H., He, M., Yin, J., Zhu, X., Li, Y., et al. (2018b). A single transcription factor promotes both yield and immunity in rice. Science *361*, 1026-1028.

Wang, S., Wu, K., Qian, Q., Liu, Q., Li, Q., Pan, Y., Ye, Y., Liu, X., Wang, J., Zhang, J., et al. (2017). Non-canonical regulation of SPL transcription factors by a human OTUB1-like deubiquitinase defines a new plant type rice associated with higher grain yield. Cell Res. *27*, 1142-1156.

Wang, Y., and Jiao, Y. (2018). Axillary meristem initiation-a way to branch out. Curr. Opin. Plant Biol. *41*, 61-66.

Wang, Y., and Li, J. (2008). Molecular basis of plant architecture. Annu. Rev. Plant Biol. *59*, 253-279.

Wang, Y., and Li, J. (2011). Branching in rice. Curr. Opin. Plant Biol. *14*, 94-99.

Wang, Y., Wang, D., Gan, T., Liu, L., Long, W., Wang, Y., Niu, M., Li, X., Zheng, M., Jiang, L., et al. (2016). CRL6, a member of the CHD protein family, is required for crown root development in rice. Plant Physiol. Biochem. *105*, 185-194.

Woll, K., Borsuk, L.A., Stransky, H., Nettleton, D., Schnable, P.S., and Hochholdinger, F. (2005). Isolation, characterization, and pericycle-specific transcriptome analyses of the novel maize lateral and seminal root initiation mutant *rum1*. Plant Physiol. *139*, 1255-1267.

Wu, K., Wang, S., Song, W., Zhang, J., Wang, Y., Liu, Q., Yu, J., Ye, Y., Li, S., Chen, J., et al. (2020a). Enhanced sustainable green revolution yield via nitrogen-responsive chromatin modulation in rice. Science *367*, eaaz2046.

Wu, Q., Xu, F., Liu, L., Char, S.N., Ding, Y., Je, B.I., Schmelz, E., Yang, B., and Jackson D. (2020b). The maize heterotrimeric G protein β subunit controls shoot meristem development and immune responses. Proc. Natl. Acad. Sci. USA *117*, 1799-1805.

Yao, R., Li, J., and Xie, D. (2018). Recent advances in molecular basis for strigolactone action. Sci. China Life Sci. *61*, 277-284.

Yu, B., Lin, Z., Li, H., Li, X., Li, J., Wang, Y., Zhang, X., Zhu, Z., Zhai, W., Wang, X., et al. (2007). *TAC1*, a major quantitative trait locus controlling tiller angle in rice. Plant J. *52*, 891-898.

Zhang, D., and Yuan, Z. (2014). Molecular control of grass inflorescence development. Annu. Rev. Plant Biol. *65*, 553-578.

Zhang, N., Yu, H., Yu, H., Cai, Y., Huang, L., Xu, C., Xiong, G., Meng, X., Wang, J., Chen, H., et al. (2018). A core regulatory pathway controlling rice tiller angle mediated by the *LAZY1*-dependent asymmetric distribution of auxin. Plant Cell *30*, 1461-1475.

Zhao, Y., Hu, Y., Dai, M., Huang, L., and Zhou, D.X. (2009). The WUSCHEL-related homeobox gene *WOX11* is required to activate shoot-borne crown root development in rice. Plant Cell *21*, 736-748.

Zhou, F., Lin, Q., Zhu, L., Ren, Y., Zhou, K., Shabek, N., Wu, F., Mao, H., Dong, W., Gan, L., et al. (2013). D14-SCFD3-dependent degradation of D53 regulates strigolactone signalling. Nature *504*, 406-410.

Zuo, J., and Li J. (2014). Molecular genetic dissection of quantitative trait loci regulating rice grain size. Annu. Rev. Genet. *48*, 99-118.

生育期、育性与杂种优势

程祝宽 刘耀光

1 国内外研究进展

1.1 生育期

生育期是农作物最重要的农艺性状之一，决定了农作物品种的生态适应性、品质和产量潜力，是农作物栽培和育种工作中优先考虑的关键因素。农作物的生长发育主要由营养生长和生殖生长构成，而开花决定着营养生长向生殖生长阶段的转变，因此开花期（抽穗期）是作物生长发育过程中最为重要的时期，对作物开花期调控的研究是实现我国农作物高产、稳产的重要前提和保障。

植物开花是由内在的遗传因素和外在的环境因素（光周期、温度等）协同调节的复杂过程。目前，对双子叶模式植物拟南芥开花的调控机制研究已比较清楚，不同途径包括光周期途径、春化途径、自主途径、温度途径、赤霉素途径和年龄途径等完成对拟南芥开花的精准调控（Fornara et al.，2010）。与拟南芥相比，农作物的开花期调控机制既有相似性也有特异性。

水稻属于典型的短日作物。当今栽培稻的分布范围为北纬55°到南纬36°（Khush，1997），远超野生稻分布的地理范围，表明水稻在驯化育种中，生育期受到选择，使水稻栽种范围向北扩展到长日照区域。水稻的抽穗期受感光性、感温性和基本营养生长性决定。研究认为光周期途径是调控水稻抽穗期的关键途径，同时，相对温度等其他因素而言，光周期更具稳定性，研究也较为深入。成花素是一种可移动的促进植物开花的信号蛋白，在开花调控过程中处于核心地位。在水稻中，Hd3a和RFT1是两个主要的成花素蛋白，前者在短日照条件下起主要作用，后者在长日照条件下发挥功能（Komiya et al.，2009），两者在很大程度上受到表观遗传调控（Sun et al.，2012）。与成花素关系最为密切的调控因子是Hd1和Ehd1两个蛋白质，它们同时介导了水稻重要的两条开花调控途径，包括保守的 OsGI-Hd1-Hd3a/RFT1 途径（Hayama et al., 2003）和特异的 Ghd7-Ehd1-Hd3a/RFT1 途径（Xue et al., 2008）。*Hd1* 是拟南芥 *CONSTANS (CO)* 的同源基因，是水稻开花调控途径中另一个关键基因，具有双向调节功能（Yano et al., 2000）。在不同日照条件下，Hd1 通过调节成花素 Hd3a 的表达，从而调控水稻开花周期（Yano et al., 2000；Izawa et al., 2002；Hayama et al., 2003）。在 Ghd7-Ehd1-Hd3a/RFT1 途径中，*Ehd1* 编码一个 B 类应答调控因子，是水稻特异的开花调控基因，受到多个开花抑制因子（如 DTH8、Hd16 等）及促进因子（如 Ehd2、Ehd3、Ehd4、Hd17、Hd18 等）的调控（Matsubara et al., 2012；Sun et al., 2014；Shibaya et al., 2016）。此外，一系列新的水稻生育期相关调控因子 HAF1、OsHAL3、DHD1 的发现与功能研究，最终形成了目前水稻生育期复杂的调控网络和作用途径（Yang et al., 2015；Su et al., 2016；Zhu et al., 2018；Zhang et al., 2019）。生育期基因在水稻的驯化和人工选育过程中尤为重要，而 *Hd1*、*Ehd1*、*Ghd7* 及 *Hd3a/RFT1* 等基因的等位分化对水稻适应更广大地区的不同光温条件发挥了重要作用（Tsuji et al., 2013；Zhao et al., 2015b）。

"高产不早熟、早熟不高产"即所谓"优而不早、早而不优"现象，是杂交稻品种培育上遇到的重大难题。最近人们克隆了一个长片段非编码 RNA *Ef-cd* 基因，可通过介导该位点组蛋白甲基化水平，正向调控另一重要开花基因 *OsSOC1/OsMADS50* 的表达，从而促使水稻早熟。*Ef-cd* 能显著提早水稻抽穗期 7～20 天，但对产量没有不利影响，甚至有不同程度的增产（Fang et al., 2019）。该基因的利用对解决直播稻和粮经、粮菜、粮油连作稻的早熟丰产，以及亚种间杂交稻"超亲晚熟"等问题具有重要价值。

大豆作为世界上种植广泛的粮食和经济作物，是人类油脂及优质植物蛋白的重要来源。与拟南芥和水稻相比，大豆生育期的研究相对滞后。现已明确的与大豆生育期相关的位点主要有 *E1*～*E4*、*E6*～*E10*、*J* 共 10 个，其中 *E1*～*E4*、*E9*、*E10*、*J* 等生育期基因已相继被克隆（Liu et al., 2008；Watanabe et al., 2009, 2011；Xia et al., 2012；Dissanayaka et al., 2016；Zhao et al., 2016；Lu et al., 2017；Samanfar et al., 2017；Liu et al., 2018），*E1* 和 *J* 基因对于大豆生育期的调控网络起最为关键的作用（Miranda et al., 2020）。

E1是豆科植物特异的转录因子，对调控大豆的开花期和成熟期贡献最大（Liu et al.，2008；Watanabe et al.，2009，2011；Xia et al.，2012；Zhao et al.，2016；Lu et al.，2017）。*J*基因在短日照条件下通过抑制*E1*基因的表达来调控开花期。突变后，*J*基因不能抑制*E1*的表达，使得大豆在低纬短日照条件下生育期延长，从而大幅度提高了大豆产量。在大豆中，*E*基因与*J*基因及*FT*基因共同组成了大豆特有的光周期调控系统PHYA(E3E4)-J-E1-FT（Lu et al.，2017）。

玉米原产于墨西哥，也是一种短日照植物。*ID1*是玉米中第一个被克隆的开花基因，与水稻中的*Ehd2*基因同源。该基因突变后将会延迟开花时间并产生异常花序（Colasanti et al.，1998）。最近的研究表明，*ID1*能够通过组蛋白甲基化来调节开花期（Mascheretti et al.，2015）。玉米中还存在与水稻*OsHd1*同源的基因*CONZ1*和水稻成花素基因*Hd3a*同源的基因*ZCN8*，它们都受到*ZmGI1*的调控（Bendix et al.，2013）。CCT结构域转录因子*ZmCCT*是水稻*OsGhd7*的同源基因，突变后导致早花，从而使得将玉米种植到更高纬度成为可能（Hung et al.，2012）。

与以上的短日照作物不同，小麦和大麦都属于长日照作物，起源于冷暖交替的地中海地区，为保护分生组织免于冷害，它们低温时不开花。冬麦必须经历春化阶段，才能正常抽穗开花，这一机制受到春化基因（*VERNALIZATION*，*VRN*）的控制（Ream et al.，2012）。*VRN1*是拟南芥*FUL*和*AP1*的同源基因，而*VRN2*是一个与水稻*Ghd7*序列同源的抑制因子，*VRN1*或*VRN2*突变后可赋予其春麦的生长习性，并将种植区域扩展到更广阔的范围。*VRN3*分别对应小麦的*TaFT*基因和大麦的*HvFT*基因，编码与拟南芥和水稻成花素的同源蛋白，在温暖和长日照条件下表达产物运输到顶端分生组织促进开花（Yan et al.，2006；Li and Dubcovsky，2008）。除了春化基因外，小麦生育期还受到光周期基因（*PHOTOPERIOD*，*PPD*）、*EPS*基因（*EARLINESS PER SE*）的调控。与小麦类似，控制大麦开花时间的基因主要包括*HvVRN-H1*、*HvVRN-H2*、*HvVRN-H3*（*HvFT1*）、*HvPPD-H1*、*HvPPD-H2*（*HvFT3*）及*EARLY MATURITY*（*EAM*）等（Hill and Li，2016）。总体来讲，相对于水稻及模式生物，对玉米、麦类作物生育期的研究仍有待深入。

1.2 育性

作物的育性既关乎作物繁殖后代的能力，也关系到人类的生产生活。作物的生殖生长从花序分化开始，经历花器官形态建成、雌雄配子体发育等一系列关键事件，任何时期发育异常都会影响作物育性，最终影响作物的产量和品质。

雌雄配子的发育首先经历复杂的减数分裂过程，其中涉及一系列重要蛋白质参与，并通过复杂的网络互作进行精确调控。对于雄配子来说，减数分裂后绒毡层的发育和细胞程序性死亡对维持小孢子正常发育至关重要，如 *TDR* (Li et al., 2006; Zhang et al., 2008)、*EAT1/DTD* (Niu et al., 2013; Ono et al., 2018)、*TIP2* (Fu et al., 2014)、*PTC1* (Li et al., 2011a) 和 *API5* (Li et al., 2011b) 等基因控制绒毡层细胞程序性死亡。*OsMADS3* 在绒毡层发育晚期清除活性氧分子，参与花药后期发育和花粉粒形成过程 (Hu et al., 2011)。*DTC1* 参与维持活性氧平衡，其突变体植株绒毡层细胞程序性死亡表现异常 (Yi et al., 2016)。

在水稻雄配子体发育的过程中，花药通过相关因子在时间和空间上有序调控来保证成熟花粉粒形成。*WDA1* (Jung et al., 2006)、*CYP703A3* (Aya et al., 2009)、*CYP704B2* (Li et al., 2010)、*DPW* (Shi et al., 2011)、*PDA1* (Zhu et al., 2013)、*OsABCG15* (Niu et al., 2013)、*OsABCG26* (Zhao et al., 2015a) 和 *OsC6* (Zhang et al., 2010) 等基因编码的蛋白质和花药表面角质层与花粉外壁形成相关，通过调控脂肪酸合成和转运影响花药和花粉发育。*DCM1* (Zhang et al., 2018) 通过影响胼胝质的降解导致花粉外壁缺陷引起花粉不育。*RIP1* 和 *AID1* 等基因在花粉成熟和花药发育后期发挥作用 (Zhu et al., 2004; Han et al., 2006)。

雌配子体发生的遗传和分子调控机制非常复杂，包括孢原细胞形成、减数分裂调节、功能大孢子选择、细胞不均等分裂及胚囊结构形成等几个发育过程。在玉米中，*AM1* 基因被认为可能同时控制大、小孢子母细胞由有丝分裂途径进入减数分裂途径 (Pawlowski et al., 2009)。*MAC1* 基因 (Wang et al., 2012) 和水稻 *MSP1* 基因 (Nonomura et al., 2003) 调控体细胞向孢原细胞转化过程。水稻 *msp1* 突变体也与玉米 *mac1* 表型相似，都产生多个孢原细胞。玉米 *IG* 基因在保持适当数目的有丝分裂周期上起作用，对应突变体的雌配子体产生更多的核，且细胞核发生不同步分裂 (Evans, 2007)。其

水稻同源基因 *OsIG1* 在调控水稻花器官、胚珠和雌配子体发育中发挥重要作用（Zhang et al., 2015）。

植物雄性不育是自然界普遍存在的现象，根据雄性不育基因的来源可将其分为细胞质雄性不育（cytoplasmic male sterility, CMS）和细胞核雄性不育（genetic male sterility, GMS）。作物雄性不育系的发现与利用，使得杂种优势获得大规模的推广和应用，对作物产量的提高有非常显著的贡献（Li et al., 2007）。

细胞质雄性不育性（CMS）指在特异的细胞质条件下，细胞器功能异常而表现败育的现象。CMS 通常是由线粒体嵌合基因引起，是线粒体基因组发生频繁的重组导致重排、插入或者缺失的结果。在水稻、小麦、玉米、高粱、油菜等作物的线粒体 CMS 相关位点中，均发现不同长度和排列方式的线粒体基因组序列和来源未知序列组成的嵌合基因，导致雄性不育；而核恢复基因（*Rf*）能抑制 CMS 表型，恢复植株的育性（Hanson and Bentolila, 2004）。野败型细胞质雄性不育基因 *WA352* 的克隆，首次阐明了植物细胞质雄性不育系统通过线粒体不育基因和核基因互作控制雄性不育的分子机制（Luo et al., 2013），随后恢复基因 *Rf4* 也被成功克隆，*Rf4* 编码一个线粒体定位的 PPR 蛋白，其介导降解 *WA352* mRNA 而恢复花粉育性（Tang et al., 2014）。

光温敏核不育水稻的发现为杂种优势的利用开辟了新途径。光温敏雄性不育由隐性核基因控制，育性恢复不受胞质基因恢保关系限制，因而比 CMS 恢复源更广泛。*pms3/p/tms12-1* 是一个长链非编码 RNA，光敏感核不育水稻与正常水稻品种在此位点存在一个碱基突变。该基因表达受光周期调控，在长日条件下，野生型水稻的 *PMS3* 基因的表达量能保证花粉正常发育（Zhu and Deng, 2012）。*tms5* 是在另一籼型温敏不育水稻'安农 S-1'中发现的一个隐性温敏不育位点。和正常水稻品种相比，温敏水稻 *tms5* 位点的核糖核酸酶 RNase Z^{S1} 基因发生功能缺失突变，导致在高温条件下高表达的泛素核糖核蛋白 Ub_{L40} 基因的 mRNA 不能被 RNase Z^{S1} 降解，进而引起花粉发育异常（Zhou et al., 2014）。*TMS10* 是近年来发现的一个温敏雄性不育基因，该基因编码一个富含亮氨酸重复序列的类受体激酶，表现为高温雄性不育和低温雄性可育。研究表明，高温条件下 *TMS10* 对维持水稻花药绒毡层的发育和小孢子形成起重要作用（Yu et al.，2017）。

1.3 杂种优势

杂种优势是指遗传基础不同的亲本杂交子一代在某些性状或综合性状上优于双亲的现象。杂种优势是生物界普遍现象，也是迄今为止人类在农业生产上运用的最成功的生命现象之一。20世纪30年代，玉米是第一个在商业化生产上大规模利用杂种优势并获极大成功的农作物。随后由于作物雄性不育系的发现，高粱、水稻、油菜、向日葵等主要农作物杂种优势的利用获得了飞跃性发展。尤其是20世纪70年代，水稻的"三系"配套在中国得以实现，为确保世界粮食安全做出了重要贡献。

杂种优势是一个非常复杂的生物学现象。近年来，大量研究表明在杂交子一代中，显性、超显性及上位性作用模式往往同时存在，杂种优势可能是这3种遗传效应其中的一种或多种效应共同作用的结果（Luo et al.，2001；Hua et al.，2003；Semel et al.，2006；Li et al.，2008）。作物的杂种优势可体现在某个单基因或位点的效应上，如玉米细胞数目调控基因 *ZmCNR1* 和 *ZmCNR2*，是玉米长势和生活力的负调控因子，可能与玉米的杂种优势有关（Guo et al.，2010）。在水稻中，Li 等（2016）通过对一个产量杂种优势的主效位点进行克隆，发现该位点实际上是一个抽穗期相关基因 *DTH8/Ghd8/LHD8*，该基因同时控制开花期、株高、分蘖数和每穗粒数等诸多性状。Wang 等（2019b）克隆到了杂交稻'广两优676'中与产量性状相关的杂种优势基因 *GW3p6*，该基因能显著提高水稻产量和千粒重。在转录组水平上，Song 等（2010）对一杂交稻及其亲本转录组进行分析表明，光合作用以及碳固定等代谢途径基因在 F_1 杂种中表达上调，这可能与杂交稻的生活力提升有关。另外，利用高通量测序和全基因组关联分析（GWAS）对杂种优势机制的研究具有明显优势。Huang 等（2015，2016）对1000多个杂交稻组合的38个性状进行了全基因组关联分析，发现对杂种优势贡献最大的是诸多有利显性等位基因的积累，通过进一步鉴定得到了控制水稻杂种优势的主要基因位点，发现这些优势基因型重新整合，在杂交水稻中实现了花期、株型及产量各要素的理想组合，形成产量杂种优势。Li 等（2016）分析了'两优培九'的杂种优势产量基础，发现了'两优培九'产量杂种优势效应主要来自优势亲本穗粒数和有效穗数的贡献。

2 未来发展趋势与关键突破口

2.1 生育期

作物的生育期是决定品种地区与季节适应性的重要性状，选育早熟高产农作物一直是育种家的目标，这给科研工作者提出了新的要求和挑战。就目前发展现状，未来农作物生育期的研究和利用可能有以下发展趋势。

主要农作物生育期研究均衡发展 目前，在各主要农作物中，对水稻生育期基因的研究相对比较清楚，进展也较快，这得益于它相对简单且较小的基因组。对其他作物，如小麦、大麦、大豆等研究则相对比较滞后。随着各作物基因组参考序列的公布，高通量测序技术及数据分析技术的发展，小麦、大麦等基因组比较庞大的作物生育期相关基因研究将会协同发展，为未来应用奠定基础。

建立农作物品种生育期数据库 随着对各种农作物生育期基因的研究，建立农作物品种生育期数据库将成为可能。通过分析我国目前已有各农作物主栽品种的相关生育期基因，考察主基因的地理分布特点，建立它们之间的联系，将为育种工作提供便利。

实现农作物生育期定向育种 通过确定我国各农作物主栽地区的光温等生态条件建立各地生态数据库，进而根据各地生态条件特点匹配农作物品种生育期数据库，筛选出具有特定生育期类型、适合当地栽种的农作物，从而实现农作物定向育种。

实现农作物生育期定制育种 进一步明确各生育期主效和微效基因的调控通路和遗传效应，通过分子设计，利用基因编辑等手段对农作物生育期进行精准控制和调节，解决品种引种，以及诸如水稻籼粳杂交稻生育期推迟的难题，实现农作物定制育种。

2.2 育性

作物的育性调控是一个极其复杂的过程，涉及花器官发育、雌雄配子发育、减数分裂等诸多过程，因此，未来的发展也必将呈现多元化趋势，但从农作物应用角度分析，可能有以下趋势。

重组频率的遗传调控 染色体的交换和重组是物种遗传多样化的基础，对作物来讲，控制了重组和交换频率就控制了育种的进程。例如，提高作物的重组频率可以大大加快育种进程，缩短有利基因的利用周期。目前在水稻减数分裂研究中，人们发现了大量重组促进因子，已鉴定出的重组促进因子包括 Zip4、Msh4、Msh5、Mer3 及 Hei10 等。研究表明，这些基因的单突变均表现为突变体中交叉结数目显著减少，并且这些突变体中 DSB 均能正常形成，同源染色体配对和联会亦能很好进行。另一类是重组抑制因子，生物体进化出了一系列机制，来抑制过多交叉的发生。目前已发现的重组抑制因子包括 FANCM、FIGL1 及 RECQ4 等，虽然它们抑制重组的机制并不相同，但是均起到抑制重组的作用。另外，目前研究还发现 ZEP1 除了作为水稻联会复合体的横丝元件外，在抑制交叉形成过程中也发挥重要作用，这些重组促进和抑制因子在不久的将来可用于作物重组频率的控制和调节。

作物倍性控制 在对作物基因功能的研究或育种过程中，经常会用到单倍体或多倍体。目前这些材料的获得，通常通过花药培养或者化学药剂处理的方式。在减数分裂研究中，先后发现了两个重要基因 *OSD1* 和 *MTL*。*OSD1* 功能缺失后，会导致减数分裂只进行第一次分裂，不进行第二次减数分裂，导致生物体染色体加倍（d'Erfurth et al.，2010）。*MTL* 是最先在玉米中发现的诱导玉米单倍体形成的基因，该基因突变后可获得一定频率的单倍体植株（Kelliher et al.，2017）。*OSD1* 和 *MTL* 基因的发现为作物获得多倍体和单倍体提供了极大便利。

人工智能不育杂交育种技术 雄性不育系的发现极大地推动了杂种优势的利用。目前杂交稻生产上通过"三系法"和"两系法"进行杂交稻制种，面临的问题是"三系法"受恢保关系制约，制种周期长，难度大。"两系法"制种受环境因素影响较大，容易发生制种不纯。这些都制约着杂交稻的进一步发展。为了解决以上问题，智能不育杂交育种技术应运而生。此技术通过获得隐性的孢子体雄性不育突变体，利用转基因技术将含有野生型孢子体育性基因、花粉配子体败育基因及标记筛选基因导入到雄性不育突变体中，产生保持系，并通过荧光分选技术生产不育系和保持系。智能不育杂交育种相关技术在玉米和水稻上已获得了成功（Chang et al.，2016）。智能不育系配组自由，不育性稳定，遗传方式简单且杂交种不携带转基因成分，为未成功应用"三系法"和"两系法"的作物开展杂种优势利用提供可能，但目前转基因植物安全性管理政策对这类技术的管理还不明确。

2.3 杂种优势

优良杂交种选育是提高粮食产量，保障我国粮食安全的重要举措。迄今为止，虽然杂种优势的理论在农作物生产中成功实践多年，但杂种优势形成的分子机制和调控机制仍然没有获得解析，使得杂种优势进一步应用受限，提升空间趋窄，难以解决作物杂种优势利用过程中显现的问题。因此未来发展的趋势之一就是明确杂种优势形成机制。随着基因组学、蛋白质组学、转录组学、代谢组学及高通量测序技术的发展，可真正从网络视角来剖析杂种优势机制，为最终解析杂种优势的分子机制提供可能。对于水稻杂种优势利用，随着更多的杂种不育基因被克隆，以分子育种技术尤其是基因编辑技术克服亚种间和种间杂种不育性，将使水稻产量有更大提升。

杂种优势利用的另一个发展趋势就是作物杂种优势的固定或者实现作物的无融合生殖。由于杂种优势主要表现在子一代上，其后代会发生性状分离，出现优势衰减现象，必须代代制种，耗费大量人力、物力及资源，同时还存在着杂种不纯风险。利用无融合生殖来固定杂种优势，是杂种优势利用的最好方式。而要实现无融合生殖，一是通过发掘无融合生殖基因，但目前主要农作物中相关基因报道和利用很少；二是通过间接方式，因为无融合生殖的本质就是让同源染色体之间不发生重组和交换。目前在植物减数分裂研究中，发现了大量包括染色体不重组，或减数分裂转变为有丝分裂等相关基因，利用这些基因资源和基因编辑技术完全可实现无融合生殖。可喜的是，通过上述途径已经在水稻杂种优势固定和无融合生殖创制方面获得突破（Wang et al., 2019a）。

3 阶段性目标

3.1 2035 年目标

至 2035 年的发展目标包括：完成主要农作物重要生育期基因的挖掘，

基本明确相关性状调控网络；完成育性控制过程基因的挖掘工作，对具有重要应用价值基因，如光温敏不育基因、杂种不育基因、重组频率调控基因等，在多作物间开展相关应用研究，实现对各农作物重组频率、染色体倍性的遗传控制，建立各作物智能不育杂交育种技术体系；通过对无融合生殖基因的挖掘，结合减数分裂相关基因的改良与应用，完成主要农作物无融合生殖材料的创制，实现杂种优势的固定。

3.2 2050年目标

至2050年的发展目标包括：完成生育期各基因的功能、效应研究，明确各基因通路和调控网络，实现生育期基因定向应用；全面完成各发育过程基因的挖掘，实现对各农作物染色体重组频率和倍性的遗传控制，建立各作物智能不育杂交育种技术体系，全面解析农作物育性控制的分子机制和调控网络；完成对杂种优势形成机理的解析，实现各主要农作物杂种优势的高效利用。

参考文献

Aya, K., Ueguchi-Tanaka, M., Kondo, M., Hamada, K., Yano, K., Nishimura, M., and Matsuoka, M. (2009). Gibberellin modulates anther development in rice via the transcriptional regulation of GAMYB. Plant Cell 21, 1453-1472.

Bendix, C., Mendoza, J.M., Stanley, D.N., Meeley, R., and Harmon, F.G. (2013). The circadian clock-associated gene gigantea1 affects maize developmental transitions. Plant, Cell & Environment 36, 1379-1390.

Chang, Z., Chen, Z., Wang, N., Xie, G., Lu, J., Yan, W., Zhou, J., Tang, X., and Deng, X.W. (2016). Construction of a male sterility system for hybrid rice breeding and seed production using a nuclear male sterility gene. Proc. Natl. Acad. Sci. USA 113, 14145-14150.

Colasanti, J., Yuan, Z., and Sundaresan, V. (1998). The indeterminate gene encodes a zinc finger protein and regulates a leaf-generated signal required for the transition to flowering in maize. Cell 93, 593-603.

d'Erfurth, I., Cromer, L., Jolivet, S., Girard, C., Horlow, C., Sun, Y., To, J.P., Berchowitz, L.E.,

Copenhaver, G.P., and Mercier, R. (2010). The cyclin—a CYCA1; 2/TAM is required for the meiosis I to meiosis II transition and cooperates with OSD1 for the prophase to first meiotic division transition. PLoS Genet. *6*, e1000989.

Dissanayaka, A., Rodriguez, T.O., Di, S., Yan, F., Githiri, S.M., Rodas, F.R., Abe, J., and Takahashi, R. (2016). Quantitative trait locus mapping of soybean maturity gene *E5*. Breeding Sci. *66*, 407-415.

Evans, M.M.S. (2007). The indeterminate gametophyte1 gene of maize encodes a LOB domain protein required for embryo sac and leaf development. Plant Cell *19*, 46-62.

Fang, J., Zhang, F., Wang, H., Wang, W., Zhao, F., Li, Z., Sun, C., Chen, F., Xu, F., Chang, S., et al. (2019). *Ef-cd* locus shortens rice maturity duration without yield penalty. Proc. Natl. Acad. Sci. USA *116*, 18717-18722.

Fornara, F., de Montaigu, A., and Coupland, G. (2010). SnapShot: control of flowering in *Arabidopsis*. Cell *141*, 550-550.

Fu, Z., Yu, J., Cheng, X., Zong, X., Xu, J., Chen, M., Li, Z., Zhang, D., and Liang, W. (2014). The rice basic helix-loop-helix transcription factor TDR INTERACTING PROTEIN2 is a central switch in early anther development. Plant Cell *26*, 1512-1524.

Guo, M., Rupe, M.A., Dieter, J.A., Zou, J., Spielbauer, D., Duncan, K.E., Howard, R.J., Hou, Z., and Simmons, C.R. (2010). Cell Number Regulator1 affects plant and organ size in maize: implications for crop yield enhancement and heterosis. Plant Cell *22*, 1057-1073.

Han, M.J., Jung, K.H., Yi, G., Lee, D.Y., and An, G. (2006). Rice Immature Pollen 1 (RIP1) is a regulator of late pollen development. Plant Cell Physiol. *47*, 1457-1472.

Hanson, M.R., and Bentolila, S. (2004). Interactions of mitochondrial and nuclear genes that affect male gametophyte development. Plant Cell *16 Suppl.*, S154-S169.

Hayama, R., Yokoi, S., Tamaki, S., Yano, M., and Shimamoto, K. (2003). Adaptation of photoperiodic control pathways produces short-day flowering in rice. Nature *422*, 719.

Hill, C.B., and Li, C. (2016). Genetic architecture of flowering phenology in cereals and opportunities for crop improvement. Front. Plant Sci. *7*, 1906.

Hu, L., Liang, W., Yin, C., Cui, X., Zong, J., Wang, X., Hu, J., and Zhang, D. (2011). Rice MADS3 regulates ROS homeostasis during late anther development. Plant Cell *23*, 515-533.

Hua, J., Xing, Y., Wu, W., Xu, C., Sun, X., Yu, S., and Zhang, Q. (2003). Single-locus heterotic effects and dominance by dominance interactions can adequately explain the genetic basis of heterosis in an elite rice hybrid. Proc. Natl. Acad. Sci. USA *100*, 2574-2579.

Huang, X., Yang, S., Gong, J., Zhao, Q., Feng, Q., Zhan, Q., Zhao, Y., Li, W., Cheng, B., and Xia, J. (2016). Genomic architecture of heterosis for yield traits in rice. Nature *537*, 629-633.

Huang, X., Yang, S., Gong, J., Zhao, Y., Feng, Q., Gong, H., Li, W., Zhan, Q., Cheng, B., and Xia, J. (2015). Genomic analysis of hybrid rice varieties reveals numerous superior alleles that contribute to heterosis. Nat. Commun. *6*, 6258.

Hung, H.Y., Shannon, L.M., Tian, F., Bradbury, P.J., Chen, C., Flint-Garcia, S.A., McMullen, M.D., Ware, D., Buckler, E.S., Doebley, J.F., et al. (2012). ZmCCT and the genetic basis of day-length adaptation underlying the postdomestication spread of maize. Proc. Natl. Acad. Sci. USA *109*, E1913-E1921.

Izawa, T., Oikawa, T., Sugiyama, N., Tanisaka, T., Yano, M., and Shimamoto, K. (2002). Phytochrome mediates the external light signal to repress FT orthologs in photoperiodic flowering of rice. Genes Dev. *16*, 2006-2020.

Jung, K.H., Han, M.J., Lee, D.Y., Lee, Y.S., Schreiber, L., Franke, R., Faust, A., Yephremov, A., Saedler, H., Kim, Y.W., et al. (2006). Wax-deficient anther1 is involved in cuticle and wax production in rice anther walls and is required for pollen development. Plant Cell *18*, 3015-3032.

Kelliher, T., Starr, D., Richbourg, L., Chintamanani, S., Delzer, B., Nuccio, M.L., Green, J., Chen, Z., McCuiston, J., and Wang, W. (2017). MATRILINEAL, a sperm-specific phospholipase, triggers maize haploid induction. Nature *542*, 105.

Khush, G.S. (1997). Origin, dispersal, cultivation and variation of rice. Plant Mol. Biol. *35*, 25-34.

Komiya, R., Yokoi, S., and Shimamoto, K. (2009). A gene network for long-day flowering activates RFT1 encoding a mobile flowering signal in rice. Development *136*, 3443-3450.

Li, C., and Dubcovsky, J. (2008). Wheat FT protein regulates VRN1 transcription through interactions with FDL2. Plant J. *55*, 543-554.

Li, D., Huang, Z., Song, S., Xin, Y., Mao, D., Lv, Q., Zhou, M., Tian, D., Tang, M., and Wu, Q. (2016). Integrated analysis of phenome, genome, and transcriptome of hybrid rice uncovered multiple heterosis-related loci for yield increase. Proc. Natl. Acad. Sci. USA *113*, E6026-E6035.

Li, H., Pinot, F., Sauveplane, V., Werck-Reichhart, D., Diehl, P., Schreiber, L., Franke, R., Zhang, P., Chen, L., Gao, Y., et al. (2010). Cytochrome P450 family member CYP704B2 catalyzes the {omega}-hydroxylation of fatty acids and is required for anther cutin biosynthesis and pollen exine formation in rice. Plant Cell *22*, 173-190.

Li, H., Yuan, Z., Vizcay-Barrena, G., Yang, C.Y., Liang, W.Q., Zong, J., Wilson, Z.A., and Zhang, D.B. (2011a). *PERSISTENT TAPETAL CELL1* encodes a PHD-finger protein that is required for tapetal cell death and pollen development in rice. Plant Physiol. *156*, 615-630.

Li, L., Lu, K., Chen, Z., Mu, T., Hu, Z., and Li, X. (2008). Dominance, over-dominance and epistasis condition the heterosis in two heterotic rice hybrids. Genetics *180*, 1725-1742.

Li, N., Zhang, D.S., Liu, H.S., Yin, C.S., Li, X.X., Liang, W.Q., Yuan, Z., Xu, B., Chu, H.W., Wang, J., et al. (2006). The rice tapetum degeneration retardation gene is required for tapetum degradation and anther development. Plant Cell *18*, 2999-3014.

Li, S.Q., Yang, D.C., and Zhu, Y.G. (2007). Characterization and use of male sterility in hybrid rice breeding. J. Integr. Plant Biol. *49*, 791-804.

Li, X., Gao, X., Wei, Y., Deng, L., Ouyang, Y., Chen, G., Li, X., Zhang, Q., and Wu, C. (2011b). Rice APOPTOSIS INHIBITOR5 coupled with two DEAD-box adenosine 5′-triphosphate-dependent RNA helicases regulates tapetum degeneration. Plant Cell *23*, 1416-1434.

Liu, B., Kanazawa, A., Matsumura, H., Takahashi, R., Harada, K., and Abe, J. (2008). Genetic redundancy in soybean photoresponses associated with duplication of phytochrome A gene. Genetics *180*, 995-1007.

Liu, W., Jiang, B.J., Ma, L.M., Zhang, S.W., Zhai, H., Xu, X., Hou, W.S., Xia, Z.J., Wu, C.X., Sun, S., et al. (2018). Functional diversification of Flowering Locus T homologs in soybean: GmFT1a and GmFT2a/5a have opposite roles in controlling flowering and maturation. New Phytol. *217*, 1335-1345.

Lu, S., Zhao, X., Hu, Y., Liu, S., Nan, H., Li, X., Fang, C., Cao, D., Shi, X., and Kong, L. (2017). Natural variation at the soybean *J* locus improves adaptation to the tropics and enhances yield. Nat. Genet. *49*, 773-779.

Luo, D.P., Xu, H., Liu, Z.L., Guo, J.X., Li, H.Y., Chen, L.T., Fang, C., Zhang, Q.Y., Bai, M.,

Yao, N., et al. (2013). A detrimental mitochondrial-nuclear interaction causes cytoplasmic male sterility in rice. Nat. Genet. *45*, 573-577.

Luo, L., Li, Z.K., Mei, H., Shu, Q., Tabien, R., Zhong, D., Ying, C., Stansel, J., Khush, G., and Paterson, A. (2001). Overdominant epistatic loci are the primary genetic basis of inbreeding depression and heterosis in rice. II. Grain yield components. Genetics *158*, 1755-1771.

Mascheretti, I., Turner, K., Roberta, S.B., Hand, A., Colasanti, J., and Rossi, V. (2015). Florigen-encoding genes of day-neutral and photoperiod-sensitive maize are regulated by different chromatin modifications at the floral transition. Plant Physiol. *168*, 1351-1363.

Matsubara, K., Ogiso-Tanaka, E., Hori, K., Ebana, K., Ando, T., and Yano, M. (2012). Natural variation in Hd17, a homolog of *Arabidopsis* ELF3 that is involved in rice photoperiodic flowering. Plant Cell Physiol. *53*, 709-716.

Miranda, C., Scaboo, A., Cober, E., Denwar, N., and Bilyeu, K. (2020). The effects and interaction of soybean maturity gene alleles controlling flowering time, maturity, and adaptation in tropical environments. BMC Plant Biol. *20*, 65.

Niu, N., Liang, W., Yang, X., Jin, W., Wilson, Z.A., Hu, J., and Zhang, D. (2013). EAT1 promotes tapetal cell death by regulating aspartic proteases during male reproductive development in rice. Nat. Commun. *4*, 1445.

Nonomura, K.I., Miyoshi, K., Eiguchi, M., Suzuki, T., Miyao, A., Hirochika, H., and Kurata, N. (2003). The *MSP1* gene is necessary to restrict the number of cells entering into male and female sporogenesis and to initiate anther wall formation in rice. Plant Cell *15*, 1728-1739.

Ono, S., Liu, H., Tsuda, K., Fukai, E., Tanaka, K., Sasaki, T., and Nonomura, K.I. (2018). EAT1 transcription factor, a non-cell-autonomous regulator of pollen production, activates meiotic small RNA biogenesis in rice anther tapetum. PLoS Genet. *14*, e1007238.

Pawlowski, W.P., Wang, C.J.R., Golubovskaya, I.N., Szymaniak, J.M., Shi, L., Hamant, O., Zhu, T., Harper, L., Sheridan, W.F., and Cande, W.Z. (2009). Maize AMEIOTIC1 is essential for multiple early meiotic processes and likely required for the initiation of meiosis. Proc. Natl. Acad. Sci. USA *106*, 3603-3608.

Ream, T.S., Woods, D.P., and Amasino, R.M. (2012). The Molecular Basis of Vernalization in Different Plant Groups. Cold Spring Harbor Symposia on Quantitative Biology *77*, 105-115.

Samanfar, B., Molnar, S.J., Charette, M., Schoenrock, A., Dehne, F., Golshani, A., Belzile, F., and Cober, E.R. (2017). Mapping and identification of a potential candidate gene for a novel maturity locus, *E10*, in soybean. Theor. Appl. Genet. *130*, 377-390.

Semel, Y., Nissenbaum, J., Menda, N., Zinder, M., Krieger, U., Issman, N., Pleban, T., Lippman, Z., Gur, A., and Zamir, D. (2006). Overdominant quantitative trait loci for yield and fitness in tomato. Proc. Natl. Acad. Sci. USA *103*, 12981-12986.

Shi, J., Tan, H.X., Yu, X.H., Liu, Y.Y., Liang, W.Q., Ranathunge, K., Franke, R.B., Schreiber, L., Wang, Y.J., Kai, G.Y., et al. (2011). Defective Pollen Wall is required for anther and microspore development in rice and encodes a fatty acyl carrier protein reductase. Plant Cell *23*, 2225-2246.

Shibaya, T., Hori, K., Ogiso-Tanaka, E., Yamanouchi, U., Shu, K., Kitazawa, N., Shomura, A., Ando, T., Ebana, K., and Wu, J. (2016). *Hd18*, encoding histone acetylase related to *Arabidopsis* FLOWERING LOCUS D, is involved in the control of flowering time in rice. Plant Cell Physiol. *57*, 1828-1838.

Song, G.S., Zhai, H.L., Peng, Y.G., Zhang, L., Wei, G., Chen, X.Y., Xiao, Y.G., Wang, L., Chen, Y.J., Wu, B., Chen, B., et al. (2010). Comparative transcriptional profiling and preliminary study on heterosis mechanism of super-hybrid rice. Mol Plant *3*, 1012-1025.

Su, L., Shan, J.X., Gao, J.P., and Lin, H.X. (2016). OsHAL3, a blue light-responsive protein, interacts with the floral regulator Hd1 to activate flowering in rice. Mol. Plant *9*, 233-244.

Sun, C., Chen, D., Fang, J., Wang, P., Deng, X., and Chu, C. (2014). Understanding the genetic and epigenetic architecture in complex network of rice flowering pathways. Protein Cell *5*, 889-898.

Sun, C., Fang, J., Zhao, T., Xu, B., Zhang, F., Liu, L., Tang, J., Zhang, G., Deng, X., Chen, F., Qian, Q., Cao, X., and Chu, C. (2012). The histone methyltransferase SDG724 mediates H3K36me2/3 deposition at *MADS50* and *RFT1*, and promotes flowering in rice. Plant Cell *24*, 3235-3247.

Tang, H., Luo, D., Zhou, D., Zhang, Q., Tian, D., Zheng, X., Chen, L., and Liu, Y.G. (2014). The rice restorer *Rf4* for wild-abortive cytoplasmic male sterility encodes a mitochondrial-localized PPR protein that functions in reduction of *WA352* transcripts. Mol. Plant *9*, 1497-1500.

Tsuji, H., Taoka, K., and Shimamoto, K. (2013). Florigen in rice: complex gene network for florigen transcription, florigen activation complex, and multiple functions. Curr. Opin. Plant Biol. *16*, 228-235.

Wang, C., Liu, Q., Shen, Y., Hua, Y., Wang, J., Lin, J., Wu, M., Sun, T., Cheng, Z., Mercier, R., and Wang, K. (2019a). Clonal seeds from hybrid rice by simultaneous genome engineering of meiosis and fertilization genes. Nat. Biotechnol. *37*, 283-286.

Wang, C.J., Nan, G.L., Kelliher, T., Timofejeva, L., Vernoud, V., Golubovskaya, I.N., Harper, L., Egger, R., Walbot, V., and Cande, W.Z. (2012). Maize *multiple archesporial cells 1* (*mac1*), an ortholog of rice *TDL1A*, modulates cell proliferation and identity in early anther development. Development *139*, 2594-2603.

Wang, C.S., Tang, S.C., Zhan, Q.L., Hou, Q.Q., Zhao, Y., Zhao, Q., Feng, Q., Zhou, C.C., Lyu, D.F., Cui, L.L., et al. (2019b). Dissecting a heterotic gene through GradedPool-Seq mapping informs a rice-improvement strategy. Nat. Commun. *10*, 2982.

Watanabe, S., Hideshima, R., Xia, Z., Tsubokura, Y., Sato, S., Nakamoto, Y., Yamanaka, N., Takahashi, R., Ishimoto, M., and Anai, T. (2009). Map-based cloning of the gene associated with soybean maturity locus *E3*. Genetics *182*, 1251-1262.

Watanabe, S., Xia, Z., Hideshima, R., Tsubokura, Y., Sato, S., Yamanaka, N., Takahashi, R., Anai, T., Tabata, S., and Kitamura, K. (2011). A map-based cloning strategy employing a residual heterozygous line reveals that the *GIGANTEA* gene is involved in soybean maturity and flowering. Genetics *188*, 395-407.

Xia, Z., Watanabe, S., Yamada, T., Tsubokura, Y., Nakashima, H., Zhai, H., Anai, T., Sato, S., Yamazaki, T., and Lv, S. (2012). Positional cloning and characterization reveal the molecular basis for soybean maturity locus *E1* that regulates photoperiodic flowering. Proc. Natl. Acad. Sci. USA *109*, E2155-E2164.

Xue, W., Xing, Y., Weng, X., Zhao, Y., Tang, W., Wang, L., Zhou, H., Yu, S., Xu, C., and Li, X. (2008). Natural variation in *Ghd7* is an important regulator of heading date and yield potential in rice. Nat. Genet. *40*, 761-767.

Yan, L., Fu, D., Li, C., Blechl, A., Tranquilli, G., Bonafede, M., Sanchez, A., Valarik, M., Yasuda, S., and Dubcovsky, J. (2006). The wheat and barley vernalization gene *VRN3* is an orthologue of *FT*. Proc. Natl. Acad. Sci. USA *103*, 19581-19586.

Yang, Y., Fu, D., Zhu, C., He, Y., Zhang, H., Liu, T., Li, X., and Wu, C. (2015). The ring-finger ubiquitin ligase HAF1 mediates Heading Date 1 degradation during photoperiodic flowering in rice. Plant Cell *27*, 2455-2468.

Yano, M., Katayose, Y., Ashikari, M., Yamanouchi, U., Monna, L., Fuse, T., Baba, T., Yamamoto, K., Umehara, Y., and Nagamura, Y. (2000). *Hd1*, a major photoperiod sensitivity quantitative trait locus in rice, is closely related to the *Arabidopsis* flowering time gene *CONSTANS*. Plant Cell *12*, 2473-2483.

Yi, J., Moon, S., Lee, Y.S., Zhu, L., Liang, W., Zhang, D., Jung, K.H., and An, G. (2016). Defective Tapetum Cell Death 1 (DTC1) regulates ROS levels by binding to metallothionein during tapetum degeneration. Plant Physiol. *170*, 1611-1623.

Yu, J.P., Han, J.J., Kim, Y.J., Song, M., Yang, Z., He, Y., Fu, R.F., Luo, Z.J., Hu, J.P., Liang, W.Q., and Zhang, D.B. (2017). Two rice receptor-like kinases maintain male fertility under changing temperatures. Proc. Natl. Acad. Sci. USA *114*, 12327-12332.

Zhang, C., Shen, Y., Tang, D., Shi, W.Q., Zhang, D.M., Du, G.J., Zhou, Y.H., Liang, G.H., Li, Y.F., and Cheng, Z.K. (2018). The zinc finger protein DCM1 is required for male meiotic cytokinesis by preserving callose in rice. PLoS Genet. *14*, e1007769.

Zhang, D.S., Liang, W.Q., Yin, C.S., Zong, J., Gu, F.W., and Zhang, D.B. (2010). *OsC6*, encoding a lipid transfer protein, is required for postmeiotic anther development in rice. Plant Physiol. *154*, 149-162.

Zhang, D.S., Liang, W.Q., Yuan, Z., Li, N., Shi, J., Wang, J., Liu, Y.M., Yu, W.J., and Zhang, D.B. (2008). Tapetum degeneration retardation is critical for aliphatic metabolism and gene regulation during rice pollen development. Mol. Plant *1*, 599-610.

Zhang, H., Zhu, S.S., Liu, T.Z., Wang, C.M., Cheng, Z.J., Zhang, X., Chen, L.P., Sheng, P.K., Cai, M.H., Li, C.N., et al. (2019). DELAYED HEADING DATE1 interacts with OsHAP5C/D, delays flowering time and enhances yield in rice. Plant Biotechnol. J. *17*, 531-539.

Zhang, J.R., Tang, W., Huang, Y.L., Niu, X.L., Zhao, Y., Han, Y., and Liu, Y.S. (2015). Down-regulation of a LBD-like gene, *OsIG1*, leads to occurrence of unusual double ovules and developmental abnormalities of various floral organs and megagametophyte in rice. J. Exp. Bot. *66*, 99-112.

Zhao, C., Takeshima, R., Zhu, J., Xu, M., Sato, M., Watanabe, S., Kanazawa, A., Liu, B., Kong, F., and Yamada, T. (2016). A recessive allele for delayed flowering at the soybean maturity locus E9 is a leaky allele of FT2a, a FLOWERING LOCUS T ortholog. BMC Plant Biol. *16*, 20.

Zhao, G., Shi, J., Liang, W., Xue, F., Luo, Q., Zhu, L., Qu, G., Chen, M., Schreiber, L., and Zhang, D. (2015a). Two ATP binding cassette G transporters, rice ATP binding cassette G26 and ATP binding cassette G15, collaboratively regulate rice male reproduction. Plant Physiol. *169*, 2064-2079.

Zhao, J., Chen, H.Y., Ren, D., Tang, H.W., Qiu, R., Feng, J.L., Long, Y.M., Niu, B.X., Chen, D.P., Zhong, T.Y., Liu, Y.G., and Guo, J.X. (2015b). Genetic interactions between diverged alleles of *Early heading date 1* (*Ehd1*) and *Heading date 3a* (*Hd3a*)/*RICE FLOWERING LOCUS T1* (*RFT1*) control differential heading and contribute to regional adaptation in rice (*Oryza sativa*). New Phytol. *208*, 936-948.

Zhou, H., Zhou, M., Yang, Y.Z., Li, J., Zhu, L.Y., Jiang, D.G., Dong, J.F., Liu, Q.J., Gu, L.F., Zhou, L.Y., et al. (2014). RNase Z^{S1} processes Ub^{L40} mRNAs and controls thermosensitive genic male sterility in rice. Nat. Commun. *5*, 4884.

Zhu, C., Peng, Q., Fu, D., Zhuang, D., Yu, Y., Duan, M., Xie, W., Cai, Y., Ouyan, Y., Lian, X., and Wu, C. (2018). The E3 ubiquitin ligase HAF1 modulates circadian accumulation of EARLY FLOWERING3 to control heading date in rice under long-day conditions. Plant Cell *30*, 2352-2367.

Zhu, D., and Deng, X.W. (2012). A non-coding RNA locus mediates environment-conditioned male sterility in rice. Cell Res. *22*, 791-792.

Zhu, L., Shi, J.X., Zhao, G.C., Zhang, D.B., and Liang, W.Q. (2013). *Post-meiotic deficient anther1* (*PDA1*) encodes an ABC transporter required for the development of anther cuticle and pollen exine in rice. J. Plant Biol. *56*, 59-68.

Zhu, Q.H., Ramm, K., Shivakkumar, R., Dennis, E.S., and Upadhyaya, N.M. (2004). The *ANTHER INDEHISCENCE1* gene encoding a single MYB domain protein is involved in anther development in rice. Plant Physiol. *135*, 1514-1525.

光 合 作 用

朱新广　林荣呈

1 国内外研究进展

20世纪六七十年代，我国集中开展了植物光合作用光能磷酸化机制研究，提出在ATP合成过程中需要有高能态存在，证明了ATP合成过程中的电化学势梯度学说（Shen and Shen，1962）。中国科学院殷宏章等科学家在20世纪60年代就认识到冠层光合作用效率对产量有重大贡献，并系统开展定量研究。进入21世纪，我国科研工作者继续对光合作用开展研究，在光反应色素蛋白复合体的结构与功能（Liu et al.，2004；Pan et al.，2018，2020；Pi et al.，2018；Qin et al.，2015，2019；Su et al.，2017；Wei，2016；Yu et al.，2018；Zhang et al.，2017，2020）、C_4光合作用（Chen et al.，2014；Jiang et al.，2016；Wang et al.，2014）、Rubisco结构与功能（Gao et al.，2016）、光合作用光系统功能调控及建成（Jin et al.，2018；Liu et al.，2012；Ouyang et al.，2017；Pi et al.，2018；Zhang et al.，2014）、叶绿体信号转导（Chi et al.，2013；Ouyang et al.，2020）、光合作用系统生物学（Xiao et al.，2017）、光合作用合成生物学（Shen et al.，2019；Zhang et al.，2015）等研究领域均获得了较大进展。

近年来，在与创制未来作物相关的光合作用改造领域，国际相关研究团队发展快速。以比尔-梅琳达·盖茨基金会（Bill & Melinda Gates Foundation）支持的国际C_4水稻项目（http://c4rice.com）及以C_3改良为核心的RIPE（Realizing Improved Photosynthetic Efficiency）项目（http://ripe.uiuc.edu）为支点，国际上已经建立了多个光合作用研究高地以及高度合作的国际研究团队，创建了开展光合作用改造研究的关键工具、平台和资源。值得一提的是，美国伊利诺伊大学及加州大学伯克利分校的RIPE项目研究团队验证了光恢复速度的关键基因可以有效提高光能利用效率，启动了利用该基因的育种实践（Kromdijk et al.，2016）。同时，科研工作者在创建光呼吸支路（South et al.，2019）、构建包含仅 4 个蛋白质的最小羧体（Long et al.，2018）等方面取得重大进展，这标志着利用合成生物学手段，改造植物光能利用效率从理论及实践上都是可行的。在 C_4 水稻项目中，目前已经明确可以激活维管束鞘细胞的关键转录因子 GOLDEN-2 LIKE 基因，这为开展 C_4 代谢改造提供了底盘（Wang et al.，2017）。与此同时，以色列威兹曼研究院的 Ron Milo 教授领导的研究团队在构建光合自养大肠杆菌创建方面获得长足进展，实现了利用 CO_2 支持大肠杆菌自养生存（Antonovsky et al.，2016；Gleizer et al.，2019），为构建光合自养型工业微生物跨出实质性一步。这些进展表明，光合合成生物学正在成为当前光合作用研究国际竞争的焦点。

在光合作用改造研究领域，我国科研工作者经过多年努力，在国际上占据领先或并跑地位。在国家相关经费支持下，尤其是在中国科学院战略性先导科技专项支持下，开展了光合作用系统模型构建工作，建立了从分子、细胞器、细胞、叶片、冠层乃至生态系统水平上的系统模型，这对光合作用系统改造具有重要的支持作用，奠定了我国在该领域的领先地位（Xiao et al.，2017）。另外，我国科研工作者与国际同行创办了专业学术杂志 *in silico Plants*，用以支持该领域的可持续性发展。在光合作用系统改造实践方面，迄今我国科研工作者在光呼吸支路改造、藻胆体重建、光合代谢途径改造等方面也开展了系统研究（Feng et al.，2007；Tang et al.，2015；Xin et al.，2015；Xu et al.，2009；Zhao et al.，2017；Shen et al.，2019）。

2 未来发展趋势及关键突破口

2.1 我国在光合作用改造研究方面存在短板

我国在光合作用改造相关领域目前存在明显短板。首先，我国光合作用研究跟实际应用间距离仍较大。值得一提的是，过去我国光合作用研究与实际生产联系比较紧密。例如，基于光合作用光合磷酸化的基础理论研究成果，我国科研工作者在 20 世纪 80 年代就发现喷洒亚硫酸氢钠，可以有效提高叶片环式电子传递，增加 ATP 供给，从而有效提高冠层光合作用效率及作物产量（魏家绵等，1989）。我国科研工作者发现籼稻和粳稻的不同品种间光抑制特性存在着差异，且粳稻对光抑制更耐受，其杂交一代的光抑制特性偏向母本（匡廷云，2004），因此提出培育耐受光抑制的籼粳杂交稻、筛选强耐受光抑制的品种为母本是籼粳亚种配组的重要原则。屠曾平（1997）利用抗光氧化的美国水稻品种'Lemont'和光抑制严重的广东品种'七桂早'杂交育出了整体光合能力比其亲本都有很大改进的杂交稻品种。中国科学院李振声院士领导的科研团队将抗光氧化能力强的小麦品种'小偃54'和'京411'杂交，培育出了光合效率高、适应范围广、品质优良的'小偃81'小麦新品种。这些都是历史上光合作用研究与生产紧密联系成功的例子。然而，从1997年至今，尽管我国在光合作用基础理论研究方面取得的进展很多，但是这些研究成果在农业生产等实用领域的应用较少。未来光合作用研究急需发展能支持作物高光效育种的新理论、技术及手段。

2.2 未来发展趋势

未来光合作用改造研究的一个主要趋势是利用合成生物学技术，人工干预、优化及改造光合系统，提高光能利用效率，从而为人类更好地提供粮食及能源（Long et al.，2015；Bailey-Serres et al.，2019；Zhu et al.，2020）。当前植物遗传学、作物基因组学、合成生物学等领域发展迅速，为开展光合作用改造研究提供了关键基因、研究技术及改造平台（Huang and Han，2014；Shao et al.，2018）。同时，当前对光合作用的新认识也为大规模开展光合作用改造研究提供了坚实的理论基础。

光合作用需要CO_2、水和光能，这些因子的变化影响光合效率。在长期自然选择过程中，为适应特定的CO_2、光、温等环境条件，植物需要在蛋白复合体结构、代谢结构与调控、细胞结构、叶片结构等不同层次形成特定的特征，以实现光能利用效率的最大化或最优化。然而，在作物驯化过程中，农田环境中的作物生长，一般处于群体环境下，与其祖先所处个体生长环境差别巨大（Bailey-Serres et al.，2019；Long et al.，2015）；近200年来，全球气候变化剧烈，空气CO_2浓度从200 ppm增加到约400 ppm，气温升高$1\sim 2$℃（Pachauri et al.，2014），这些外界环境快速变化远超于光合作用的适应或进化速度，使得现有植物或作物光合器官不具备实现最高光能利用效率的最优光合器官特征。设计并创造更高效光合器官，使其更加适合于当前环境条件，是作物高光效改良及育种的重要方向。进一步讲，为了适应特定 CO_2、光及温度环境，各类植物（包括高等植物和低等植物）也进化出多种多样的高效光能吸收、传递、转化途径。例如，蓝藻进化出具有高效光能捕获能力的藻胆体，蓝藻、真核藻类乃至高等 C_4 植物进化出高效 CO_2 浓缩机制等。这些途径如果在 C_3 作物中得以重建，必将大幅度提高作物产量潜力（Kubis and Bar-Even，2019；Zhu et al.，2020）。

最后，利用合成生物学手段，设计并创造当前自然界不存在的光合天线系统、CO_2 固定系统，也成为当前光合作用改造研究的重要前沿领域之一。以色列科学家 Ron Milo 团队利用人工进化方法，创造了具有光合自养能力的大肠杆菌（Antonovsky et al.，2016；Gleizer et al.，2019）；该团队还利用计算机模拟手段，设计出具备 CO_2 固定能力但不需要 Rubisco 参与的全新 CO_2 固定通路（Bar-Even et al.，2010）。当前材料科学的发展也为创造全新光能吸收、转化材料提供了新思路。例如，上转光（upper-conversion）纳米材料能够将短波长、高能量的光转化成波长为 400～700 nm 的可见光，从而使得光合作用更加有效地利用环境光（Dong et al.，2019）。纳米材料与生物系统的耦联则能够实现全新的自然界尚不存在的杂化系统。例如，利用无机材料硫化镉向不具备光合作用能力的细菌传递光能，从而使细菌能够利用光能固定 CO_2 并转化为醋酸（Sakimoto et al.，2016）。可以预期，随着具有特定光子吸收及释放能力、物质吸附及释放特性、氧化还原特性的各类新材料的发展及其与光合作用器官的耦合，具有更高的光能转化能力的生物-无机杂化系统将被研发出来，从而大大增强光合作用提供人类粮食及能源的能力。

2.3 关键突破口

基于以上讨论，光合作用的改造及优化研究将在以下几个方面集中突破。

现有光合自然变异的改造及优化。利用自然界现有光合系统，挖掘当前系统中控制光能利用效率的关键遗传变异，进而通过基因工程或者定向进化技术，改造、优化现有光合系统，提高植物光能利用率。

光合系统与植物基本代谢及源-库过程的耦合优化。光合作用并不是一个孤立的过程,而是与基本代谢、源-库过程紧密互作。研究光合作用与基本代谢、源-库过程的互作及调控关系,实现对这些互作及调控关系的有效改造,是提高光能利用率的重要手段。

自然界已有高光效途径的跨物种重建。将自然界已有的高效光合系统在作物中进行重建,这包括创建具有更广光谱吸收特征的天线系统、创建各类 CO_2 浓缩机制,包括基于羧体的 CO_2 蓝藻浓缩机制、基于蛋白质体的真核藻类 CO_2 浓缩机制及基于叶片花环结构的 C_4 CO_2 浓缩机制等。

自然界未有光合途径的设计及再造。充分利用当前合成生物学手段及材料科学领域的研究进展,设计自然界尚不存在的全新光合元件、通路及系统,增强植物利用、吸收及转化光能的能力,以及对 CO_2 的固定能力,大幅提高作物光合作用效率。

3 阶段性目标

3.1 2035 年目标

至 2035 年,光合作用相关领域的研究目标主要如下。

实现针对现有光合途径自然变异的改造及优化。优化 Rubisco 的动力学参数、ATP 合成酶结构、光合系统天线大小、叶片叶肉细胞的大小及分布、NPQ 动态变化、气孔分布及动态变化、碳代谢酶的含量及调控模式等。

优化光合作用与植物基本代谢及源-库过程的互作。优化光合代谢与呼吸代谢、氮代谢的互作,优化光合产物的运输、存储、分配及利用模式,优化光合产物对光合机器的抑制效应等。

实现针对自然界已有光合代谢通路的改造。针对 C_3、C_4 及景天酸代谢途径,设计并创造新的代谢通路,实现对这些代谢通路的优化改造,提高光能、CO_2 及各类营养元素的利用效率;在高等植物中建立叶绿素 d 及叶绿素 f 的合成途径。

构建人工合成材料与光合器官的有机整合,提高光能及 CO_2 利用效率。创建具备吸收利用非可见光谱的全新材料,创建光敏型 CO_2 吸附材料,实现与光合器官的有机整合,构建具有更高光能利用率的无机-有机杂合光合系统。

3.2 2050年目标

至2050年，光合作用相关领域的研究目标主要如下。

实现自然界已有高效光合元件或途径的跨物种重建。实现自然界中各类高光效光合元件、途径及系统在C_3作物中的重建及整合。

设计并创造材料-生物杂化系统。创制新材料，实现光能吸收、传递、转化等功能，并将该材料与光合生物材料结合，建立全新、高效的材料-生物器官混化系统，提升光合效率。

设计并构建全新光合途径。设计并实现全新光合元件、途径及系统，实现高效光能吸收、传递及转化，有效捕捉太阳能生产化学能，最大化能源及粮食生产潜力。

参考文献

匡廷云 (2004). 作物光能利用效率与调控. 济南：山东科学技术出版社.

屠曾平 (1997). 水稻光合特性研究与高光效育种. 中国农业科学 3, 28-35.

魏家绵, 沈允钢, 李德耀, 徐春和 (1989). 亚硫酸氢钠在低光强下对叶绿体循环光合磷酸化的促进作用. 植物生理学报 1, 101-104.

Antonovsky, N., Gleizer, S., Noor, E., Zohar, Y., Herz, E., Barenholz, U., Zelcbuch, L., Amram, S., Wides, A., Tepper, N., et al. (2016). Sugar synthesis from CO_2 in *Escherichia coli*. Cell 166, 115-125.

Bailey-Serres, J., Parker, J.E., Ainsworth, E.A., Oldroyd, G.E.D., and Schroeder, J.I. (2019). Genetic strategies for improving crop yields. Nature 575, 109-118.

Bar-Even, A., Noor, E., Lewis, N.E., and Milo, R. (2010). Design and analysis of synthetic carbon fixation pathways. Proc. Natl. Acad. Sci. USA 107, 8889-8894.

Chen, Y., Lu, T., Wang, H., Shen, J., Bu, T., Chao, Q., Gao, Z., Zhu, X., Wang, Y., and Wang, B. (2014). Posttranslational modification of maize chloroplast pyruvate orthophosphate dikinase reveals the precise regulatory mechanism of its enzymatic activity. Plant Physiol. 165, 534-549.

Chi, W., Sun, X., and Zhang, L. (2013). Intracellular signaling from plastid to nucleus. Annu. Rev. Plant Biol. 64, 559-582.

Dong, R., Li, Y., Li, W., Zhang, H., Liu, Y., Ma, L., Wang, X., and Lei, B. (2019). Recent developments in luminescent nanoparticles for plant imaging and photosynthesis. J. Rare Earths *37*, 903-915.

Feng, L., Wang, K., Li, Y., Tan, Y., Kong, J., Li, H., Li, Y., and Zhu, Y. (2007). Overexpression of SBPase enhances photosynthesis against high temperature stress in transgenic rice plants. Plant Cell Rep. *26*, 1635-1646.

Gao, X., Hong, H., Li, W., Yang, L., Huang, J., Xiao, Y., Chen, X., and Chen, G. (2016). Downregulation of rubisco activity by non-enzymatic acetylation of RbcL. Mol. Plant *9*, 1018-1027.

Gleizer, S., Ben-Nissan, R., Bar-On, Y.M., Antonovsky, N., Noor, E., Zohar, Y., Jona, G., Krieger, E., Shamshoum, M., Bar-Even, A., et al. (2019). Conversion of *Escherichia coli* to generate all biomass carbon from CO_2. Cell *179*, 1255-1263.

Huang, X., and Han, B. (2014). Natural variations and genome-wide association studies in crop plants. Annu. Rev. Plant Biol. *65*, 531-551.

Jiang, L., Chen, Y., Zheng, J., Chen, Z., Liu, Y., Tao, Y., Wu, W., Chen, Z., and Wang, B. (2016). Structural basis of reversible phosphorylation by maize pyruvate orthophosphate dikinase regulatory protein. Plant Physiol. *170*, 732-741.

Jin, H., Fu, M., Duan, Z., Duan, S., Li, M., Dong, X., Liu, B., Feng, D., Wang, J., Peng, L., et al. (2018). LOW PHOTOSYNTHETIC EFFICIENCY 1 is required for light-regulated photosystem II biogenesis in *Arabidopsis*. Proc. Natl. Acad. Sci. USA *115*, E6075-E6084.

Kromdijk, J., Glowacka, K., Leonelli, L., Gabilly, S.T., Iwai, M., Niyogi, K.K., and Long, S.P. (2016). Improving photosynthesis and crop productivity by accelerating recovery from photoprotection. Science *354*, 857-861.

Kubis, A., and Bar-Even, A. (2019). Synthetic biology approaches for improving photosynthesis. J. Exp. Bot. *70*, 1425-1433.

Liu, J., Yang, H., Lu, Q., Wen, X., Chen, F., Peng, L., Zhang, L., and Lu, C. (2012). PSBP-DOMAIN PROTEIN1, a nuclear-encoded thylakoid lumenal protein, is essential for photosystem i assembly in *Arabidopsis*. Plant Cell *24*, 4992-5006.

Liu, Z., Yan, H., Wang, K., Kuang, T., Zhang, J., Gui, L., An, X., and Chang, W. (2004). Crystal structure of spinach major light-harvesting complex at 2.72 angstrom resolution. Nature *428*, 287-292.

Long, B., Hee, W.Y., Sharwood, R., Rae, B., Kaines, S., Lim, Y. L., Nguyen, N., Massey, B., Bala, S., von Caemmerer, S., et al. (2018). Carboxysome encapsulation of the CO_2-fixing enzyme Rubisco in tobacco chloroplasts. Nat. Commun. *9*, 3570.

Long, S.P., Marshall, A.M., and Zhu, X. (2015). Engineering crop photosynthesis and yield potential to meet global food demand of 2050. Cell *161*, 56-66.

Ouyang, M., Li, X., Zhang, J., Feng, P., Pu, H., Kong, L., Bai, Z., Rong, L., Xu, X., Chi, W., et al. (2020). Liquid-liquid phase transition drives intra-chloroplast cargo sorting. Cell *180*, 1144-1159.

Ouyang, M., Li, X., Zhao, S., Pu, H., Shen, J., Adam, Z., Clausen, T., and Zhang, L. (2017). The crystal structure of Deg9 reveals a novel octameric-type HtrA protease. Nat. Plants *3*, 973-982.

Pachauri, R.K., Allen, M.R., Barros, V.R., Broome, J., Cramer, W., Christ, R., Church, J.A., Clarke, L., Dahe, Q.D., Dasqupta, P., et al. (2014). IPCC, 2014: climate change 2014 synthesis report. Contribution of working groups I, II and III to the fifth assessment report of the intergovernmental panel on climate Change. IPCC. http://www.mendeley.com/research/climate-change-2014-synthesis-report-contribution-working-groups-i-ii-iii-fifth-assessment-report-in-20 [2020-7-23].

Pan, X., Cao, D., Xie, F., Xu, F., Su, X., Mi, H., Zhang, X., and Li, M. (2020). Structural basis for electron transport mechanism of complex I-like photosynthetic NAD(P)H dehydrogenase. Nat. Commun. *11*, 610-610.

Pan, X., Ma, J., Su, X., Cao, P., Chang, W., Liu, Z., Zhang, X., and Li, M. (2018). Structure of the maize photosystem I supercomplex with light-harvesting complexes I and II. Science *360*, 1109-1112.

Pi, X., Tian, L., Dai, H., Qin, X., Cheng, L., Kuang, T., Sui, S., and Shen, J. (2018). Unique organization of photosystem I-light-harvesting supercomplex revealed by cryo-EM from a red alga. Proc. Natl. Acad. Sci. USA *115*, 4423-4428.

Qin, X., Pi, X., Wang, W., Han, G., Zhu, L., Liu, M., Cheng, L., Shen, J., Kuang, T., and Sui, S. (2019). Structure of a green algal photosystem I in complex with a large number of light-harvesting complex I subunits. Nat. Plants *5*, 263-272.

Qin, X., Suga, M., Kuang, T., and Shen, J. (2015). Structural basis for energy transfer pathways in the plant PSI-LHCI supercomplex. Science *348*, 989-995.

Sakimoto, K.K., Wong, A.B., and Yang, P. (2016). Self-photosensitization of nonphotosynthetic bacteria for solar-to-chemical production. Science *351*, 74-77.

Shao, Y., Lu, N., Wu, Z., Cai, C., Wang, S., Zhang, L., Zhou, F., Xiao, S.B., Liu, L., Zeng, X., et al. (2018). Creating a functional single-chromosome yeast. Nature *560*, 331-335.

Shen, B., Wang, L., Lin, X., Yao, Z., Xu, H., Zhu, C., Teng, H., Cui, L., Liu, E., Zhang, J., et al. (2019). Engineering a new chloroplastic photorespiratory bypass to increase photosynthetic efficiency and productivity in rice. Mol. Plant *12*, 199-214.

Shen, Y.K., and Shen, G.M. (1962). Studies on photophosphorylation II. The "light intensity effect" and the intermediate steps of photophosphorylation. Acta Biochim. Biophys. Sin. *2*, 60-68.

South, P.F., Cavanagh, A.P., Liu, H.W., Ort, D.R. (2019) Synthetic glycolate metabolism pathways stimulate crop growth and productivity in the field. Science *363*, eaat9077.

Su, X., Ma, J., Wei, X., Cao, P., Zhu, D., Chang, W., Liu, Z., Zhang, X., and Li, M. (2017). Structure and assembly mechanism of plant C2S2M2-type PSII-LHCII supercomplex. Science *357*, 815-820.

Tang, K., Ding, W.L., Hoppner, A., Zhao, C., Zhang, L., Hontani, Y., Kennis, J.T.M., Gartner, W., Scheer, H., Zhou, M., et al. (2015). The terminal phycobilisome emitter, L-CM: a light-harvesting pigment with a phytochrome chromophore. Proc. Natl. Acad. Sci. USA *112*, 15880-15885.

Wang, P., Khoshravesh, R., Karki, S., Tapia, R., Balahadia, C.P., Bandyopadhyay, A., Quick, W.P., Furbank, R., Sage, T.L., and Langdale, J.A. (2017). Re-creation of a key step in the evolutionary switch from C3 to C4 leaf anatomy. Curr. Biol. *27*, 3278-3287.

Wang, Y., Long, S.P., Zhu, X. (2014). Elements required for an efficient NADP-Malic enzyme type C_4 photosynthesis. Plant Physiol. *164*, 2231-2246.

Wei, X., Su, X., Cao, P., Liu, X., Chang, W., Li, M., Zhang, X., Liu, Z. (2016). Structure of spinach photosystem II-LHCII supercomplex at 3.2 Å resolution. Nature *534*, 69-74.

Xiao, Y., Chang, T., Song, Q., Wang, S., Tholen, D., Wang, Y., Xin, C., Zheng, G., Zhao, H., and Zhu, X. (2017). ePlant for quantitative and predictive plant science research in the big data era—lay the foundation for the future model guided crop breeding, engineering and agronomy. Quant. Biol. *5*, 260-271.

Xin, C., Tholen, D., Devloo, V., and Zhu, X. (2015). The benefits of photorespiratory bypasses: how can they work? Plant Physiol. *167*, 574-585.

Xu, H., Zhang, J., Zeng, J., Jiang, L., Liu, E., Peng, C., He, Z., and Peng, X. (2009). Inducible antisense suppression of glycolate oxidase reveals its strong regulation over photosynthesis in rice. J. Exp. Bot. *60*, 1799-1809.

Yu, L., Suga, M., Wang-Otomo, Z.Y., and Shen, J. (2018). Structure of photosynthetic LH1-RC supercomplex at 1.9 Å resolution. Nature *556*, 209-213.

Zhang, C., Chen, C., Dong, H., Shen, J., Dau, H., and Zhao, J. (2015). A synthetic Mn4Ca-cluster mimicking the oxygen-evolving center of photosynthesis. Science *348*, 690-693.

Zhang, C., Shuai, J., Ran, Z., Zhao, J., Wu, Z., Liao, R., Wu, J., Ma, W., and Lei, M. (2020). Structural insights into NDH-1 mediated cyclic electron transfer. Nat. Commun. *11*, 888.

Zhang, F., Tang, W., Hedtke, B., Zhong, L., Liu, L., Peng, L., Lu, C., Grimm, B., and Lin, R. (2014). Tetrapyrrole biosynthetic enzyme protoporphyrinogen IX oxidase 1 is required for plastid RNA editing. Proc. Natl. Acad. Sci. USA *111*, 2023-2028.

Zhang, J., Ma, J., Liu, D., Qin, S., Sun, S., Zhao, J., and Sui, S. (2017). Structure of phycobilisome from the red alga *Griffithsia pacifica*. Nature *551*, 57-63.

Zhao, C., Hoppner, A., Xu, Q., Gartner, W., Scheer, H., Zhou, M., and Zhao, K. (2017). Structures and enzymatic mechanisms of phycobiliprotein lyases CpcE/F and PecE/F. Proc. Natl. Acad. Sci. USA *114*, 13170-13175.

Zhu, X., Ort, D.R., Parry, M., and von Caemmerer, S. (2020). A wish list for synthetic biology in photosynthesis research. J. Exp. Bot. *71*, 2219-2225.

水资源利用效率与抗旱

熊立仲

水资源匮乏和作物生产用水之间的矛盾随着全球气候变化的加剧和人口不断增加而愈加突出。解决这一矛盾的有效办法之一是提高作物水资源利用效率，培育节水抗旱的作物新品种。提高作物水资源利用效率可以节约农业用水，为粮食生产提供保障。应用节水抗旱作物品种则可以有效减少极端环境下农作物的损失，尤其是保障缺水地区的粮食安全。

1 国内外研究进展

1.1 作物抗旱性的遗传学研究

作物叶片形态和生理特征常被作为评价避旱性的指标，渗透调节、细胞膜稳定性、脯氨酸和可溶性糖含量等一般被作为评价抗旱性的生理指标（Hu and Xiong, 2014）。Guo 等（2018）首次利用表型组平台对 529 份水稻品种开展无损表型鉴定并进行全基因组关联分析，一共扫描到 470 个关联位点。这些位点包括一些已报道的抗旱相关基因，也发现一个新的与抗旱性密切相关的基因，如 *OsPP15*。在玉米中，Wang 等（2016）利用全基因组关联分析鉴定到一个与抗旱相关基因 *ZmVPP1*，编码液泡型 H^+ 型焦磷酸酶，通过提高光合效率和根系发育来增强玉米的抗旱性。

根系在抗旱中起着至关重要的作用，根性状包括根体积、最大根长、根粗、深根比例、根的穿透力和根茎比等（Qu et al., 2008）。Li 等（2017a）通过全基因组关联分析对水稻成熟期干旱条件下的根系性状的遗传基础进行解析，并验证了 *Nal1* 和 *OsJAZ1* 等代表性候选基因控制根系相关性状的功能。目前通过正向遗传学手段克隆抗旱基因的一个代表成功例子是水稻 *DRO1* 基因。*DRO1* 控制根的生长角度，参与根尖细胞的伸长并引起根的向地性弯曲，在中度和深度干旱胁迫条件下，*DRO1* 近等基因系的单株产量显著高于受体亲本 IR64（Uga et al., 2013）。Li 等（2015）用 Yuefu/IL392 的 $F_{2:3}$ 群体定位到的 QTL（*qRT9*）分别解释了 32.5% 和 28.1% 的根粗和根长变异，其候选基因被推测是一个 bHLH 转录因子 OsbHLH120。

1.2 作物抗旱功能基因的鉴定

当植物面临干旱胁迫时，多种功能型蛋白和调控型蛋白参与多条路径来抵御干旱胁迫。其中功能蛋白涉及渗透物质合成、活性氧清除及物质运输等；调控型蛋白通过调控基因表达量、转录后的修饰、信号转导等途径参与抗旱过程。本节仅对其中一些代表性基因做简要介绍。

作物干旱应答中很多基因的激活依赖脱落酸（abscisic acid，ABA）信号，bZIP/AREB/ABF 类转录因子在依赖 ABA 的调控植物抗旱性的信号通路中起着非常关键的作用。在水稻中，bZIP 家族基因 *OsbZIP23* 超量表达植株对 ABA 的敏感性及抗旱性和耐盐性都显著提高（Xiang et al., 2008）。超量表达 *OsbZIP46CA1*（编码一个具有组成型活性的 bZIP 转录因子）也能够显著增强水稻的抗旱性（Tang et al., 2012a）。OsbZIP46 的互作蛋白 MODD（Mediator of OsbZIP46 Deactivation and Degradation）负调控 ABA 信号和耐旱性，其通过与 OsTPR3-HDA702 抑制因子复合物的相互作用，下调 OsbZIP46 靶基因的组蛋白乙酰化水平来抑制 OsbZIP46 活性，同时通过与 U-box 型 E3 泛素连接酶 OsPUB70 相互作用促进 OsbZIP46 降解，从而达到对 ABA 信号转导和干旱胁迫响应的精细调控（Tang et al., 2016）。超量表达 *OsbZIP16*（Chen et al., 2012）、*OsbZIP71*（Liu et al., 2014）、*OsbZIP12*（Zhang et al., 2017）的转基因植株的抗旱性也明显增强。在玉米中，*ZmbZIP4* 参与玉米根系的发育，提高植物对非生物胁迫的耐受能力（Ma et al., 2018）。

不依赖 ABA 的逆境应答途径主要由 AP2/ERF 家族成员 DREB1/CBF 和 DREB2 类转录因子介导（Todaka et al., 2015; Yoshida et al., 2014）。在水稻中，分别超量表达 *OsDREB1* 和 *DREB1* 基因均能提高耐旱性（Ito et al., 2006）。超量表达玉米 *ZmCBF3* 的转基因水稻对干旱、盐胁迫、低温的抗性都有显著增强（Xu et al., 2011）。Liu 等（2013）发现 *ZmDREB2.7* 与玉米苗期抗旱显著关联，*ZmDREB2.7* 的 DNA 多态性在基因的启动子区而非编码区，表明 *ZmDREB2.7* 可能是通过不同的表达模式来响应干旱；超量表达 *ZmDREB2.7*

的转基因拟南芥植株耐旱性增强。

NAC 转录因子是植物特有的一类转录因子。Hu 等（2006）分离出水稻第一个逆境响应的 NAC 转录因子 *SNAC1*，超量表达植株对 ABA 敏感性增强，干旱来临时超量表达植株通过关闭更多的气孔从而减少失水速率，但是光合效率却未受影响。此外，超量表达 *SNAC1* 的转基因小麦表现出耐旱和耐盐的能力显著增强（Saad et al.，2013）。在水稻中超量表达 *SNAC3*、*OsNAC5*、*OsNAC9* 和 *OsNAC10* 都可以一定程度上增强抗旱性和产量（Fang et al.，2015；Jeong et al.，2010，2013；Redillas et al.，2012）。然而，超量表达 *OsNAC6* 的转基因水稻虽表现出耐脱水性和耐盐胁迫增强的表型，但是植株生长迟缓、产量降低（Nakashima et al.，2007）。

Mao 等（2015）在玉米中发现 *ZmNAC111* 基因启动子区的 MITE 插入与玉米抗旱性的自然变异显著关联，提高 *ZmNAC111* 表达量可以增强苗期抗旱性和水分利用效率。在拟南芥中超量表达 *ZmSNAC1* 的转基因植株对 ABA 敏感性增强，在苗期耐旱性增强（Lu et al.，2012）。*ZmNAC55* 的表达受到干旱、高浓度的盐和 ABA 的诱导，超量表达 *ZmNAC55* 的转基因拟南芥植株展现出较强的抗旱能力（Mao et al.，2016）。超量表达 *TaNAC69* 增强了普通小麦中与胁迫相关基因的转录水平，进而增强小麦的抗旱性（Xue et al.，2011）。在烟草中超量表达 *TaNAC2a* 基因也增强了转基因植株的耐旱性（Tang et al.，2012b）。

多个来自 MYC/MYB 家族的转录因子也被证实是重要的抗旱调控基因。在水稻中超量表达 *MYB2* 使抗旱、耐盐、耐低温能力都有不同程度的增强（Yang et al.，2012）。*OsMYB48-1* 作为一个新的 MYB 相关转录因子，通过调控干旱胁迫下的 ABA 合成正调控抗旱（Xiong et al.，2014）。Chen 等（2015）报道了海岛棉中 R2R3-type MYB 转录因子基因 *GbMYB5*，超量表达该基因的转基因烟草表现出抗旱性增强。Shen 等（2012）发现超量表达了其他转录因子基因（*WRKY30*）的水稻植株的抗旱能力也显著增强。

OsSKIPa 是人类 SKIP 在水稻中的同源蛋白。组成型超量表达 *OsSKIPa* 的植株具有更强的活性氧清除能力及更高的逆境相关基因转录水平，从而在生育期表现较强的耐旱性（Hou et al.，2009）。

受体类激酶（receptor-like kinase，RLK）也参与植物抗旱性调控。水

稻中 *OsSIK1* 超量表达能增强抗旱性和耐盐性，而基因敲出突变体 *sik1* 和 RNAi 植株对干旱和盐胁迫表现敏感（Ouyang et al.，2010）。水稻中另外一个 RLK 基因 *OsSIK2* 超量表达也能增加植株对干旱和盐的耐受性（Chen et al.，2013）。超量表达 *LRK2* 增强植物对干旱胁迫的耐受性，这可能与营养生长期侧根数目增加有关（Kang et al.，2017）。

植物钙依赖性蛋白激酶（CPK/CDPK）在多种应激反应中发挥作用。超量表达 *OsCPK4* 的转基因水稻植株的抗旱能力显著增强，*OsCPK4* 可能参与保护细胞膜免受胁迫诱导的氧化损伤（Campo et al.，2014）。*OsCPK10* 通过调节过氧化氢蛋白酶的积累，提高水稻植株的抗氧化能力，从而产生耐旱性（Bundó and Coca，2017）。超量表达 *OsCDPK1* 的转基因植株表现出一定程度的耐旱性，并能激活 *GF14c* 的表达；而且超量表达 *GF14c* 的转基因植株也表现出一定程度的耐旱性（Ho et al.，2013）。

PP2C 蛋白磷酸酶家族成员广泛分布于植物生长发育及逆境应答过程中（Lammers and Lavi，2007）。其中 A 亚家族成员多参与到 ABA 信号途径的 PYL-SnRK-PP2C 信号转导复合体中，负调控植物的逆境应答（Bai et al.，2013；Fujii et al.，2009；Komatsu et al.，2013；Soon et al.，2012）。最近在水稻中发现 F2 亚家族成员 *OsPP18* 在超量表达时可以引起 ROS 清除相关基因的表达量及 ROS 清除酶活性上调，导致植株在逆境下 ROS 积累量下降，从而增强水稻的抗氧化胁迫能力（You et al.，2014）。Xiang 等（2017）通过候选基因关联分析在玉米中鉴定到一个 PP2C-A 亚家族成员 *ZmPP2C-A10* 通过负调控 ABA 信号转导从而降低玉米苗期的抗旱性；该基因 5′-UTR 的顺式元件 ERSE（endoplasmic reticulum stress response element）的缺失导致了表型的变异。

1.3 作物节水抗旱遗传改良

尽管目前作物新品种审定时并没有抗旱性这一指标，但提高作物抗旱性一直是育种工作者关注的重要目标。以水稻为例，上海农业基因中心等单位近十多年来致力于培育节水抗旱稻（water-saving and drought-resistance rice，WDR）并取得了可喜的进展。节水抗旱稻主要特征是具备较高抗旱性（包括避旱性和耐旱性）和较高的水分利用效率，同时保证水稻的高产优质特性。

WDR在灌溉条件下，在保证米质和产量的同时可节水50%以上，从而大量节约农业用水（Luo，2010）。我国已制定了严格的WDR鉴定指标，近年来多个单位培育了一批具有WDR特征的新品种。以'旱优73'为例，该品种节水抗旱性强，产量高，适应性广，稻米品质经农业部（现农业农村部）稻米及制品质量监督检验测试中心检测，达到国家二级优质米标准（罗利军，2018）。2016年4月，中华人民共和国农业部正式发布实施《节水抗旱稻术语》行业标准（NY/T 2862—2015），为WDR选育确定了明确指标。

从事作物抗逆基础和应用基础研究的人员一直在尝试利用现代分子生物学技术克隆和挖掘作物节水抗旱基因，开发节水抗旱基因位点紧密连锁的分子标记，或利用转基因技术进行作物节水抗旱性遗传改良。随着研究的深入和技术的发展，生物技术与常规育种相结合将成为培育节水抗旱作物新品种的重要手段。

2 未来发展趋势与关键突破口

人口的压力和气候环境的变化是粮食生产面临的巨大挑战，因此作物对不利环境适应性的遗传改良显得尤为重要。得益于快速发展的现代生物技术，作物水分利用效率和抗旱特性研究不断取得进展，节水抗旱新品种（如节水抗旱稻）也展示出广阔的应用前景。针对国家绿色农业和可持续发展的需求，以及人们对餐桌食物品质追求的逐步提高，未来作物育种的主要发展趋势是整合绿色性状与优质性状，培育"资源节约、优质高产、环境广适"的新型作物。

2.1 作物水分利用效率研究趋势与突破

光合速率、气孔导度、叶肉导度、冠层结构等多种性状直接影响着水分利用效率（water use efficiency，WUE），已经证明通过育种和现代生物技术可以提高作物WUE，科学家们在作物表型、数量遗传、分子生理和建模研究方面已经取得了很多进展（Leakey et al.，2019）。作物叶片水平的WUE变化主要归因于叶片的净光合速率和气孔导度的变化，与作物生理生态和叶片结构性状有关，受气候因子（光照、水分、温度）等影响，同时作物WUE

是一个可以遗传的复杂性状，受多基因控制，并表现出一定的遗传变异性状（Flexas et al.，2016）。未来作物水资源利用效率有望在以下几个方面获得更多的进展和突破。

生理生态和遗传调控 作物叶片的光合器官对环境变化较为敏感，对作物 WUE 具有指示作用。首先，气孔密度、气孔大小、保卫细胞形态、气孔导度是调节作物 WUE 的关键（Dunn et al.，2019）。C_3 和 C_4 植物内源 WUE（iWUE）的遗传变异主要是由气孔导度的变化驱动的，近期高通量表型分析和 WUE 组分性状建模的发展为不同环境基因型与表型关系的研究创造新的机会（Leakey et al.，2019）。因此，寻找气候变化条件下可以有效增加光合作用和 WUE 的关键因素是以后研究的重要方向。其次，在气孔导度和光合羧化酶等方面，需要结合已有研究成果，通过遗传变异挖掘和生理生化机制的研究，从叶绿体和叶肉细胞的结构特性、Rubisco 大小亚基活性等方面来探讨作物 WUE 的提高。

多时空尺度及研究方法 作物 WUE 的研究随着现代技术的发展，在叶片尺度、单株尺度、群体尺度，甚至农田区域尺度均可进行直接观测研究。气体交换法常用于叶片尺度 WUE 的瞬时测定，便于研究作物的短期水分利用状况。近年来，涡度相关技术快速发展使得农田尺度 WUE 的观测难度逐渐减少；结合遥感监测，尤其是无人机技术监测 WUE 各种组分性状，农田尺度 WUE 的研究将是近期研究的热点。稳定性碳同位素技术能够定量跟踪各个生态系统中不同生态过程中碳同化效率，并且能够排除干扰。该技术虽然很早被提出可以作为一种可靠的研究手段，但该技术成本投入过高，一些学者正努力寻找其替代方法。因此，利用现代技术对不同尺度条件下作物 WUE 研究方法的开发依然有很大的提升空间。

尺度传递 尺度传递是指作物不同尺度下 WUE 相互表征的方法。目前针对各种尺度作物 WUE 的观测方法、评估模型和调控机制等均有较为成熟的理论研究成果，但是各尺度之间的传递与拓展未有深入研究。作物各个尺度 WUE 的研究方法并不能相互代替使用，各个尺度之间的相关性变化较大，严重限制了大田条件下的大批量材料筛选和鉴定的准确性和效率。进行尺度传递和拓展上的研究能深入了解作物因环境变化的响应机制，对于提高农业 WUE 意义重大。但由于作物 WUE 受到诸多环境因子影响，目前仍缺少被广泛认可的拓展方法。同时还应完善作物 WUE 测定方法，在各个尺度试验观测的数据基础上建立作物叶片、单株、群体乃至区域多尺度联系的模型将是未来作物 WUE 尺度传递研究的发展方向。

2.2 作物节水抗旱研究发展趋势和突破

在水资源短缺条件下，高 WUE 可以提高作物产量，但 WUE 和抗旱性是显著不同但彼此相关的性状（Leakey et al., 2019）。未来作物要实现节水抗旱特性培育，系统开展作物节水抗旱性状的功能基因组研究可能是突破口之一。目前作物节水抗旱特性面临诸多科学问题需要进行深入研究，如作物如何感受缺水或渗透胁迫信号，是否存在早期缺水或渗透胁迫感受器；干旱应答调控和植物激素以及生长发育信号之间的交叉（crosstalk），特别是作物如何"判断形势"（如干旱形成的强度、快慢、发生时期等），从而调整生长发育进程，实现繁种或把产量损失降到最低；作物干旱逆境与其他非生物逆境以及生物逆境信号之间的关系；抗旱性与节水的关系。

作物抗旱性的鉴定方法和指标　文献报道了大量的不同作物抗旱性的鉴定方法和指标，其至同一作物在不同发育阶段或不同种植环境条件下也有很多不同的鉴定方法和指标。因此，针对特定作物、特定环境，以及特定研究目的（如基因功能鉴定、遗传分析和育种应用等），建立可以通用的鉴定方法和指标体系显得非常迫切。同时，研发建立高通量、高精度的抗旱表型性状自动化测定技术和平台也是开展抗旱功能基因组研究非常重要的支撑。

作物抗旱性的遗传基础和分子机制　人工构建分离群体的遗传连锁分析和自然变异群体的全基因组关联分析已被广泛应用于抗逆主效 QTL/基因的定位与克隆。挖掘重要抗旱基因、解析抗旱性遗传调控网络、开展优异等位基因型的发掘和利用是未来的研究重点。特别是要鉴定出更多的作物在驯化过程中和不同生态条件下产生变异的重要抗逆基因，这对于揭示关键的抗逆机制和寻找抗逆遗传改良途径都非常重要。另外，多种组学技术（包括转录组、代谢组、蛋白组、表观组、离子组等）已越来越多地被广泛用到作物非生物逆境抗性研究，但如何分析和整合这些"组学"大数据，获取作物干旱应答和适应的关键途径，从而全面揭示作物抗旱性的分子机制已成为一个新的挑战。

基于功能基因组学的作物抗旱分子育种　当前作物抗旱遗传改良主要依赖于传统育种。虽然基于抗旱 QTL 的分子辅助选择或转基因抗旱（如转基因抗旱玉米）也有成功报道，但总体上看，抗旱功能基因组研究成果在作物抗旱新品种培育的应用还非常有限。因此，如何将抗旱功能基因组研究与遗传改良有机结合是下一阶段需要重点关注的问题。

3 阶段性目标

3.1 2035 年目标

至 2035 年，作物 WUE 研究有望能突破目前单一尺度的研究范畴，并进行有效的多尺度关联和尺度传递；同时建立通用的抗旱特性的鉴定方法和指标体系，揭示作物节水抗旱的机理并应用于新品种的培育。

基本建立作物 WUE 多时空尺度关联分析模型　以目前各种尺度作物 WUE 较为成熟的理论研究成果为基础，结合稳定碳同位素技术或替代技术，在多种时空尺度上整合作物生理生态过程对环境因子变化的响应，建立多个时空尺度的作物 WUE 关联模型，对不同尺度下作物 WUE 相关性状进行快速分析。

初步建立作物 WUE 尺度传递和拓展方法　在关联分析模型的基础上，将叶片尺度、单株尺度、群体尺度和农田区域尺度作物 WUE 研究成果相互表征和验证，初步建立较为统一的尺度传递和拓展方法，并指导实际生产。

建立作物抗旱特性的鉴定方法和指标　对特定作物、特定逆境及特定研究目的（如基因功能鉴定、遗传分析、育种应用等），建立业内可以通用的鉴定方法和指标体系。

基本完整地揭示主要作物节水抗旱性的遗传和分子机制　根据报道的作物抗旱性遗传基础以及与作物抗旱性相关基因的功能看出，抗旱性涉及多种不同的机制，遗传基础复杂，成百上千的基因直接或间接地参与了干旱应答与适应过程。因此，根据特定作物挖掘关键的途径或基因仍然是未来一段时间内主要工作。在此基础上整合多学科知识和技术开展研究，到 2035 年前基本完整地揭示主要作物节水抗旱性的遗传和分子机制。

应用分子设计、基因组选择和基因组编辑技术培育节水抗旱性增强的作物新品种 到 2035 年，预期水稻新品种比当前品种生理节水（同等产量前提下少用水）30% 以上，小麦、玉米等旱地作物比当前品种生理节水 30% 以上；抗旱性平均提高 20% 以上（即相对于当前的优良抗旱品种在相同发育时期同等干旱胁迫条件下少减产 50% 以上）。

3.2 2050 年目标

至 2050 年，作物节水抗旱的研究应该建立了完善的标准、方法和模型等，能够结合快速发展的现代化生物技术揭示作物节水抗旱机制，培育的新型抗旱作物应该具备高效的节水抗旱特性。

建成完善的作物多尺度传递的 WUE 分析方法和平台，满足分析不同作物和不同尺度下 WUE 和作物性状的分析，建成快速的 WUE 分析模型。

完整地揭示主要作物节水抗旱性与其他重要农艺性状相互关系的内在分子机制，从不同尺度揭示作物适应气候和水资源变化的机制。

通过模块组装或从头合成节水抗旱性主要通路的基因组，精准高效地创制适应未来气候变化的新型作物，其节水抗旱能力进一步大幅提高。到 2050 年，预期水稻新品种比当前品种生理节水 30% 以上，小麦、玉米等旱地作物比当前品种生理节水 40% 以上；抗旱性平均提高 60% 以上。

致谢：本章在撰写过程中得到了白宝伟和涂海甫的协助，特此致谢！

参考文献

罗利军 (2018). 节水抗旱稻的培育与应用. 生命科学 *30*, 1108-1112.

Bai, G., Yang, D.H., Zhao, Y., Ha, S., Yang, F., Ma, J., Gao, X.S., Wang, Z.M., and Zhu, J.K. (2013). Interactions between soybean ABA receptors and type 2C protein phosphatases. Plant Mol. Biol. *83*, 651-664.

Bundó, M., and Coca, M. (2017). Calcium-dependent protein kinase OsCPK10 mediates both drought tolerance and blast disease resistance in rice plants. J. Exp. Bot. *68*, 2963-2975.

Campo, S., Baldrich, P., Messeguer, J., Lalanne, E., Coca, M., and San Segundo, B. (2014). Overexpression of a calcium-dependent protein kinase donfers salt and drought tolerance in rice by preventing membrane lipid peroxidation. Plant Physiol. *165*, 688-704.

Chen, H., Chen, W., Zhou, J.L., He, H., Chen, L.B., Chen, H.D., and Deng, X.W. (2012). Basic leucine zipper transcription factor OsbZIP16 positively regulates drought resistance in rice. Plant Sci. *193*, 8-17.

Chen, L.J., Wuriyanghan, H., Zhang, Y.Q., Duan, K.X., Chen, H.W., Li, Q.T., Lu, X., He, S.J., Ma, B., Zhang, W.K., et al. (2013). An S-domain receptor-like kinase, OsSIK2, confers abiotic stress tolerance and delays dark-induced leaf senescence in rice. Plant Physiol. *163*, 1752-1765.

Chen, T.Z., Li, W.J., Hu, X.H., Guo, J.R., Liu, A.M., and Zhang, B.L. (2015). A cotton MYB transcription factor, GbMYB5, is positively involved in plant adaptive response to drought stress. Plant Cell Physiol. *56*, 917-929.

Dunn, J., Hunt, L., Afsharinafar, M., Al Meselmani, M., Mitchell, A., Howells, R., Wallington, E., Fleming, A.J., and Gray, J.E. (2019). Reduced stomatal density in bread wheat leads to increased water-use efficiency. J. Exp. Bot. *70*, 4737-4748.

Fang, Y.J., Liao, K.F., Du, H., Xu, Y., Song, H.Z., Li, X.H., and Xiong, L.Z. (2015). A stress-responsive NAC transcription factor SNAC3 confers heat and drought tolerance through modulation of reactive oxygen species in rice. J. Exp. Bot. *66*, 6803-6817.

Ferrero-Serrano, A., and Assmann, S.M. (2016). The alpha-subunit of the rice heterotrimeric G protein, RGA1, regulates drought tolerance during the vegetative phase in the dwarf rice mutant *d1*. J. Exp. Bot. *67*, 3433-3443.

Flexas, J., Díaz-Espejo, A., Conesa, M.A., Coopman, R.E., Douthe, C., Gago, J., Gallé, A., Galmés, J., Medrano, H., Ribas-Carbo, M., et al. (2016). Mesophyll conductance to CO_2 and rubisco as targets for improving intrinsic water use efficiency in C3 plants. Plant Cell Environ. *39*, 965-982.

Fujii, H., Chinnusamy, V., Rodrigues, A., Rubio, S., Antoni, R., Park, S.Y., Cutler, S.R., Sheen, J., Rodriguez, P.L., and Zhu, J.K. (2009). *In vitro* reconstitution of an abscisic acid signalling pathway. Nature *462*, 660-664.

Guo, Z., Yang, W., Chang, Y., Ma, X., Tu, H., Xiong, F., Jiang, N., Feng, H., Huang, C., Yang, P., et al. (2018). Genome-wide association studies of image traits reveal genetic architecture of drought resistance in rice. Mol. Plant *11*, 789-805.

Ho, S.L., Huang, L.F., Lu, C.A., He, S.L., Wang, C.C., Yu, S.P., Chen, J.C., and Yu, S.M. (2013). Sugar starvation- and GA-inducible calcium-dependent protein kinase 1 feedback regulates GA biosynthesis and activates a 14-3-3 protein to confer drought tolerance in rice seedlings. Plant Mol. Biol. *81*, 347-361.

Hou, X., Xie, K.B., Yao, J.L., Qi, Z.Y., and Xiong, L.Z. (2009). A homolog of human ski-interacting protein in rice positively regulates cell viability and stress tolerance. Proc. Natl. Acad. Sci. USA *106*, 6410-6415.

Hu, H., and Xiong, L. (2014). Genetic engineering and breeding of drought-resistant crops. Annu. Rev. Plant Biol. *65*, 715-741.

Hu, H.H., Dai, M.Q., Yao, J.L., Xiao, B.Z., Li, X.H., Zhang, Q.F., and Xiong, L.Z. (2006). Overexpressing a NAM, ATAF, and CUC (NAC) transcription factor enhances drought resistance and salt tolerance in rice. Proc. Natl. Acad. Sci. USA *103*, 12987-12992.

Ito, Y., Katsura, K., Maruyama, K., Taji, T., Kobayashi, M., Seki, M., Shinozaki, K., and Yamaguchi-Shinozaki, K. (2006). Functional analysis of rice DREB1/CBF-type transcription factors involved in cold-responsive gene expression in transgenic rice. Plant Cell Physiol. *47*, 141-153.

Jeong, J.S., Kim, Y.S., Baek, K.H., Jung, H., Ha, S.H., Do Choi, Y., Kim, M., Reuzeau, C., and Kim, J.K. (2010). Root-specific expression of OsNAC10 improves drought tolerance and grain yield in rice under field drought conditions. Plant Physiol. *153*, 185-197.

Jeong, J.S., Kim, Y.S., Redillas, M.C.F.R., Jang, G., Jung, H., Bang, S.W., Choi, Y.D., Ha, S.H., Reuzeau, C., and Kim, J.K. (2013). OsNAC5 overexpression enlarges root diameter in rice plants leading to enhanced drought tolerance and increased grain yield in the field. Plant Biotechnol. J. *11*, 101-114.

Kang, J., Li, J., Gao, S., Tian, C., and Zha, X. (2017). Overexpression of the leucine-rich receptor-like kinase gene *LRK2* increases drought tolerance and tiller number in rice. Plant Biotechnol. J. *15*, 1175-1185.

Komatsu, K., Suzuki, N., Kuwamura, M., Nishikawa, Y., Nakatani, M., Ohtawa, H., Takezawa, D., Seki, M., Tanaka, M., Taji, T., et al. (2013). Group A PP2Cs evolved in land plants as key regulators of intrinsic desiccation tolerance. Nat. Commun. *4*, 2219.

Lammers, T., and Lavi, S. (2007). Role of type 2C protein phosphatases in growth regulation and in cellular stress signaling. Crit. Rev. Biochem. Mol. *42*, 437-461.

Leakey, A.D.B., Ferguson, J.N., Pignon, C.P., Wu, A., Jin, Z., Hammer, G.L., and Lobell, D.B. (2019). Water use efficiency as a constraint and target for improving the resilience and productivity of C_3 and C_4 crops. Annu. Rev. Plant Biol. *70*, 781-808.

Li, H.W., Zang, B.S., Deng, X.W., and Wang, X.P. (2011). Overexpression of the trehalose-6-phosphate synthase gene *OsTPS1* enhances abiotic stress tolerance in rice. Planta *234*, 1007-1018.

Li, J., Li, Y., Yin, Z., Jiang, J., Zhang, M., Guo, X., Ye, Z., Zhao, Y., Xiong, H., Zhang, Z., et al. (2017b). OsASR5 enhances drought tolerance through a stomatal closure pathway associated with ABA and H_2O_2 signalling in rice. Plant Biotechnol. J. *15*, 183-196.

Li, J.Z., Han, Y.C., Liu, L., Chen, Y.P., Du, Y.X., Zhang, J., Sun, H.Z., and Zhao, Q.Z. (2015). *qRT9*, a quantitative trait locus controlling root thickness and root length in upland rice. J. Exp. Bot. *66*, 2723-2732.

Li, L., Du, Y., He, C., Dietrich, C.R., Li, J., Ma, X., Wang, R., Liu, Q., Liu, S., Wang, G., et al. (2019). The maize *glossy6* gene is involved in cuticular wax deposition and drought tolerance. J. Exp. Bot. *70*, 3089-3099.

Li, X., Guo, Z., Lv, Y., Cen, X., Ding, X., Wu, H., Li, X., Huang, J., and Xiong, L. (2017a). Genetic control of the root system in rice under normal and drought stress conditions by genome-wide association study. PLoS Genet. *13*, e1006889.

Liu, C.T., Mao, B.G., Ou, S.J., Wang, W., Liu, L.C., Wu, Y.B., Chu, C.C., and Wang, X.P. (2014). OsbZIP71, a bZIP transcription factor, confers salinity and drought tolerance in rice. Plant Mol. Biol. *84*, 19-36.

Liu, S.X., Wang, X.L., Wang, H.W., Xin, H.B., Yang, X.H., Yan, J.B., Li, J.S., Tran, L.S.P., Shinozaki, K., Yamaguchi-Shinozaki, K., et al. (2013). Genome-wide analysis of *ZmDREB* genes and their association with natural variation in drought tolerance at seedling stage of *Zea mays* L. PLoS Genet. *9*, e1003790.

Lu, M., Ying, S., Zhang, D.F., Shi, Y.S., Song, Y.C., Wang, T.Y., and Li, Y. (2012). A maize stress-responsive NAC transcription factor, ZmSNAC1, confers enhanced tolerance to dehydration in transgenic *Arabidopsis*. Plant Cell Rep. *31*, 1701-1711.

Luo, L.J. (2010). Breeding for water-saving and drought-resistance rice (WDR) in China. J. Exp. Bot. *61*, 3509-3517.

Ma, H., Liu, C., Li, Z., Ran, Q., Xie, G., Wang, B., Fang, S., Chu, J., and Zhang, J. (2018). ZmbZIP4 contributes to stress resistance in maize by regulating ABA synthesis and root development. Plant Physiol. *178*, 753-770.

Mao, H., Yu, L., Han, R., Li, Z., and Liu, H. (2016). ZmNAC55, a maize stress-responsive NAC transcription factor, confers drought resistance in transgenic *Arabidopsis*. Plant Physiol. Biochem. *105*, 55-66.

Mao, H.D., Wang, H.W., Liu, S.X., Li, Z., Yang, X.H., Yan, J.B., Li, J.S., Tran, L.S.P., and Qin, F. (2015). A transposable element in a *NAC* gene is associated with drought tolerance in maize seedlings. Nat. Commun. *6*, 8326.

Nakashima, K., Tran, L.S., Van Nguyen, D., Fujita, M., Maruyama, K., Todaka, D., Ito, Y., Hayashi, N., Shinozaki, K., and Yamaguchi-Shinozaki, K. (2007). Functional analysis of a NAC-type transcription factor OsNAC6 involved in abiotic and biotic stress-responsive gene expression in rice. Plant J. *51*, 617-630.

Ouyang, S.Q., Liu, Y.F., Liu, P., Lei, G., He, S.J., Ma, B., Zhang, W.K., Zhang, J.S., and Chen, S.Y. (2010). Receptor-like kinase OsSIK1 improves drought and salt stress tolerance in rice (*Oryza sativa*) plants. Plant J. *62*, 316-329.

Qu, Y., Mu, P., Zhang, H., Chen, C.Y., Gao, Y., Tian, Y., Wen, F., and Li, Z. (2008). Mapping QTLs of root morphological traits at different growth stages in rice. Genetica *133*, 187-200.

Redillas, M.C.F.R., Jeong, J.S., Kim, Y.S., Jung, H., Bang, S.W., Choi, Y.D., Ha, S.H., Reuzeau, C., and Kim, J.K. (2012). The overexpression of *OsNAC9* alters the root architecture of rice plants enhancing drought resistance and grain yield under field conditions. Plant Biotechnol. J. *10*, 792-805.

Saad, A.S.I., Li, X., Li, H.P., Huang, T., Gao, C.S., Guo, M.W., Cheng, W., Zhao, G.Y., and Liao, Y.C. (2013). A rice stress-responsive *NAC* gene enhances tolerance of transgenic wheat to drought and salt stresses. Plant Sci. *203*, 33-40.

Shen, H.S., Liu, C.T., Zhang, Y., Meng, X.P., Zhou, X., Chu, C.C., and Wang, X.P. (2012). OsWRKY30 is activated by MAP kinases to confer drought tolerance in rice. Plant Mol.

Biol. *80*, 241-253.

Soon, F.F., Ng, L.M., Zhou, X.E., West, G.M., Kovach, A., Tan, M.H.E., Suino-Powell, K.M., He, Y.Z., Xu, Y., Chalmers, M.J., et al. (2012). Molecular mimicry regulates ABA signaling by SnRK2 kinases and PP2C phosphatases. Science *335*, 85-88.

Tang, N., Ma, S., Zong, W., Yang, N., Lv, Y., Yan, C., Guo, Z., Li, J., Li, X., Xiang, Y., et al. (2016). MODD mediates deactivation and degradation of OsbZIP46 to negatively regulate ABA signaling and drought resistance in rice. Plant Cell *28*, 2161-2177.

Tang, N., Zhang, H., Li, X., Xiao, J., and Xiong, L. (2012a). Constitutive activation of transcription factor OsbZIP46 improves drought tolerance in rice. Plant Physiol. *158*, 1755-1768.

Tang, Y.M., Liu, M.Y., Gao, S.Q., Zhang, Z., Zhao, X.P., Zhao, C.P., Zhang, F.T., and Chen, X.P. (2012b). Molecular characterization of novel *TaNAC* genes in wheat and overexpression of TaNAC2a confers drought tolerance in tobacco. Physiol. Plantarum *144*, 210-224.

Todaka, D., Shinozaki, K., and Yamaguchi-Shinozaki, K. (2015). Recent advances in the dissection of drought-stress regulatory networks and strategies for development of drought-tolerant transgenic rice plants. Front. Plant Sci. *6*, 84.

Uga, Y., Sugimoto, K., Ogawa, S., Rane, J., Ishitani, M., Hara, N., Kitomi, Y., Inukai, Y., Ono, K., Kanno, N., et al. (2013). Control of root system architecture by *DEEPER ROOTING 1* increases rice yield under drought conditions. Nat. Genet. *45*, 1097-1102.

Wang, X., Wang, H., Liu, S., Ferjani, A., Li, J., Yan, J., Yang, X., and Qin, F. (2016). Genetic variation in *ZmVPP1* contributes to drought tolerance in maize seedlings. Nat. Genet. *48*, 1233-1241.

Xiang, Y., Sun, X., Gao, S., Qin, F., and Dai, M. (2017). Deletion of an endoplasmic reticulum stress response element in a *ZmPP2C-A* gene facilitates drought tolerance of maize seedlings. Mol. Plant *10*, 456-469.

Xiang, Y., Tang, N., Du, H., Ye, H.Y., and Xiong, L.Z. (2008). Characterization of OsbZIP23 as a key player of the basic leucine zipper transcription factor family for conferring abscisic acid sensitivity and salinity and drought tolerance in rice. Plant Physiol. *148*, 1938-1952.

Xiong, H., Yu, J., Miao, J., Li, J., Zhang, H., Wang, X., Liu, P., Zhao, Y., Jiang, C., Yin, Z., et al. (2018). Natural variation in increases drought tolerance in rice by inducing ROS scavenging.

Plant Physiol. *178*, 451-467.

Xiong, H.Y., Li, J.J., Liu, P.L., Duan, J.Z., Zhao, Y., Guo, X., Li, Y., Zhang, H.L., Ali, J., and Li, Z.C. (2014). Overexpression of OsMYB48-1, a novel MYB-related transcription factor, enhances drought and salinity tolerance in rice. PLoS One *9*, e92913.

Xu, M.Y., Li, L.H., Fan, Y.L., Wan, J.M., and Wang, L. (2011). *ZmCBF3* overexpression improves tolerance to abiotic stress in transgenic rice (*Oryza sativ*a) without yield penalty. Plant Cell Rep. *30*, 1949-1957.

Xue, G.P., Way, H.M., Richardson, T., Drenth, J., Joyce, P.A., and McIntyre, C.L. (2011). Overexpression of *TaNAC69* leads to enhanced transcript levels of stress up-regulated genes and dehydration tolerance in bread wheat. Mol. Plant *4*, 697-712.

Yang, A., Dai, X.Y., and Zhang, W.H. (2012). A R2R3-type MYB gene, *OsMYB2*, is involved in salt, cold, and dehydration tolerance in rice. J. Exp. Bot. *63*, 2541-2556.

Yao, L., Cheng, X., Gu, Z., Huang, W., Li, S., Wang, L., Wang, Y.F., Xu, P., Ma, H., and Ge, X. (2018). The AWPM-19 family protein OsPM1 mediates abscisic acid influx and drought response in rice. Plant Cell *30*, 1258-1276.

Yoshida, T., Mogami, J., and Yamaguchi-Shinozaki, K. (2014). ABA-dependent and ABA-independent signaling in response to osmotic stress in plants. Curr. Opin. Plant Biol. *21*, 133-139.

You, J., Zong, W., Hu, H.H., Li, X.H., Xiao, J.H., and Xiong, L.Z. (2014). A STRESS-RESPONSIVE NAC1-regulated protein phosphatase gene rice protein phosphatase18 modulates drought and oxidative stress tolerance through abscisic acid-independent reactive oxygen species scavenging in rice. Plant Physiol. *166*, 2100-2114.

Zhang, C., Li, C., Liu, J., Lv, Y., Yu, C., Li, H., Zhao, T., and Liu, B. (2017). The OsABF1 transcription factor improves drought tolerance by activating the transcription of COR413-TM1 in rice. J. Exp. Bot. *68*, 4695-4707.

Zhang, H., Xiang, Y., He, N., Liu, X., Liu, H., Fang, L., Zhang, F., Sun, X., Zhang, D., Li, X., et al. (2020). Enhanced vitamin C production mediated by an ABA-induced PTP-like nucleotidase improves drought tolerance of *Arabidopsis* and maize. Mol. Plant *13*, 760-776.

Zhu, D., Chang, Y., Pei, T., Zhang, X., Liu, L., Li, Y., Zhuang, J., Yang, H., Qin, F., Song, C., et al. (2020). MAPK-like protein 1 positively regulates maize seedling drought sensitivity by suppressing ABA biosynthesis. Plant J. *102*, 747-760.

Zhu, X., and Xiong, L. (2013). Putative megaenzyme DWA1 plays essential roles in drought resistance by regulating stress-induced wax deposition in rice. Proc. Natl. Acad. Sci. USA *110*, 17790-17795.

养分资源高效利用

徐国华 廖 红 陈彩艳 龚继明

1 国内外研究进展

我国土壤类型繁多,不同土壤类型的酸碱度(pH)、养分含量和有效性也不同。总体上,我国土壤 pH 从南向北递增,南方多酸性土壤,北方多碱性土壤,其中西北地区土壤碱性最强。在中碱性旱地土壤中,有效氮的形态主要为硝态氮,而在酸性土壤和稻田等淹水土壤中,有效氮的形态主要为铵态氮和硝态氮。土壤中的磷含量则从南到北、从东到西呈现逐渐增加的趋势。此外,酸性土壤由于 pH 低,常伴有缺磷及有毒离子镉、铝等生物有效性增加的问题,从而对作物产量和食品安全造成影响;而中碱性土壤由于盐基饱和度过高,常发生盐碱害及铁、锰、锌等微量元素缺乏的问题(Teng et al., 2014)。因此,多样化的土壤类型决定了作物养分效率改良策略的多样性。在土壤环境趋于恶化的背景下,综合考虑不同营养元素和非必需重金属元素间的相互作用,是植物养分效率改良必须考虑的问题之一。

养分高效利用是一个非常复杂的性状,所以相当长时期内的研究都主要在模式植物拟南芥中开展。但目前一个显著的趋势是向水稻、玉米等农作物转移。近年来,在氮磷钾营养信号调控、共生固氮、营养与器官发育、重金属吸收转运等方面均取得了突破性进展,鉴定了多个重要的养分和重金属转运蛋白,并且从细胞和植物个体水平深入解析了调控养分吸收转运的信号网络。根系与根际微生物互作机制、养分调控器官发育的机理及矿质元素之间互作的离子组学研究等新领域也取得了重要进展。

1.1 养分资源的吸收利用

农作物的高产优质离不开充分和平衡的养分供应,但是肥料的过量施用加剧了环境的负担,导致土壤 pH 下降、温室气体排放增加、水体富营养化、重金属污染等一系列生态环境问题。未来农业强调资源节约、环境友好，寄希望于通过大幅提高养分资源利用效率，在减少施肥量的同时保持和提高作物产量。因此，揭示植物对养分吸收利用的生理和分子机制，通过分子遗传学手段提高其利用效率，一直是植物营养生物学的重要研究内容，也是未来作物遗传改良的重要方向及实现农业可持续性发展的重要途径。科学家们对氮、磷、钾这三种重要的大量营养元素展开了深入的生理和分子机制研究，发现植物进化出了两套运输系统，分别负责土壤中营养元素充足和缺乏时的吸收（Luan et al., 2017; Puga et al., 2017; Tsay et al., 2007; Wu et al., 2013）。

在氮素吸收利用和氮与其他元素互作方面，科学家们已建立起从细胞膜硝酸盐受体及信号感知到细胞核的信号响应等主信号通路，阐明了植物协同利用氮磷实现营养平衡的分子机制（Hu and Chu, 2020; Hu et al., 2019; Maeda et al., 2018; Medici et al., 2015; Zhang et al., 2020a）。植物对氮、钾的吸收具有非常强的协同性（Roy et al., 2014; Xia et al., 2015），首先表现在根系氮和钾吸收的关键转运体受同一套蛋白磷酸化机制调控（Cubero-Font et al., 2016; Ho et al., 2009; Liu and Tsay, 2003; Xu et al., 2006），其次一些硝酸盐转运蛋白（如NRT1.5）能够同时参与对钾离子的分配（Drechsler et al., 2015; Li et al., 2017a; Lin et al., 2008; Meng et al., 2016）。此外，硝酸盐在植物体内的分配甚至和重金属镉具有正相关关系（Li et al., 2010; Luo et al., 2012; Mao et al., 2014）。

在磷素营养方面，克隆了磷代谢的中心调控因子，构建了磷信号调控的基本网络（Hu et al., 2011; Lv et al., 2014; Puga et al., 2017; Rubio et al., 2001; Wang et al., 2014; Wild et al., 2016）。找到了两条主要的信号转导途径，一条是miR399-PHO2-PHO1/Pht1途径，其作用在于增加磷吸收、木质部装载及往地上部运输；另一条是PHR1-PHF1-Pht1途径或者PHR1-miR827-NLA1-Pht1/SPX-MFS途径，可以增加植物磷吸收、转运及重新分配，并且诱导植物其他一系列缺磷响应基因的表达（Bari et al., 2006; Lin et al., 2010; Wang et al., 2012）。克隆了具有不同磷亲和力和生理功能的磷酸盐转运蛋白编码基因，揭示了这些转运蛋白的翻译后修饰调控（Ai et al., 2009; Chen et al., 2015, 2019; Chang et al., 2019; Li et al., 2015; Xie et al.,

2013)。在水稻中,发现了类似硫转运体的负责磷分配转运的蛋白(SPDT),它控制着磷在籽粒和茎叶中的分配,其突变之后可显著降低种子中的磷含量,但不影响产量和种子活力,为通过作物遗传改良减少土壤磷损失、降低水体富营养化提供了一个新途径(Hu and Chu,2017;Yamaji et al.,2017)。

科学家们在微量营养元素(如 Fe)的吸收转运和利用方面也做了大量工作,发现在单子叶植物和双子叶植物中分别存在两种不同的铁吸收策略(Curie and Briat,2003)——螯合策略和还原策略,即通过分泌麦根酸螯合土壤中不易吸收的三价铁,或者分泌氢离子并通过还原酶将三价铁活化并还原为易溶的二价铁。水稻及部分单子叶植物中同时拥有两套吸收系统(Ishimaru et al.,2006)。自第一个 Fe 营养调控转录因子 *Fer* 被鉴定以来(Ling et al.,2002),利用拟南芥和水稻作为模式植物,人们已经初步解析出植物中 Fe 营养的调控网络(Gao et al.,2020;Lei et al.,2020;Zhang et al.,2020b)。通过基因工程方法调控这些信号分子的表达能够增加作物的 Fe 利用效率,提高其营养价值。在野生二粒小麦(*Triticum trrgidum* var. *dicoccoides*)基因组中,科学家发现了一个 NAC 转录因子能调控 Fe 和 Zn 从叶片向籽粒迁移;将该转录因子从野生二粒小麦中导入到栽培小麦中能提高小麦的蛋白质、Fe 和 Zn 含量(Uauy et al.,2006),有意思的是,水稻中的一个 NAC 转录因子 *OsNAP* 也具有提高营养元素包括大量元素和微量元素由叶片向籽粒转运的功能(Liang et al.,2014)。在水稻中的研究发现两个在剑叶特异表达的液泡铁转运蛋白基因 *OsVIT1* 和 *OsVIT2* 参与调控铁的液泡区隔化,当这两个基因被敲除时,铁和锌向水稻籽粒迁移大幅增加,预示着一种新型的铁锌生物强化策略。但值得注意的是,通过这些策略籽粒铁锌富集的水稻在大田种植时也富集重金属镉,再次提醒铁锌生物强化必须考虑到多元素互作问题(Zhang et al.,2012)。

1.2 营养高效与环境互作

我国农业面临着可用耕地面积少、土壤质量差、极端气象灾害频发等一系列环境问题,极大地限制了作物的稳产高产。虽然目前以耐盐和耐旱为目标的分子育种已经选育出不少优异抗逆作物品种,但是在全球气候变化背景下,如何提高作物养分利用效率,选育出抗多重逆境且养分高效的新品种是未来可持续发展农业的关键。例如，大气二氧化碳浓度上升对 C_3 和 C_4 植物氮素的同化利用及其生长的影响取决于植物种类和氮素供应形态（Bloom et al.，2012；Rubio-Asensio and Bloom，2017）。植物氮素利用中有一个重要的生理现象，就是氮素吸收后一般会被运输到植物地上部位，进入叶绿体，通过 GS/GOGAT 循环直接利用光合作用提供的碳骨架、能量和还原力，将无机氮转化为有机氮，从而将光合作用和氮素同化耦联起来（Xu et al.，2012），这个生理现象被命名为光合氮素同化（nitrogen photo assimilation）。但在逆境情况下，包括低温、弱光、盐胁迫甚至重金属胁迫，光合作用和氮素同化的耦联关系会被解开，这个生理现象被命名为逆境诱导的硝酸盐再分配（stress-induced nitrate allocation to roots，SINAR）。研究表明，NRT1.5 和 NRT1.8 在此过程中起着非常重要作用，它们通过在多种逆境条件下的联动，精细调控硝酸盐在植物地上和地下部分的分配，从而调控植物在生长和逆境耐受间的能量分配和平衡（Chen et al.，2012；Li et al.，2010）。进一步研究还发现，乙烯和茉莉酸信号通路通过协同 NRT1.5/NRT1.8 的表达，有效整合了环境信号和养分信号间的协同（Chen et al.，2012；Zhang et al.，2014）。另有研究发现，一种南方油菜品种体内 SINAR 机制的弱化使其获得了更高的氮素利用效率。以上研究说明，通过新的技术手段，开展 SINAR 改良可能成为多变环境条件下提高氮素利用效率的一个有效手段（Han et al.，2016）。

1.3 有害元素的吸收分配

近年来，在有害元素的吸收、运输和解毒机制，以及其与必需元素互作机制等方面研究取得了一些突破性进展。锰离子吸收通道 OsNRAMP5 是镉进入水稻根部的主要通道，OsCd1 也参与水稻镉吸收和转运，将这两个基因敲除或者整合低表达等位基因都能显著降低稻米镉含量（Ishikawa et al.，2012；Sasaki et al.，2012；Yan et al.，2019）。HMA3 是一个液泡膜上的转运蛋白，能够将根系吸收的镉滞留在根液泡中从而降低籽粒和地上部位的镉含量（Chao et al.，2012；Maccaferri et al.，2019；Ueno et al.，2010）。CAL1 是定向调控镉向细胞外分泌的类防御素蛋白，专一地控制镉向水稻地上部迁移而不增加水稻籽粒中镉积累，为培育兼具安全生产和重金属植物修复能力的新型水稻品种奠定了理论基础（Luo et al.，2018）。在砷的研究方面，*HAC1*

基因能够调控砷酸盐还原为亚砷酸盐，从而排出体外；当 $HAC1$ 功能缺失时，砷酸盐则利用磷酸盐运输通路被运送到植物的地上部位，对植物造成极大的毒害作用（Chao et al.，2014）。对于酸性土壤中生长的植物来说，铝毒害是常见的逆境因子，因此植物进化出排斥与忍耐两套机制应对铝毒害，通过有机酸、酚类和多糖类物质将 Al^{3+} 螯合在体外（排斥）或体内形成非毒性的复合物（忍耐）（Kochian et al.，2004；Ma et al.，2001）。因此，部分作物的抗铝性与 MATE（multidrug and toxic compound extrusion）转运蛋白有关，该转运蛋白能够将柠檬酸从根尖分泌出来，螯合根际的 Al^{3+} 并降低其毒性，在大麦、高粱和玉米等作物中由于等位变异或拷贝数变化所导致的 $MATE$ 表达量变化是造成其铝毒害耐受性差异的重要因素（Fujii et al.，2012；Magalhaes et al.，2007；Maron et al.，2013）。此外，在拟南芥中也克隆了调控多种矿质元素积累的 $SIC1$ 基因，其通过调控磷脂酰胆碱的分配从而调节多种矿质元素转运蛋白的亚细胞定位，最终达到必需营养元素和非必需重金属元素等多元素复合调控，为植物修复、生物强化，以及狭义的营养高效提供了一个离子组学角度的新研究思路（Gao et al.，2017）。

2 未来发展趋势与关键突破口

2.1 养分利用效率的研究向农作物倾斜

近年来，农作物养分利用效率的研究受到越来越多的重视。第一次绿色革命的发源地 Centro Internacional de Mejoramientode Maizy Trigo（CIMMYT）的全球小麦育种计划（Global Wheat Program）和英国洛桑研究站（Rothamsted Research）的 20：20 Wheat® 战略研究计划都高度重视提高小麦的养分利用效率。欧盟第七框架计划（7th Framework Programme，FP7）在 2009～2014 年资助多个欧盟国家和中国科学家开展作物氮利用计划，进行小麦、玉米、油菜等农作物养分高效利用的土壤肥料学和遗传学研究，旨在通过合理施肥和品种改良提高农作物肥料利用效率。杜邦先锋（DuPont Pioneer）、先正达（Syngenta）、巴斯夫（BASF）等跨国公司也非常重视研发提高小麦、水稻、玉米、油菜等农作物养分效率的生物技术。比尔 - 梅琳达·盖茨基金会（Bill & Melinda Gates Foundation）提供大量经费支持逐步建立非豆科植物（玉米、大麦等）的共生固氮体系。从 20 世纪 90 年代起，国际上开展了 HarvestPlus 计划，旨在通过生物强化（biofortification）措施提高粮食作物中微量矿质元素和维生素的含量，进而提高人类营养和健康。

我国对提高作物养分资源利用效率的研究也同样十分重视。2016 年，在首批国家重点研发计划七大作物育种项目中设置了"主要农作物养分高效利用性状形成的遗传与分子基础"项目。随着分子生物学技术的发展，养分高效利用基因的克隆和相应作物育种材料的培育进程也大大加快（Hu et al.，2015；Kopriva and Chu，2018；Li et al.，2017b，2018）。以水稻为例，对 *NRT1.1B*（Hu et al.，2015）、*NRT2.3b*（Fan et al.，2016）、*ARE1*（Wang et al.，2018a）、*NRT1.1A*（Wang et al.，2018b）、*GRF4*（Li et al.，2018）、*NPF6.1*（Tang et al.，2019）、*OsNR2*（Gao et al.，2019b）、*NGR5*（Wu et al.，2020）等重要基因的克隆，为培育氮素高效水稻及其他农作物提供了十分重要的分子遗传基础。

2.2 精准调控实现养分的动态和高效循环利用

氮、磷、钾是维持植物生长发育最主要的物质基础，但目前对作物养分效率的提高主要集中单位施肥量下产量的提高上，较少考虑对花期（成熟期）、株型（株高和分蘖等）及品质指标等协同效应的影响。由于作物器官的用途不尽相同（如饲料、果实、蔬菜、药用等），以及健康需求（如富含特定氨基酸、高蛋白质、低糖等）的变化，未来对作物的需求会从对作物整体产量和生物量的提高，逐渐转向特定生育期内特定器官生物量和特定营养与代谢产物的增加。因此，利用分子手段实现氮磷钾在作物体内的定向流动，充分利用液泡中储存的养分，从而有选择性地调节氮磷钾等营养物质在不同器官中的比例，提高养分在作物不同生长时期和不同器官（如种子、叶片、果实等）的富集；通过有针对性地提高以氮磷钾为主要成分的代谢产物的定向合成，从而实现作物器官水平的氮磷钾高效吸收、分配和再利用的人工选择，降低氮磷钾养分在冗余器官中的浪费。

微量元素和有毒重金属元素的积累不仅关系到作物产量，还关系到农产品的营养和安全品质。提高必需微量元素的利用效率，强化其在食用器官的积累并降低有毒重金属含量成为作物微量元素相关研究的主要目标。另外，各微量元素的调控及离子平衡不是孤立存在的,植物体内存在复杂的互作调控网络,如 Fe 和 Zn 具有拮抗效应,然而 Fe 和 Zn 能缓解 Cd 对植物的毒性,还能影响植物 P 的利用效率。因此,如何提高植物协同利用各种微量元素的效率是急需攻克的科学难题。

为实现这一目标,挖掘养分高效利用的植物遗传资源,从中克隆调控基因是关键。基于营养高效相关表型遗传差异,构建极端品种间的重组自交系、剩余杂合群体等遗传材料,结合 QTL 和 GWAS 挖掘优良等位变异(基因)是获得养分高效基因、培育养分高效作物品种的重要途径(Zhang et al., 2020c)。同时,通过具有不同优势性状品种之间的杂交和基因编辑技术也可以实现优异性状的组合,培育出营养高效的优质新品种。而且 CRISPR/Cas9 基因编辑技术近年来取得了快速发展(Lin et al., 2020),使精准改造养分高效利用相关基因成为可能。最近研究发现,细胞分裂素在水稻锌的吸收及转运中起重要调控作用。通过对根中细胞分裂素代谢相关基因的精准调控,使水稻籽粒胚乳中锌含量大幅度增加,产量也有显著提高,且重金属镉等没有变化。这项研究为培育富锌水稻及其他作物提供了全新的思路(Gao et al., 2019a)。

2.3 作物与微生物互作提升养分利用效率

影响作物对氮、磷、钾等养分吸收能力的主要位置在根际。一方面,植物通过激活根系趋向性,形成理想的根系构型,以便从土壤中快速感应和高效吸收养分;另一方面,根系也通过与根系微生物的协作增强土壤和肥料中养分的有效性,如侵染和定殖于根系表层细胞中的内生细菌以及菌根真菌可以促进作物对氮、磷等养分的吸收利用以供作物生长所需;根际土壤中硝化菌类能将铵态氮转化为硝态氮供给作物根系吸收;根际有益菌(如促生菌等)可以通过其代谢产生的化学物质促进作物根系发生等(Ditengou et al., 2015)。最近的研究还发现,水稻籼粳亚种间根系微生物组与其氮肥利用效率有着密切关系,*NRT1.1B* 通过调控水稻根系微生物组从而调控氮素利用效率,该发现为通过调控根际微环境进而减少氮肥的施用提供了全新思路(Zhang et al., 2019)。在大豆中也发现控制根构型的 QTL 直接影响根际微生物的结构和组成。具有良好根构型的大豆基因型,不仅能招募解磷、解钾等有益细菌，还能调控根际真菌组分，使之更有利于氮磷的吸收（Xu et al.，2020；Zhong et al.，2019）。

通过提高根系对氮磷养分的敏感性和可塑性，构建更有利于作物吸收养分的理想根系构型；同时利用和改善根际微生物群落，建立根系与微生物共生互惠体系，从而促进植物在根系与土壤相互作用区域内对养分的感知和吸收，最终提高植物的竞争优势和营养效率。

2.4 养分高效利用机制与生态环境进一步协调

植物在根际互作区中感应和吸收养分的过程中会与多种外部信号，如根际pH、根际微生物组成、盐碱、干旱、重金属毒害及微量元素缺乏等互作，这些信号可通过影响根系对土壤局部养分感知与吸收的变化影响植物的生长发育（Verbon and Liberman，2016；Xuan et al.，2017）。例如，植物根系在吸收铵和硝酸盐时也会导致根表面或根际的pH变化，根际pH对硝酸盐转运蛋白的转运活性具有调节作用，从而反作用于根系对硝酸盐的感知和吸收（Fan et al.，2016；Xuan et al.，2017）；作物的养分高效离不开稳定的水源供给，根际土壤中养分溶解性和移动性受到土壤中水分影响，缺水条件会显著抑制根系对氮、磷养分的吸收和利用（He and Dijkstra，2014）；中国沿海有大量滩涂和盐碱地，提高作物对钾、硝酸盐的选择性吸收，有可能抑制钠和氯离子的吸收，从而提高耐盐能力和养分的利用效率。此外，在全球大气CO_2浓度升高的条件下，喜铵作物和喜硝作物对此响应呈相反趋势（Bloom et al.，2012；Rubio-Asensio and Bloom，2017）。有毒重金属对氮素吸收同化具有强烈的抑制作用，而且其吸收积累往往与氮素间具有正相关。因此，作物品种设计过程中对此要有不同的育种选择策略。

为了应对全球气候变化造成的极端气象条件加剧（如水资源分配不均），以及土壤环境恶化（酸化、盐碱化、重金属污染等）所导致的可耕作土地资源减少，提高作物在多变环境条件下养分效率是未来培育抗逆且资源高效作物新品种的关键。

2.5 养分稳态的实时监测——智慧农业

智慧农业是未来农业的发展方向。在标准化农业种植管理中，快速判断不同生育期下作物体内主要器官养分状态是实现精确化田间施肥管理的重要途径。随着作物养分吸收利用相关基因的挖掘和功能解析，尤其是各种矿质元素（如氮、磷、钾、微量元素及有害元素）响应基因的鉴定，在未来农作物中通过基因编辑和转基因技术导入相关特异报告基因，结合遥感和农业信息工程技术综合分析作物的生长表型（表型组分析）与养分供应状态之间的关联性（Yendrek et al.，2017），实现对作物不同器官内养分的快速判断，确定作物和土壤中养分的供应和利用状态，从而实时、精确地控制田间肥料施用。

3 阶段性目标

3.1 2035 年目标

至 2035 年的目标包括：培育出单产提高、主要养分需耗显著降低、优质多抗的主要作物品种；分离鉴定典型土壤类型中促进矿质养分吸收利用的微生物菌群；完善主要作物氮、磷、钾及微量元素的信号调控网络；针对不同土壤类型，提出改良作物养分效率的主要途径；依赖作物表型组和农业信息工程技术实现对作物体内各器官养分状况的快速判断，确定大田作物养分利用状态和土壤中养分供应状态，实现大田实时、精确施肥目标。

3.2 2050 年目标

至 2050 年的目标包括：人工智能技术常规性介入，高效、精准提取土壤化学、植物生理、转录组、蛋白组、离子组、代谢组等大数据中蕴含的信息，明确主要作物氮磷钾及代表性微量元素的稳态及信号调控网络，完成相应数字化模型构建，指导生产中综合不同作物特性、土壤类型、土壤理化状态、根际微生物互作状态、温度、光、水分等多重因素条件下的最佳和动态肥料施用模式，最大限度降低无效施肥；培育智能化、多功能性新型作物应对未来复杂土壤环境；同时实现作物栽培条件的全人工监控、调控和最终的标准化、设施化。

参考文献

Ai, P., Sun, S., Zhao, J., Fan, X., Xin, W., Guo, Q., Yu, L., Shen, Q., Wu, P., Miller, A.J., et al. (2009). Two rice phosphate transporters, OsPht1; 2 and OsPht1; 6, have different functions and kinetic properties in uptake and translocation. Plant J. *57*, 798-809.

Bari, R., Datt Pant, B., Stitt, M., and Scheible, W.R. (2006). PHO2, microRNA399, and PHR1 define a phosphate-signaling pathway in plants. Plant Physiol. *141*, 988-999.

Bloom, A.J., Asensio, J.S., Randall, L., Rachmilevitch, S., Cousins, A.B., and Carlisle, E.A. (2012). CO_2 enrichment inhibits shoot nitrate assimilation in C3 but not C4 plants and slows growth under nitrate in C_3 plants. Ecology *93*, 355-367.

Chang, M.X., Gu, M., Xia, Y., Dai, X.L., Dai, C.R., Zhang, J., Wang, S.C., Qu, H.Y., Yamaji, N., Ma, F.J., et al. (2019). OsPHT1; 3 mediates uptake, translocation, and remobilization of phosphate under extremely low phosphate regimes. Plant Physiol. *179*, 656-670.

Chao D. Y, Silva A, Baxter I, Huang Y. S, Nordborg M, Danku J, Lahner, B., Yakubova, E., and Salt, D.E. (2012). Genome-wide association atudies identify heavy metal ATPase3 as the primary determinant of natural variation in leaf cadmium in *Arabidopsis thaliana*. PLoS Genet. *8*, e1002923.

Chao, D.Y., Chen, Y., Chen, J., Shi, S., Chen, Z., Wang, C., Danku, J.M., Zhao, F.J., and Salt, D.E. (2014). Genome-wide association mapping identifies a new arsenate reductase enzyme critical for limiting arsenic accumulation in plants. PLoS Biol. *12*, e1002009.

Chen, C.Z., Lv, X.F., Li, J.Y., Yi, H.Y., and Gong, J.M. (2012). *Arabidopsis* NRT1.5 is another essential component in the regulation of nitrate reallocation and stress tolerance. Plant Physiol. *159*, 1582-1590.

Chen, J., Wang, Y., Wang, F., Yang, J., Gao, M., Li, C., Liu, Y., Yamaji, N., Ma, J.F., Paz-Ares, J., et al. (2015). The rice CK2 kinase regulates trafficking of phosphate transporters in response to phosphate levels. Plant Cell *27*, 711-723.

Chen, L., Qin, L., Zhou, L., Li, X., Chen, Z., Sun, L., Wang, W., Lin, Z., Zhao, J., Yamaji, N., et al. (2019). A nodule-localized phosphate transporter GmPT7 plays an important role in enhancing symbiotic N_2 fixation and yield in soybean. New Phytol. *221*, 2013-2025.

Cubero-Font, P., Maierhofer, T., Jaslan, J., Rosales, M.A., Espartero, J., Diaz-Rueda, P., Muller,

H.M., Hurter, A.L., Al-Rasheid, K.A., Marten, I., et al. (2016). Silent S-type anion channel subunit SLAH1 gates SLAH3 open for chloride root-to-shoot translocation. Curr. Biol. *26*, 2213-2220.

Curie, C., and Briat, J.F. (2003). Iron transport and signaling in plants. Annu. Rev. Plant Biol. *54*, 183-206.

Ditengou, F.A., Muller, A., Rosenkranz, M., Felten, J., Lasok, H., van Doorn, M.M., Legue, V., Palme, K., Schnitzler, J.P., and Polle, A. (2015). Volatile signalling by sesquiterpenes from ectomycorrhizal fungi reprogrammes root architecture. Nat. Commun. *6*, 6279.

Drechsler, N., Zheng, Y., Bohner, A., Nobmann, B., von Wiren, N., Kunze, R., and Rausch, C. (2015). Nitrate-dependent control of shoot K homeostasis by the nitrate transporter1/peptide transporter family member NPF7.3/NRT1.5 and the stelar K^+ outward rectifier SKOR in *Arabidopsis*. Plant Physiol. *169*, 2832-2847.

Fan, X., Tang, Z., Tan, Y., Zhang, Y., Luo, B., Yang, M., Lian, X., Shen, Q., Miller, A.J., and Xu, G. (2016). Overexpression of a pH-sensitive nitrate transporter in rice increases crop yields. Proc. Natl. Acad. Sci. USA *113*, 7118-7123.

Fujii, M., Yokosho, K., Yamaji, N., Saisho, D., Yamane, M., Takahashi, H., Sato, K., Nakazono, M., and Ma, J.F. (2012) Acquisition of aluminium tolerance by modification of a single gene in barley. Nat. Commun. *3*, 713.

Gao, F., Robe, K., Bettembourg, M., Navarro, N., Rofidal, V., Santoni, V., Gaymard, F., Vignols, F., Roschzttardtz, H., Izquierdo, E., et al. (2020). The transcription factor bHLH121 interacts with bHLH105 (ILR3) and its closest homologs to regulate iron homeostasis in *Arabidopsis*. Plant Cell. *32*, 508-524.

Gao, S., Xiao, Y., Xu, F., Gao, X., Cao, S., Zhang, F., Wang, G., Sanders, D., and Chu, C. (2019a). Cytokinin-dependent regulatory module underlies the maintenance of zinc nutrition in rice. New Phytol. *224*, 202-215.

Gao, Y.Q., Chen, J.G., Chen, Z.R., An, D., Lv, Q.Y., Han, M.L., Wang, Y.L., Salt, D.E., and Chao, D.Y. (2017). A new vesicle trafficking regulator CTL1 plays a crucial role in ion homeostasis. PLoS Biol. *15*, e2002978.

Gao, Z., Wang, Y., Chen, G., Zhang, A., Yang, S., Shang, L., Wang, D., Ruan, B., Liu, C., Jiang, H., et al. (2019b). The indica nitrate reductase gene *OsNR2* allele enhances rice yield potential and nitrogen use efficiency. Nat. Commun. *10*, 5207.

Han, Y.L., Song, H.X., Liao, Q., Yu, Y., Jian, S.F., Lepo, J.E., Liu, Q., Rong, X.M., Tian, C., Zeng, J., et al. (2016). Nitrogen use efficiency is mediated by vacuolar nitrate sequestration capacity in roots of *Brassica napus*. Plant Physiol. *170*, 1684-1698.

He, M., and Dijkstra, F.A. (2014). Drought effect on plant nitrogen and phosphorus: a meta-analysis. New Phytol. *204*, 924-931.

Ho, C.H., Lin, S.H., Hu, H.C., and Tsay, Y.F. (2009). CHL1 functions as a nitrate sensor in plants. Cell *138*, 1184-1194.

Hu, B., and Chu, C. (2017). Node-based transporter: Switching phosphorus distribution. Nat Plants *3*, 17002.

Hu, B., and Chu, C. (2020). Nitrogen-phosphorus interplay: old story with molecular tale. New Phytol. *225*, 1455-1460.

Hu, B., Jiang, Z., Wang, W., Qiu, Y., Zhang, Z., Liu, Y., Gao, X., Liu, L., Qian, Y., Huang, X., et al. (2019). Nitrate-NRT1.1B-SPX4 cascade integrates nitrogen and phosphorus signaling networks in plants. Nat. Plants *5*, 401-413.

Hu, B., Wang, W., Ou, S., Tang, J., Li, H., Che, R., Zhang, Z., Chai, X., Wang, H., Wang, Y., et al. (2015). Variation in NRT1.1B contributes to nitrate-use divergence between rice subspecies. Nat. Genet. *47*, 834-838.

Hu, B., Zhu, C., Li, F., Tang, J., Wang, Y., Lin, A., Liu, L., Che, R., and Chu, C. (2011). LEAF TIP NECROSIS1 plays a pivotal role in the regulation of multiple phosphate starvation responses in rice. Plant Physiol. *156*, 1101-1115.

Ishikawa, S., Ishimaru, Y., Igura, M., Kuramata, M., Abe, T., Senoura, T., Hase, Y., Arao, T., Nishizawa, N.K., and Nakanishi, H. (2012) Ion-beam irradiation, gene identification, and marker-assisted breeding in the development of low-cadmium rice. Proc. Natl. Acad. Sci. USA *109*, 19166-19171.

Ishimaru, Y., Suzuki, M., Tsukamoto, T., Suzuki, K., Nakazono, M., Kobayashi, T., Wada, Y., Watanabe, S., Matsuhashi, S., Takahashi, M., et al. (2006). Rice plants take up iron as an Fe^{3+}-phytosiderophore and as Fe^{2+}. Plant J. *45*, 335-346.

Kochian, L.V., Hoekenga, O.A., and Piñeros, M.A. (2004). How do crop plants tolerate acid soils? Mechanisms of aluminum tolerance and phosphorous efficiency. Annu. Rev. Plant

Biol. *55*, 459-493.

Kopriva, S. and Chu, C. (2018). Are we ready to improve phosphorus homeostasis in rice? J. Exp. Bot. *69*, 3515-3522.

Lei, R., Li, Y., Cai, Y., Li, C., Pu, M., Lu, C., Yang, Y., and Liang, G. (2020). bHLH121 functions as a direct link that facilitates the activation of FIT by bHLH IVc transcription factors for maintaining Fe homeostasis in *Arabidopsis*. Mol. Plant *13*, 634-649.

Li, H., Hu, B. and Chu, C. (2017b). Nitrogen use efficiency in crops: lessons from *Arabidopsis* and rice. J. Exp. Bot. *68*, 2477-2488.

Li, H., Yu, M., Du, X.Q., Wang, Z.F., Wu, W.H., Quintero, F.J., Jin, X.H., Li, H.D., and Wang, Y. (2017a). NRT1.5/NPF7.3 functions as a proton-coupled H^+/K^+ antiporter for K^+ loading into the sylem in *Arabidopsis*. Plant Cell *29*, 2016-2026.

Li, J.Y., Fu, Y.L., Pike, S.M., Bao, J., Tian, W., Zhang, Y., Chen, C.Z., Li, H.M., Huang, J., Li, L.G., et al. (2010). The *Arabidopsis* nitrate transporter NRT1.8 functions in nitrate removal from the xylem sap and mediates cadmium tolerance. Plant Cell *22*, 1633-1646.

Li, S., Tian, Y., Wu, K., Ye, Y., Yu, J., Zhang, J., Liu, Q., Hu, M., Li, H., Tong, Y., et al. (2018). Modulating plant growth-metabolism coordination for sustainable agriculture. Nature *560*, 595-600.

Li, Y., Zhang, J., Zhang, X., Fan, H., Gu, M., Qu, H., and Xu, G. (2015). Phosphate transporter OsPht1;8 in rice plays an important role in phosphorus redistribution from source to sink organs and allocation between embryo and endosperm of seeds. Plant Sci. *230*, 23-32.

Liang, C., Wang, Y., Zhu, Y., Tang, J., Hu, B., Liu, L., Ou, S., Wu, H., Sun, X., Chu, J., and Chu, C. (2014). OsNAP connects absisic acid and leaf senescence by fine tuning absisic acid biosynthesis and directly targeting senescence-associated genes in rice. Proc. Natl. Acad. Sci. USA *111*, 10013-10018.

Lin, Q., Zong, Y., Xue, C., Wang, S., Jin, S., Zhu, Z., Wang, Y., Anzalone A.V., Raguram, A., Doman, J.L. et al. (2020). Prime genome editing in rice and wheat. Nat. Biotechnol. *38*, 582-585.

Lin, S.H., Kuo, H.F., Canivenc, G., Lin, C.S., Lepetit, M., Hsu, P.K., Tillard, P., Lin, H.L., Wang, Y.Y., Tsai, C.B., et al. (2008). Mutation of the *Arabidopsis NRT1.5* nitrate transporter causes defective root-to-shoot nitrate transport. Plant Cell *20*, 2514-2528.

Lin, S.I., Santi, C., Jobet, E., Lacut, E., El Kholti, N., Karlowski, W.M., Verdeil, J.L., Breitler, J.C., Perin, C., Ko, S.S., et al. (2010). Complex regulation of two target genes encoding SPX-MFS proteins by rice miR827 in response to phosphate starvation. Plant Cell Physiol. *51*, 2119-2131.

Ling, H. Q., Bauer, P., Keller, B., and Ganal., M. (2002). The *fer* gene encoding a bHLH transcriptional regulator controls development and physiology in response to iron in tomato. Proc. Natl. Acad. Sci. USA *99*, 13938-13943.

Liu, K.H., and Tsay, Y.F. (2003). Switching between the two action modes of the dual-affinity nitrate transporter CHL1 by phosphorylation. EMBO J. *22*, 1005-1013.

Luan, M., Tang, R.J., Tang, Y., Tian, W., Hou, C., Zhao, F., Lan, W., and Luan, S. (2017). Transport and homeostasis of potassium and phosphate: limiting factors for sustainable crop production. J. Exp. Bot. *68*, 3091-3105.

Luo, B.F., Du, S.T., Lu, K.X., Liu, W.J., Lin, X.Y., and Jin, C.W. (2012). Iron uptake system mediates nitrate-facilitated cadmium accumulation in tomato (*Solanum lycopersicum*) plants. J. Exp. Bot. *63*, 3127-3136.

Luo, J.S., Huang, J., Zeng, D.L., Peng, J.S., Zhang, G.B., Ma, H.L., Guan, Y., Yi, H.Y., Fu, Y.L., Han, B., et al. (2018). A defensin-like protein drives cadmium efflux and allocation in rice. Nat. Commun. *9*, 645.

Lv, Q., Zhong, Y., Wang, Y., Wang, Z., Zhang, L., Shi, J., Wu, Z., Liu, Y., Mao, C., Yi, K., et al. (2014). SPX4 negatively regulates phosphate signaling and homeostasis through its interaction with PHR2 in rice. Plant Cell *26*, 1586-1597.

Ma, J.F., Ryan, P.R., and Delhaize, E. (2001). Aluminium tolerance in plants and the complexing role of organic acids. Trends Plant Sci. *6*, 273-278.

Maccaferri, M., Harris, N.S., Twardziok, S.O., Pasam, R.K., Gundlach, H., Spannagl, M., Ormanbekova, D., Lux, T., Prade, V.M., Milner, S.G., et al. (2019). Durum wheat genome highlights past domestication signatures and future improvement targets. Nat. Genet. *51*, 885-895.

Maeda, Y., Konishi, M., Kiba, T., Sakuraba, Y., Sawaki, N., Kurai, T., Ueda, Y., Sakakibara, H., and Yanagisawa, S. (2018). A NIGT1-centred transcriptional cascade regulates nitrate signalling and incorporates phosphorus starvation signals in *Arabidopsis*. Nat. Commun. *9*, 1376.

Magalhaes, J.V., Liu, J., Guimaraes, C.T., Lana, U.G., Alves, V.M., Wang, Y.H., Schaffert, R.E., Hoekenga, O.A., Pineros, M.A., Shaff, J.E., et al. (2007). A gene in the multidrug and toxic compound extrusion (MATE) family confers aluminum tolerance in sorghum. Nat. Genet. *39*, 1156-1161.

Mao, Q.Q., Guan, M.Y., Lu, K.X., Du, S.T., Fan, S.K., Ye, Y.Q., Lin, X.Y., and Jin, C.W. (2014). Inhibition of nitrate transporter 1.1-controlled nitrate uptake reduces cadmium uptake in *Arabidopsis*. Plant Physiol. *166*, 934-944.

Maron, L.G., Guimaraes, C.T., Kirst, M., Albert, P.S., Birchler, J.A., Bradbury, P.J., Buckler, E.S., Coluccio, A.E., Danilova, T.V., Kudrna, D., et al. (2013). Aluminum tolerance in maize is associated with higher *MATE1* gene copy number. Proc. Natl. Acad. Sci. USA *110*, 5241-5246.

Medici, A., Marshall-Colon, A., Ronzier, E., Szponarski, W., Wang, R., Gojon, A., Crawford, N.M., Ruffel, S., Coruzzi, G.M., and Krouk, G. (2015). AtNIGT1/HRS1 integrates nitrate and phosphate signals at the *Arabidopsis* root tip. Nat. Commun. *6*, 6274.

Meng, S., Peng, J.S., He, Y.N., Zhang, G.B., Yi, H.Y., Fu, Y.L., and Gong, J.M. (2016). *Arabidopsis* NRT1.5 mediates the suppression of nitrate starvation-induced leaf senescence by modulating foliar potassium level. Mol. Plant *9*, 461-470.

Puga, M.I., Rojas-Triana, M., de Lorenzo, L., Leyva, A., Rubio, V., and Paz-Ares, J. (2017). Novel signals in the regulation of Pi starvation responses in plants: facts and promises. Curr. Opin. Plant Biol. *39*, 40-49.

Roy, S.J., Negrao, S., and Tester, M. (2014). Salt resistant crop plants. Curr. Opin. Biotechnol. *26*, 115-124.

Rubio, V., Linhares, F., Solano, R., Martin, A.C., Iglesias, J., Leyva, A., and Pazares, J. (2001). A conserved MYB transcription factor involved in phosphate starvation signaling both in vascular plants and in unicellular algae. Genes Dev. *15*, 2122-2133.

Rubio-Asensio, J.S., and Bloom, A.J. (2017). Inorganic nitrogen form: a major player in wheat and *Arabidopsis* responses to elevated CO_2. J. Exp. Bot. *68*, 2611-2625.

Sasaki, A., Yamaji, N., Yokosho, K., and Ma, J.F. (2012). Nramp5 is a major transporter responsible for manganese and cadmium uptake in rice. Plant Cell *24*, 2155-2167.

Tang, W., Ye, J., Yao, .X, Zhao, P., Xuan, W., Tian, Y., Zhang, Y., Xu, S., An, H., Chen, G., et al. (2019). Genome-wide associated study identifies NAC42-activated nitrate transporter conferring high nitrogen use efficiency in rice. Nat. Commun. *10*, 5279.

Teng, Y., Wu, J., Lu, S., Wang, Y., Jiao, X., and Song, L. (2014). Soil and soil environmental quality monitoring in China: a review. Environ. Internation. *69*, 177-199.

Tsay, Y.F., Chiu, C.C., Tsai, C.B., Ho, C.H., and Hsu, P.K. (2007). Nitrate transporters and peptide transporters. FEBS Lett. *581*, 2290-2300.

Uauy C, Distelfeld A, Fahima T, Blechl A, Dubcovsky J. (2006). A *NAC* gene regulating senescence improves grain protein, zinc, and iron content in wheat. Science *314*, 1298-1301.

Ueno, D., Yamaji, N., Kono, I., Huang, C.F., Ando, T., Yano, M., and Ma, J.F. (2010). Gene limiting cadmium accumulation in rice. Proc. Natl. Acad. Sci. USA *107*, 16500-16505.

Verbon, E.H., and Liberman, L.M. (2016). Beneficial microbes affect endogenous mechanisms controlling root development. Trends Plant Sci. *21*, 218-229.

Wang, C., Huang, W., Ying, Y., Li, S., Secco, D., Tyerman, S., Whelan, J., and Shou, H. (2012). Functional characterization of the rice SPX-MFS family reveals a key role of OsSPX-MFS1 in controlling phosphate homeostasis in leaves. New Phytol. *196*, 139-148.

Wang, Q., Nian, J., Xie, X., Yu, H., Zhang, J., Bai, J., Dong, G., Hu, J., Bai, B., Chen, L., et al. (2018a). Genetic variations in *ARE1* mediate grain yield by modulating nitrogen utilization in rice. Nat. Commun. *9*, 735.

Wang, W., Hu, B., Yuan, D., Liu, Y., Che, R., Hu, Y., Ou, S., Zhang, Z., Wang, H., Li, H., et al. (2018b). Expression of the nitrate transporter *OsNRT1.1A/OsNPF6.3* confers high yield and early maturation in rice. Plant Cell *30*, 638-651.

Wang, Z., Ruan, W., Shi, J., Zhang, L., Xiang, D., Yang, C., Li, C., Wu, Z., Liu, Y., Yu, Y., et al. (2014). Rice SPX1 and SPX2 inhibit phosphate starvation responses through interacting with PHR2 in a phosphate-dependent manner. Proc. Natl. Acad. Sci. USA *111*, 14953-14958.

Wild, R., Gerasimaite, R., Jung, J.Y., Truffault, V., Pavlovic, I., Schmidt, A., Saiardi, A., Jessen, H.J., Poirier, Y., Hothorn, M., et al. (2016). Control of eukaryotic phosphate homeostasis by inositol polyphosphate sensor domains. Science *352*, 986-990.

Wu, K., Wang, S., Song, W., Zhang, J., Wang, Y., Liu, Q., Yu, J., Ye, Y., Li, S., Chen, J., et al. (2020). Enhanced sustainable green revolution yield via nitrogen-responsive chromatin modulation in rice. Science *367*, eaaz2046.

Wu, P., Shou, H., Xu, G., and Lian, X. (2013). Improvement of phosphorus efficiency in rice on the basis of understanding phosphate signaling and homeostasis. Curr. Opin. Plant Biol. *16*, 205-212.

Xia, X., Fan, X., Wei, J., Feng, H., Qu, H., Xie, D., Miller, A.J., and Xu, G. (2015). Rice nitrate transporter OsNPF2.4 functions in low-affinity acquisition and long-distance transport. J. Exp. Bot. *66*, 317-331.

Xie, X., Huang, W., Liu, F., Tang, N., Liu, Y., Lin, H., and Zhao, B. (2013). Functional analysis of the novel mycorrhiza-specific phosphate transporter AsPT1 and PHT1 family from Astragalus sinicus during the arbuscular mycorrhizal symbiosis. New Phytol. *198*, 836-852.

Xu, G., Fan, X., and Miller, A.J. (2012). Plant nitrogen assimilation and use efficiency. Annu. Rev. Plant Biol. *63*, 153-182.

Xu, H., Yang, Y., Tian, Y., Xu, R., Zhong, Y., and Liao, H. (2020). Rhizobium inoculation drives the shifting of rhizosphere fungal community in a host genotype dependent manner. Front. Mocrobiol. *10*, 3135.

Xu, J., Li, H.D., Chen, L.Q., Wang, Y., Liu, L.L., He, L., and Wu, W.H. (2006). A protein kinase, interacting with two calcineurin B-like proteins, regulates K^+ transporter AKT1 in *Arabidopsis*. Cell *125*, 1347-1360.

Xuan, W., Beeckman, T., and Xu, G. (2017). Plant nitrogen nutrition: sensing and signaling. Curr. Opin. Plant Biol. *39*, 57-65.

Yamaji, N., Takemoto, Y., Miyaji, T., Mitani-Ueno, N., Yoshida, K.T., and Ma, J.F. (2017). Reducing phosphorus accumulation in rice grains with an impaired transporter in the node. Nature *541*, 92-95.

Yan, H., Xu, W., Xie, J., Gao, Y., Wu, L., Sun, L., Feng, L., Chen, X., Zhang, T., Dai, C., et al. (2019). Variation of a major facilitator superfamily gene contributes to differential cadmium accumulation between rice subspecies. Nat. Commun. *10*, 2562.

Yendrek, C.R., Tomaz, T., Montes, C.M., Cao, Y., Morse, A.M., Brown, P.J., McIntyre, L.M., Leakey, A.D., and Ainsworth, E.A. (2017). High-throughput phenotyping of maize leaf physiological and biochemical traits using hyperspectral reflectance. Plant Physiol. *173*, 614-626.

Zhang, G.B., Yi, H.Y., and Gong, J.M. (2014). The *Arabidopsis* ethylene/jasmonic acid-NRT signaling module coordinates nitrate reallocation and the trade-off between growth and environmental adaptation. Plant Cell 26, 3984-3998.

Zhang, H., Li, Y., Pu, M., Xu, P., Liang, G., and Yu, D. (2020b). *Oryza sativa* POSITIVE REGULATOR OF IRON DEFICIENCY RESPONSE 2 (OsPRI2) and OsPRI3 are involved in the maintenance of Fe homeostasis. Plant Cell Environ. 43, 261-274.

Zhang, J., Liu, Y.X., Zhang, N., Hu, B., Jin, T., Xu, H., Qin, Y., Yan, P., Zhang, X., Guo, X., et al. (2019). *NRT1.1B* is associated with root microbiota composition and nitrogen use in field-grown rice. Nat. Biotechnol. 37, 676-684.

Zhang, Y., Xu, Y., Yi, H., and Gong, J. (2012). Vacuolar membrane transporters OsVIT1 and OsVIT2 modulate iron translocation between flag leaves and seeds in rice. Plant J. 72, 400-410.

Zhang, Z., Gao, S., and Chu, C. (2020c). Improvement of nutrient use efficiency in rice: current toolbox and future perspectives. Theor. Appl. Genet. 133, 1365-1384.

Zhang, Z., Hu, B., and Chu, C. (2020a). Towards the integration of hierarchical nitrogen signalling network in plants. Curr. Opin. Plant Biol. 55, 60-65.

Zhong, Y., Yang, Y., Liu, P., Xu, R., Rensing, C., Fu, X., and Liao, H. (2019). Genotype and rhizobium inoculation modulate the assembly of soybean rhizobacterial communities. Plant Cell Environ. 42, 2028-2044.

Zhu, Y., Di, T., Xu, G., Chen, X., Zeng, H., Yan, F., and Shen, Q. (2009). Adaptation of plasma membrane H^+-ATPase of rice roots to low pH as related to ammonium nutrition. Plant Cell Environ. 32, 1428-1440.

生物固氮

谢芳 李霞

1 国内外研究进展

氮是自然界重要的生命元素和植物生长发育所必需的主要营养元素，氮素也是决定作物产量的首要因素。氮素来源于大气中的氮气，尽管空气中氮含量高达78%，却不能被植物直接利用。大气中的氮气只有被转化为可溶性离子态氮和硝态氮后方能被植物同化和利用，这个过程被称为氮气的固定过程或者固氮过程。氮气的固定可通过自然、工业和生物三种途径实现。自然固氮是在闪电等高能环境下氮气转化为氨和硝酸的过程，在植物氮素供应中发挥微小作用。工业固氮是采用高温、高压和化学催化的方法将氮气转化为氨的过程。大量工业氮肥的生产和施用不仅消耗能源，还对环境造成了极大的危害。生物固氮是指固氮微生物（主要是原核微生物和固氮蓝藻）利用固氮酶将氮气还原为氨的过程，是自然界环境友好和高效的固氮方式。在生物固氮中，豆科植物与根瘤菌间的共生固氮是效率最高的生物固氮形式，是过去几十年世界各国生物固氮科学家研究的重点。科学家们在豆科植物与根瘤菌间的信号交流、根瘤菌侵染和根瘤形态建成等遗传机制方面取得了重要突破，为培育具有固氮能力的未来作物提供了科学依据。

1.1 豆科植物与根瘤菌间的共生固氮

豆科植物与根瘤菌间共生固氮效率高，基本可以满足植物生长期间对氮素的需求，还可以为间作或轮作的植物提供部分氮素。但豆科植物与根瘤菌间的共生固氮有着非常严格的专一性，所以阐明共生固氮的分子机制，不仅能提高豆科植物与根瘤菌间的共生固氮效率，还能为探索扩大宿主范围提供科学依据。

1.1.1 结瘤因子信号转导途径

豆科植物与根瘤菌间的共生互作起始于共生双方的早期信号交流，豆科植物分泌黄酮与类黄酮类物质，结合到根瘤菌的 *NodD* 基因上，NodD 激活 *Nod* 基因的表达，从而合成结瘤因子（nodulation factors，NF）（Debellé and Promé，1996）。NF 是一类脂质几丁寡糖（lipochito-oligosaccharide，LCO），不同根瘤菌分泌的 NF 的寡糖骨架是相同的，但具有不同的长度和饱和度，以及不同的侧链修饰，这些修饰是共生互作中宿主特异性的关键决定因素（Debellé and Promé，1996）。

近 20 年，科学家们对 NF 信号转导研究取得了非常大的突破。不同豆科植物结瘤因子受体相继被鉴定，这是一类具有 LysM 结构域的受体激酶（Limpens et al.，2003；Radutoiu et al.，2003）。除 NF 受体外，一些信号转导组分，包括共生受体激酶 LjSYMRK/MtDMI2，核孔复合物 LjNUP85、LjNUP133、LjNENA，离子通道蛋白 LjPOLLUX/MtDMI1、LjCASTOR、MtCNGC15 和 MtMCA8，都是 NF 激活细胞核内钙离子震荡（calcium spiking）所必需的（Charpentier et al.，2008，2016；Groth et al.，2010；Kistner et al.，2005；Saito et al.，2007；Stracke et al.，2002）。随后钙离子信号被依赖于钙离子和 CaM 的钙调激酶 LjCCaMK/MtDMI3 所解析，与 LjCYCLOPS/MtIPD3 相互作用并磷酸化 LjCYCLOPS/MtIPD3，进一步激活下游转录因子，包括 GRAS 类的转录因子 DELLA、NSP1、NSP2，AP2/ERF 类的转录因 ERN1，以及 RWP-RK 类的转录因子 NIN 等，从而将共生信号传导下去（Cerri et al.，2017；Fonouni-Farde et al.，2015，2016；Gleason et al.，2006；Horvath et al.，2011；Jin et al.，2016；Kaló et al.，2005；Marsh et al.，2007；Middleton et al.，2007；Schauser et al.，1999；Singh et al.，2014）。

NF 信号转导通路的一些组分（包括 LjSYMRK/MtDMI2、核孔复合物蛋白、离子通道和钙离子信号解码器 LjCCaMK/MtDMI3 和 LjCYCLOPS/MtIPD3），不仅是豆科植物与根瘤菌共生固氮所必需，也是植物与菌根真菌共生互作所必需，因此，称此通路为共同共生信号途径（common symbiosis-signaling pathway）（Downie，2014；Parniske，2008）。由于 80% 以上陆生植物，包括水稻、玉米等农作物，都可与菌根真菌共生互作，成为遗传操作其他作物与根瘤菌共生互作并固氮的分子基础。

1.1.2 根瘤菌侵染

根瘤菌可以通过表皮细胞间隙或者裂隙（crack entry）侵入宿主植物，但最主要还是通过在根毛内形成一个新的管状结构——侵入线（infection thread）侵入宿主植物（Jones et al.，2007；Oldroyd and Downie，2008）。根瘤菌包裹在根毛卷曲内，质膜内陷形成一向内的管状通道，即侵染线；根瘤菌通过侵染线从表皮向皮层细胞延伸，到达根瘤原基中。根瘤菌分泌的 NF 诱导植物细胞壁和细胞骨架的重排，是侵染线形成所必需的（Hossain et al.，2012；Qiu et al.，2015；Xie et al.，2012；Yokota et al.，2009）。还有一些遗传组分被分离鉴定出是侵染线形成所必需的，如 MtRPG、MtLIN/LjCERBERUS、MtVAPYRIN 等，但其分子机制尚未得到解析（Arrighi et al.，2008；Kiss et al.，2009；Murray et al.，2011；Yano et al.，2009）。根瘤菌分泌的胞外多糖（exopolysaccharide，EPS）也是侵染线起始与延伸所必需的。已知 EPS 很可能是作为一种信号分子调控侵染线的形成。在百脉根的研究中，鉴定出 EPS 的受体 LjEPR3 也是一类具有 LysM 结构域的受体激酶，而 NF 信号的转录因子 NIN 和 ERN1 可以激活 *LjEPR3* 的表达，从而提出调控根瘤菌侵染过程的"two-step"模型（Kawaharada et al.，2015，2017）。但 *epr3* 突变体能够被野生型根瘤菌正常侵染，表明很可能还有其他组分作为共受体来感知 EPS 信号；对 *LjEPR3* 在蒺藜苜蓿的同源基因 *MtLYK10* 的研究认为，根瘤菌琥珀酰聚糖相关的侵染表型不依赖于 MtLYK10（Maillet et al.，2020），因此，蒺藜苜蓿中的 LYK10 是否是 EPS 的受体还有待于进一步研究。同时，EPS 信号是如何被传递下去的，目前仍不清楚。侵染线的形成对根瘤菌能否侵入植物至关重要，目前对豆科植物根瘤菌侵染的分子机制了解的还非常有限。因此，对根瘤菌侵入宿主植物的研究，对将来探索非豆科植物共生固氮具有非常重要的作用。

1.1.3 根瘤器官的形成

根瘤菌可以诱导宿主植物皮层细胞分裂形成根瘤原基。一系列研究表明，根瘤菌侵染和根瘤器官形成，在遗传上是相互独立的，在空间上也是分开的，但是在时间上是协同调控的（Oldroyd and Downie，2008）。研究表明，植物激素，特别是生长素和细胞分裂素，对根瘤器官的形成非常重要。生长素转运抑制剂和细胞分裂素都可以诱导假瘤的形成，而细胞分裂素受体LjLHK1的功能获得（gain-of-function）突变体 *Ljsnf2* 和 *Ljsnf5* 也可以自发形成根瘤（Heckmann et al.，2011；Long and Long，2011；Tirichine et al.，2007）。结瘤因子信号对根瘤原基的形成至关重要，LjCCaMK 的功能获得性突变体 *snf1* 也能自发结瘤（Gleason et al.，2006；Tirichine et al.，2006）。这些自发结瘤都依赖于 NF 信号通路中的下游转录因子的存在，包括 NSP1、NSP2、ERN1 和 NIN。

根瘤和侧根都是植物根部胚胎后形成的侧生器官，都能为植物获取氮素。根瘤与侧根发育具有一定的相似性，但也有诸多不同之处。从进化角度看，根瘤很有可能在其形成过程中招募了侧根或其他根系发育程序（Sprent，2007）。最近英国和日本两个实验室分别以蒺藜苜蓿和百脉根为材料，发现共生特异的转录因子 NIN 激活下游基因，包括 *LBD16*、*STY* 等的表达，从而激活根瘤的发育过程（Schiessl et al.，2019；Soyano et al.，2019），这表明根瘤和侧根通过共同的基因或程序来调控其发育过程。NIN 调控根瘤发育和侧根发育基因的表达，NIN 或 *NF-YA* 过表达后形成假瘤，部分转基因植物表现出侧根发育异常表型（Sayonao et al.，2013）。但是，最初决定根瘤器官细胞命的运转变机制并未得到解析，即这些研究仍然没有揭示为什么豆科植物能响应根瘤菌分泌的信号形成根瘤。

1.1.4 氮素的固定和营养交换

根瘤菌进入根瘤原基后，从侵染线中释放出来，被共生体膜包裹形成类菌体，利用植物光合产物的能量，将氮气还原为氨，最后以氨基酸的形式被植物吸收利用（Oldroyd et al.，2011）。在此过程中，根瘤菌能否从侵染线中释放出来、共生体膜能否形成，都是根瘤菌共生固氮的前提条件。目前，已经鉴定出一些植物蛋白，如 LjFEN1、LjSST、LjSEN1、MtDNFs 等，是氮素固定所必需的（Farkas et al.，2014；Hakoyama et al.，2009；Horváth et al.，2015；Krusell et al.，2005；Pan and Wang，2017；Starker et al.，2006；

Wang et al., 2010）。固氮酶是一种对氧浓度非常敏感的酶，豆科植物需表达特异的豆血红蛋白，为固氮酶发挥固氮能力提供微氧环境（Ott et al., 2005；Wang et al., 2019）。

1.1.5 结瘤自我调控途径

共生固氮是一个非常耗能的过程，在长期的进化过程中，豆科植物进化出精细的调控系统，控制根瘤的数目"恰到好处"。研究发现，豆科植物通过根—地上—根的长距离信号转导途径，以局部和系统的方式调控根瘤的数目，这被称为根瘤自我调控途径（autoregulation of nodulation，AON）（Reid et al., 2011）。近年来，对 AON 途径的长距离信号分子认识取得了较大的突破，根部分泌的 CLE 短肽被糖基化后移到地上，与受体 LjHAR1 结合，进而抑制根瘤的形成；而地上到地下的信号分子认为有小 RNA miR2111 和细胞分裂素等（Gautrat et al., 2020；Okamoto et al., 2013；Sasaki et al., 2014；Tsikou et al., 2018）。

1.2 环境对共生固氮的调控

多种环境因素对共生固氮有着重要的调控作用。固氮酶是一种 Mo-Fe 蛋白，钼、铁、硫对固氮酶活性的发挥非常重要，硼、磷、钙、铜等调控着共生固氮过程（Weisany et al., 2013）。环境中高浓度的氮抑制共生固氮过程及固氮酶活性。研究表明，NLPs（NIN-like protein）在调控硝态氮抑制根瘤形成中有着重要作用（Lin et al., 2018；Nishida et al., 2018）。各种生物、非生物逆境对共生固氮也有着非常重要的调控作用，但目前对这些方面的研究国内外还非常薄弱。

1.3 高通量测序在共生固氮中的研究

近年来，随着测序技术的发展，一些模式豆科植物的基因组陆续被测序，另外，固氮（豆科）植物与其他植物的比较基因组学研究，为揭示豆科植物在进化过程中获得共生互作的分子基础提供了很大的帮助。研究结果表明，只有为数不多的基因，包括 NIN、RGP 和 NFP2 是固氮植物所特有的（Griesmann et al., 2018；van Velzen et al., 2018）。

1.4 对非豆科植物共生固氮的研究及自身固氮潜能的探索

山麻黄（*Parasponia andersonii*）是除豆科植物之外唯一可以与根瘤菌共生固氮的植物（Behm et al.，2014）。放线菌 *Frankia* 可以与包括 8 个科的植物（被称为 actinorhizal 植物）共生互作形成根瘤并固氮（Svistoonoff et al.，2014）。尽管大多数 actinorhizal 植物是木本植物，遗传操作困难，但近年来科学家们在 actinorhizal 植物基因组序列、转录组分析、遗传转化等方面均取得非常大的进展（Froussart et al.，2016）。研究发现，很多在豆科植物与根瘤菌共生固氮中的基因，在山麻黄与根瘤菌间共生固氮和 *Frankia* 与 actinorhizal 植物共生互作中也有相似功能，如 SYMRK、CCaMK 和 NIN 等（Clavijo et al.，2015；Svistoonoff et al.，2014）。

随着对豆科植物共生固氮分子机制研究的深入，科学家们认识到豆科植物与根瘤菌间共生互作与大部分陆生植物与菌根真菌间的共生互作享有共同的信号转导通路，因此，开始探索一些主要农作物，如玉米、小麦、水稻，与根瘤菌共生互作的可能性（Charpentier and Oldroyd，2010；Beatty and Good，2011；Mus et al.，2016；Oldroyd and Dixon，2014）。同时，随着合成生物学的发展，科学家们也试图通过合成生物学获得有活性的最小固氮基因簇，转入农作物中从而获得自身固氮的能力（Hu and Ribbe，2011，2016，2013；Yang et al.，2017）。

2 未来发展趋势与关键突破口

生物固氮性状形成的遗传学机制是培育具有高效固氮性状未来作物的基础，也有助于实现我国农业"化肥和农药减施增效"目标、保障国家粮食安全和促进绿色农业和生态文明发展。世界各国都非常重视生物固氮基础研究。

2.1 生物固氮遗传学基础

生物固氮遗传机制和调控网络是实现固氮作物遗传改良的基础。国外一直大力支持对生物固氮，特别是豆科植物共生固氮机制的研究，在认识固氮微生物和植物识别、固氮微生物共生固氮的关键过程及调控机制方面有了初步认识，已经对固氮微生物固氮、豆科植物和根瘤菌共生固氮以及非豆科植物与固氮菌联合固氮开展研究，以期全面深入的解析生物固氮分子调控网络，为固氮作物的创制提供科学依据。

此外，国内外还非常重视非豆科植物固氮及遗传机制研究。科学家们在全球广泛搜集可以共生固氮的非豆科植物资源，对山麻黄和蕨类植物满江红等植物基因组及其与根瘤菌和固氮蓝细菌（念珠藻）共生固氮过程展开了研究，并对豆科植物和非豆科共生固氮演化的分子基础开展系统研究，以期揭示非豆科植物不能结瘤固氮的根源，该项研究筛选了可用于作物固氮改良的基因资源，扩充了种质材料遗传基础，为作物固氮性状的改良奠定了基础（Griesmann et al.，2018；Li et al.，2018；van Velzen et al.，2018）。

2.2 非豆科作物结瘤固氮性状的改良

基于豆科植物共生高效固氮的特性，让非豆科粮食作物，如水稻、玉米等，能够结瘤固氮一直是科学家们的目标。基于对豆科植物共生固氮主要过程和调控分子机制及网络的认识，特别是非豆科作物与菌根真菌共生与共生固氮相似的遗传基础，科学家们一直致力于遗传改良非豆科作物结瘤固氮的研究。目前及未来的研究仍然集中在通过豆科植物共生固氮信号通路中关键基因的叠加实现根瘤菌和作物识别、定植和根瘤形态建成的研究（Beatty and Good，2011；Mus et al.，2016；Rosenblueth et al.，2018）。此外，国外也在尝试对非豆科共生固氮植物山麻黄的近源种进行遗传改良，以期为共生固氮作物的培育提供科学依据。

2.3 遗传改良与非豆科植物联合固氮菌提高固氮效率

让非豆科植物，特别是禾本科作物实现根瘤固氮是一个挑战性更大的系统工程。因此，基于在长期进化中非豆科植物和固氮菌存在联合固氮的事实，分离和鉴定与非豆科作物能够联合固氮的菌株，进而改良联合固氮菌与作物联合及固氮的能力，将是未来实现非豆科作物生物固氮的一个重要途径（Beatty and Good，2011；Li et al.，2018；van Deynze et al.，2018）。解析非豆科作物感应联合固氮菌并与其联合固氮的遗传基础，并进一步改良这些作物与联合固氮菌的识别和固氮能力，可以较快推动非豆科植物生物固氮的应用。

2.4 合成生物学与非豆科作物固氮性状

生物固氮的核心是固氮酶,让植物能够表达具有酶活力的固氮酶一直是目前研究的一个重点和热点。研究主要集中在将固氮酶基因在植物细胞器,如叶绿体、根系质体和线粒体中表达(Cheng et al.,2005;Yang et al.,2017),以期保证为固氮酶提供低氧和足够能量的作用环境,实现固氮功能(Beatty and Good,2011;Rosenblueth et al.,2018)。

2.5 利用固氮微生物及微生物组改良促进作物固氮

固氮菌和植物识别的特异性是限制植物固氮特性改良的重要因子。因此,通过遗传工程改良根瘤菌与植物识别和互作是实现固氮作物改良的关键。此外,微生物组及其功能研究蓬勃发展,使我们认识到根际微生物对作物性状的巨大影响。最近国内外开展了豆科植物微生物组研究,建立了对豆科植物根际微生物组成特点及其对共生固氮性状影响的初步认识(Hartman et al.,2017;Zgadzaj et al.,2016)。未来研究将进一步揭示微生物组成的功能及影响豆科植物共生固氮的机制,并对豆科植物根际微生物对非豆科作物的影响进行探索。基于植物在微生物组成中的主导作用,阐明决定共生固氮的重要调控机制,将有助于推动遗传改良作物代谢等提高固氮微生物与作物的互作和固氮性状的建成。

2.6 探索新的研究手段

先进的研究手段是推进固氮作物育种研究的关键因素之一。为了支撑固氮作物育种研究,国外研究不断向基因发掘、组学分析、现代信息分析、自动化(智能)技术、纳米技术等相关研发领域延伸。科学家密切跟踪生物科技最新进展,对有望取得突破的技术通过项目支持;基于已经积累的大量基因和技术储备,及时应用于固氮作物创制的研发,加快了作物固氮研究的进程。

创制固氮非豆科作物是一个大的系统工程,涉及多学科、多种技术的应用,只有组织优势团队,优势互补,联合攻关,才能保证极高的研究效率,稳定研发体系和团队。目前,美国国家科学基金会(National Science Foundation,NSF)、英国生物技术和生物学研究理事会(Biotechnology and Biological Sciences Research Council，BBSRC)、印度农业研究理事会（Indian Council of Agricultural Research，ICAR）先后启动了相关研发项目（Rosenblueth et al., 2018）。此外，比尔-梅琳达·盖茨基金会（Bill & Melinda Gates Foundation）也斥巨资支持欧美多所大学和研究所组成高水平科学家团队，对植物共生固氮、微生物、生物信息学等进行联合攻关，已经在非豆科作物响应根瘤菌信号方面取得了阶段性进展，使非豆科作物固氮创制实现了概念性突破，也拥有了大量的技术储备，为进一步研究奠定了坚实基础。

3 阶段性目标

根据国内外生物固氮基础研究和培育高效固氮作物新品种研究的发展趋势，我国未来作物生物固氮研究总体思路是，围绕培育具有高效共生固氮能力的未来作物为目标，加强对生物固氮性状形成的基础研究和相关技术方法研发支持力度，推动以优势科研单位和大学为主体的研究联盟，通过资源和技术平台共享及联合攻关，全面推进生物固氮，特别是植物共生固氮遗传机制、高效固氮种质资源创新、非豆科植物固氮研究方法与技术等基础性研究和未来高效固氮新作物培育的进程，使我国生物固氮基础研究和高效固氮新作物育种工作健康快速发展，全面提升我国生物固氮基础研发、技术研发及应用的发展水平，以满足我国未来绿色农业发展的需求。

3.1 2035年目标

生物固氮性状形成的分子调控网络是培育生物固氮新作物的重要基础。在未来15年里，通过基因组学、蛋白质组学、代谢组学、遗传学等综合研究手段和方法，发掘豆科植物和主要非豆科植物与作物共生固氮性状形成、调控的关键基因和元件，揭示豆科植物和固氮根瘤菌（特异性）识别、根瘤菌侵入、根瘤形态建成和根瘤固氮调控的分子机制和遗传调控网络；挖掘具有重要育种价值的新基因和优异等位变异，为高效和稳效固氮豆科作物分子育种提供理论基础和优异基因资源；通过转基因等手段实现对固氮菌和植物识别、根瘤形态建成、作物高效和稳效固氮重要过程的重塑；深入解析固氮微生物固氮的分子调控网络和固氮酶等重要蛋白质的结构基础，在应用合成生物学设计、构建和创制高效固氮微生物及豆科植物方面取得突破；探索非豆科重要粮食作物生物固氮路径，在固氮菌识别与侵入非豆科作物，以及根瘤发生等方面实现概念验证和技术突破，为实现固氮新作物创制提供理论指导和技术储备。总之，到2035年前，针对生物固氮和具有生物固氮能力重要农作物培育和创制中迫切需要解决的关键科学问题，重点加强理论基础研究和技术探索，为解决我国生物固氮和固氮作物新品种培育中的重大理论和技术问题提供原创性成果，为培育高效和稳效固氮作物新品种和新作物奠定基础。

3.2 2050年目标

在2035年到2050年的15年间，飞速发展的技术，包括基因技术、数据分析技术、自动化和人工智能等，将进一步带动固氮生物学研究和培育固氮作物新品种的发展及突破。到2050年，能够实现对豆科植物共生固氮、微生物固氮和生物固氮的遗传机制和调控网络的全面解析；随着对生物固氮调控基因和调控网络复杂数据规律性认识的不断深入，可以利用数据分析技术直接设计和编纂参与生物固氮过程的基因和分子调控网络；实现通过基因工程（转基因技术）和合成生物学技术，操纵和改变微生物和作物固氮相关的生物路径；在创制高效固氮豆科植物及固氮和（或）共生固氮非豆科作物方面取得理论和技术突破；利用分子设计培育高效和稳效固氮豆科作物，并基本实现具有生物固氮能力新作物的创制，推动氮高效和生态环境友好作物的个性化设计和创制。

参考文献

Arrighi J.F., Godfroy, O., de Billy, F., Saurat, O., Jauneau, A., Gough, C. (2008). The *RPG* gene of *Medicago truncatula* controls *rhizobium*-directed polar growth during infection. Proc. Natl. Acad. Sci. USA *105*, 9817-9822.

Beatty, P.H., and Good, A.G. (2011). Plant Science. Future prospects for cereals that fix nitrogen. Science *333*, 416-417.

Behm, J.E., Geurts, R., and Kiers, E.T. (2014). *Parasponia*: a novel system for studying mutualism stability. Trends Plant Sci. *19*, 757-763.

Cerri, M.R., Wang, Q., Stolz, P., Folgmann, J., Frances, L., Katzer, K., Li, X., Heckmann, A.B., Wang, T.L., Downie, J.A., et al. (2017). The ERN1 transcription factor gene is a target of the CCaMK/CYCLOPS complex and controls rhizobial infection in *Lotus japonicus*. New Phytol. *215*, 323-337.

Charpentier, M., Bredemeier, R., Wanner, G., Takeda, N., Schleiff, E., and Parniske, M. (2008). *Lotus japonicus* CASTOR and POLLUX are ion channels essential for perinuclear calcium spiking in legume root endosymbiosis. Plant Cell *20*, 3467-3479.

Charpentier, M., and Oldroyd, G. (2010). How close are we to nitrogen-fixing cereals? Curr. Opin. Plant Biol. *13*, 556-564.

Charpentier, M., Sun, J., Vaz Martins, T., Radhakrishnan, G.V., Findlay, K., Soumpourou, E., Thouin, J., Very, A.A., Sanders, D., Morris, R.J., et al. (2016). Nuclear-localized cyclic nucleotide-gated channels mediate symbiotic calcium oscillations. Science *352*, 1102-1105.

Cheng Q., Dowson-Day M., Shen G., Ray D. (2005). The *Klebsiella* pneumoniae nitrogenase Fe protein gene (*nifH*) functionally substitutes for the *chlL* gene in *Chlamydomonas reinhardtii*. Biochem. Biophy. Res. *329*, 966-975.

Clavijo, F., Diedhiou, I., Vaissayre, V., Brottier, L., Acolatse, J., Moukouanga, D., Crabos, A., Auguy, F., Franche, C., Gherbi, H., et al. (2015). The Casuarina *NIN* gene is transcriptionally activated throughout *Frankia* root infection as well as in response to bacterial diffusible signals. New Phytol. *208*, 887-903.

Debellé, F., Promé, J.C. (1996). *Rhizobium* lipo-chitooligosaccharide nodulation factors: signaling molecules mediating recognition and morphogenesis. Annu. Rev. Biochem. *65*, 503-535.

Downie, J.A. (2014). Legume nodulation. Curr. Biol. *24*, R184-R190.

Farkas, A., Maroti, G., Durgo, H., Gyorgypal, Z., Lima, R.M., Medzihradszky, K.F., Kereszt, A., Mergaert, P., and Kondorosi, E. (2014). *Medicago truncatula* symbiotic peptide NCR247 contributes to bacteroid differentiation through multiple mechanisms. Proc. Natl. Acad. Sci. USA *111*, 5183-5188.

Fonouni-Farde, C., Diet, A., and Frugier, F. (2016). Root development and endosymbioses:

DELLAs lead the orchestra. Trends Plant Sci. *21*, 898-900.

Fonouni-Farde, S.T., Baudin, W., Brault M., Wen J., Kirankumar S. M., Andreas N., Frugier, F., and Diet, A. (2015). DELLA-mediated gibberellin signalling regulates Nod factor signalling and rhizobial infection. Nat. Commun. *7*, 12636.

Froussart, E., Bonneau, J., Franche, C., and Bogusz, D. (2016). Recent advances in actinorhizal symbiosis signaling. Plant Mol. Biol. *90*, 613-622.

Gautrat, P., Laffont, C., and Frugier, F. (2020). Compact root architecture 2 promotes root competence for nodulation through the *miR2111* systemic effector. Curr. Biol. *30*, 1339-1345.

Gleason, C., Chaudhuri, S., Yang, T., Munoz, A., Poovaiah, B.W., and Oldroyd, G.E. (2006). Nodulation independent of rhizobia induced by a calcium-activated kinase lacking autoinhibition. Nature *441*, 1149-1152.

Groth, M., Takeda, N., Perry, J., Uchida, H., Dräxl, S., Brachmann, A., Sato, S., Tabata, S., Kawaguchi, M., Wang. T.L., et al. (2010). NENA, a *lotus japonicus* homolog of *Sec13*, is required for rhizodermal infection by arbuscular mycorrhiza fungi and rhizobia but dispensable for cortical endosymbiotic development. Plant Cell *22*, 2509-2526.

Griesmann, M., Chang, Y., Liu, X., Song, Y., Haberer, G., Crook, M.B., Billault-Penneteau, B., Lauressergues, D., Keller, J., Imanishi, L., et al. (2018). Phylogenomics reveals multiple losses of nitrogen-fixing root nodule symbiosis. Science *361*, 144.

Hakoyama, T., Niimi, K., Watanabe, H., Tabata, R., Matsubara, J., Sato, S., Nakamura, Y., Tabata, S., Jichun, L., Matsumoto, T., et al. (2009). Host plant genome overcomes the lack of a bacterial gene for symbiotic nitrogen fixation. Nature *462*, 514-517.

Hartman, K., van der Heijden, M.G., Roussely-Provent, V., Walser, J.C., Schlaeppi, K. (2017). Deciphering composition and function of the root microbiome of a legume plant. Microbiome *5*, 2.

Heckmann, A.B., Sandal, N., Bek, A.S., Madsen, L.H., Jurkiewicz, A., Nielsen, M.W., Tirichine, L., and Stougaard, J. (2011). Cytokinin induction of root nodule primordia in *Lotus japonicus* is regulated by a mechanism operating in the root cortex. Mol. Plant Microbe Interact. *24*, 1385-1395.

Horváth, B., Domonkos, Á., Kereszt, A., Szűcs, A., Ábrahám, E., Ayaydin, F., Bóka, K., Chen,

Y., Chen, R., Murray, J.D., et al. (2015). Loss of the nodule-specific cysteine rich peptide, NCR169, abolishes symbiotic nitrogen fixation in the *Medicago truncatula dnf7* mutant. Proc. Natl. Acad. Sci. USA *112*, 15232-15237.

Horvath, B., Yeun, L.H., Domonkos, A., Halasz, G., Gobbato, E., Ayaydin, F., Miro, K., Hirsch, S., Sun, J., Tadege, M., et al. (2011). *Medicago truncatula* IPD3 is a member of the common symbiotic signaling pathway required for rhizobial and mycorrhizal symbioses. Mol. Plant Microbe Interact. *24*, 1345-1358.

Hossain, M.S., Liao, J., James, E.K., Sato, S., Tabata, S., Jurkiewicz, A., Madsen, L.H., Stougaard, J., Ross L, and Szczyglowski, K. (2012). *Lotus japonicus* ARPC1 is required for rhizobial infection. Plant Physiol. *160*, 917-928.

Hu, Y., and Ribbe, M.W. (2011). Biosynthesis of nitrogenase FeMoco. Coord. Chem Rev. *255*, 1218-1224.

Hu, Y., and Ribbe, M. W. (2013). Biosynthesis of the iron-molybdenum cofactor of nitrogenase. J. Biol. Chem. *288*, 13173-13177.

Hu, Y., and Ribbe, M.W. (2016). Biosynthesis of the metalloclusters of nitrogenases. Annu. Rev. Biochem. *85*, 455-483.

Jin, Y., Liu, H., Luo, D., Yu, N., Dong, W., Wang, C., Zhang, X., Dai, H., Yang, J., and Wang, E. (2016). DELLA proteins are common components of symbiotic rhizobial and mycorrhizal signalling pathways. Nat. Commun. *7*, 12433.

Jones, K.M., Kobayashi, H., Davies, B.W., Taga, M.E., and Walker, G.C. (2007). How rhizobial symbionts invade plants: the *Sinorhizobium-Medicago* model. Nat. Rev. Microbiol. *5*, 619-633.

Kaló, P., Gleason, C., Edwards, A., Marsh, J., Mitra, R.M., Hirsch, S., Jakab, J., Sims, S., Long, S.R., Rogers, J., et al. (2005). Nodulation signaling in legumes requires NSP2, a member of the GRAS family of transcriptional regulators. Science *308*, 1786-1789.

Kawaharada, Y., Kelly, S., Nielsen, M.W., Hjuler, C.T., Gysel, K., Muszyński, A., Carlson, R.W., Thygesen, M.B., Sandal, N., Asmussen, M.H., et al. (2015). Receptor-mediated exopolysaccharide perception controls bacterial infection. Nature *523*, 308-312.

Kawaharada, Y., Nielsen, M.W., Kelly, S., James, E.K., Andersen, K.R., Rasmussen, S.R., Fuchtbauer, W., Madsen, L.H., Heckmann, A.B., Radutoiu, S., et al. (2017). Differential regulation of the *Epr3* receptor coordinates membrane-restricted rhizobial colonization of root nodule primordia. Nat. Commun. *8*, 14534.

Kiss, E., Oláh, B., Kaló, P., Morales, M., Heckmann, A.B., Borbola, A., Lózsa, A., Kontár, K., Middleton, P., Downie, J.A., et al. (2009). LIN, a novel type of U-Box/WD40 protein, controls early infection by rhizobia in legumes. Plant Physiol. *151*, 1239-1249.

Kistner, C., Winzer, T., Pitzschke, A., Mulder, L., Sato, S., Kaneko, T., Tabata, S., Sandal, N., Stougaard, J., Webb, K.J., et al. (2005). Seven *Lotus japonicus* genes required for transcriptional reprogramming of the root during fungal and bacterial symbiosis. Plant Cell *17*, 2217-2229.

Krusell, L., Krause, K., Ott, T., Desbrosses, G., Kramer, U., Sato, S., Nakamura, Y., Tabata, S., James, E.K., Sandal, N., et al. (2005). The sulfate transporter SST1 is crucial for symbiotic nitrogen fixation in *Lotus japonicus* root nodules. Plant Cell *17*, 1625-1636.

Li, F.W., Brouwer, P., Carretero-Paulet, L., Cheng, S., de Vries, J., Delaux, P.M., Eily, A., Koppers, N., Kuo, L.Y., Li, Z., et al. (2018). Fern genomes elucidate land plant evolution and cyanobacterial symbioses. Nat. Plants *4*, 460-472.

Limpens, C.F., Smit, P., Willemse, J., Bisseling, T., and Geurts, R. (2003). LysM domain receptor kinases regulating rhizobial Nod factor-induced infection. Science *24*, 630-633.

Lin, J.S., Li, X., Luo, Z.L., Mysore, K.S., Wen, J., and Xie, F. (2018). NIN interacts with NLPs to mediate nitrate inhibition of nodulation in *Medicago truncatula*. Nat. Plants *4*, 942-952.

Long, A.P.R., and Long, S.R. (2011). Pseudonodule formation by wild type and symbiotic mutant *Medicago truncatula* in response to auxin transport inhibitors. Mol. Plant-Microbe Inter. *11*, 1372-1384.

Maillet, F., Fournier, J., Mendis, H.C., Tadege, M., Wen, J., Ratet, P., Mysore, K.S., Gough, C., and Jones, K. (2020). *Sinorhizobium meliloti* succinylated high-molecular-weight succionglycan and the *Medicago truncatula* LysM receptor-like kinase MtLYK10 participate independently in symbiotic infection. Plant J. *102*, 311-326.

Marsh, J.F., Rakocevic, A., Mitra, R.M., Brocard, L., Sun, J., Eschstruth, A., Long, S.R., Schultze, M., Ratet, P., and Oldroyd, G.E. (2007). *Medicago truncatula* NIN is essential for rhizobial-independent nodule organogenesis induced by autoactive calcium/calmodulin-dependent protein kinase. Plant Physiol. *144*, 324-335.

Middleton, P.H., Jakab, J., Penmetsa, R.V., Starker, C.G., Doll, J., Kalo, P., Prabhu, R., Marsh, J.F., Mitra, R.M., Kereszt, A., et al. (2007). An ERF transcription factor in *Medicago truncatula* that is essential for Nod factor signal transduction. Plant Cell *19*, 1221-1234.

Murray, J.D., Torres-Jerez, I., Tang, Y., Allen, S., Andriankaja, M., Li, G., Laxmi, A., Cheng, X., Wen, J., et al. (2011). *Vapyrin*, a gene essential for intracellular progression of arbuscular mycorrhizal symbiosis, is also essential for infection by rhizobia in the nodule symbiosis of *Medicago truncatula*. Plant J. *65*, 244-252.

Mus, F., Crook, M.B., Garcia, K., Garcia Costas, A., Geddes, B.A., Kouri, E.D., Paramasivan, P., Ryu, M.H., Oldroyd, G.E.D., Poole, P.S., et al. (2016). Symbiotic nitrogen fixation and the challenges to its extension to nonlegumes. Appl. Environ. Microbiol. *82*, 3698-3710.

Nishida, H., Tanaka, S., Handa, Y., Ito, M., Sakamoto, Y., Matsunaga, S., Betsuyaku, S., Miura, K., Soyano, T., Kawaguchi, M., et al. (2018). A NIN-LIKE PROTEIN mediates nitrate-induced control of root nodule symbiosis in *Lotus japonicus*. Nat. Commun. *9*, 499.

Okamoto, S., Shinohara, H., Mori, T., Matsubayashi, Y., and Kawaguchi, M. (2013). Root-derived CLE glycopeptides control nodulation by direct binding to HAR1 receptor kinase. Nat. Commun. *4*, 2191.

Oldroyd, G.E., and Dixon, R. (2014). Biotechnological solutions to the nitrogen problem. Curr. Opin. Biotechnol. *26*, 19-24.

Oldroyd, G.E., and Downie, J.A. (2008). Coordinating nodule morphogenesis with rhizobial infection in legumes. Annu. Rev. Plant Biol. *59*, 519-546.

Oldroyd, G.E., Murray, J.D., Poole, P.S., and Downie, J.A. (2011). The rules of engagement in the legume-rhizobial symbiosis. Annu. Rev. Genet. *45*, 119-144.

Ott, T., van Dongen, J.T., Gunther, C., Krusell, L., Desbrosses, G., Vigeolas, H., Bock, V., Czechowski, T., Geigenberger, P., and Udvardi, M.K. (2005). Symbiotic leghemoglobins are crucial for nitrogen fixation in legume root nodules but not for general plant growth and development. Curr. Biol. *15*, 531-535.

Pan, H., and Wang, D. (2017). Nodule cysteine-rich peptides maintain a working balance during nitrogen-fixing symbiosis. Nat. Plants *3*, 17048.

Parniske, M. (2008). Arbuscular mycorrhiza: the mother of plant root endosymbioses. Nat. Rev. Microbiol. *6*, 763-775.

Qiu, L., Lin, J.S., Xu, J., Sato, S., Parniske, M., Wang, T.L., Downie, J.A., and Xie, F. (2015). SCARN a novel class of SCAR protein that is required for root-hair infection during legume nodulation. PLoS Genet. *11*, e1005623.

Radutoiu, M., Madsen, E., Felle, H.H., Umehara, Y., Grønlund, M., Sato, Y., Nakamura, Y., Tabata, S., Sandal, N., and Stougaard, J. (2003). Plant recognition of symbiotic bacteria requires two LysM receptor-like kinase. Nature *425*, 585-592.

Reid, D.E., Ferguson, B.J., Hayashi, S., Lin, Y.H., and Gresshoff, P.M. (2011). Molecular mechanisms controlling legume autoregulation of nodulation. Ann. Bot. *108*, 789-795.

Rosenblueth, M., Ormeño-Orrillo, E., López-López, A., Rogel, M.A., Reyes-Hernández, B.J., Martínez-Romero, J.C., Reddy, P.M., and Martínez-Romero, E. (2018). Nitrogen fixation in cereals. Front. Microbiol. *9*, 1974.

Schauser, A.R., Stiller, J., and Stougaard, J. (1999). A plant regulator controlling development of symbiotic root nodules. Nature *402*, 191-195.

Schiessl, K., Lilley, J.L.S., Lee, T., Tamvakis, I., Kohlen, W., Bailey, P.C., Thomas, A., Luptak, J., Ramakrishnan, K., Carpenter, M.D., et al. (2019). NODULE INCEPTION recruits the lateral root developmental program for symbiotic nodule organogenesis in Medicago truncatula. Curr. Biol. *29*, 3657-3668.

Soyano, T., Kouchi, H., Hirota, A., Hayashi, M. (2013). NODULE INCEPTION directly targets NF-Y subunit genes to regulate essential processes of root nodule development in *Lotus japonicus*. PLoS Genet. *9*, e1003352.

Soyano, T., Shimoda, Y., Kawaguchi, M. and Hayashi, M. (2019). A shared gene drives lateral root development and root nodule symbiosis pathways in *Lotus*. Science *366*, 1021-1023.

Sprent, J.I. (2007). Evolving ideas of legume evolution and diversity: a taxonomic perspective on the occurrence of nodulation. New Phytol. *174*, 11-25

Saito, K., Yoshikawa, M., Yano, K., Miwa, H., Uchida, H., Asamizu, E., Sato, S., Tabata, S., Imaizumi-Anraku, H., Umehara, Y., et al. (2007). NUCLEOPORIN85 is required for calcium spiking, fungal and bacterial symbioses, and seed production in *Lotus japonicus*. Plant Cell *19*, 610-624.

Sasaki, T., Suzaki, T., Soyano, T., Kojima, M., Sakakibara, H., and Kawaguchi, M. (2014). Shoot-derived cytokinins systemically regulate root nodulation. Nat. Commun. *5*, 4983.

Singh, S., Katzer, K., Lambert, J., Cerri, M., and Parniske, M. (2014). CYCLOPS, a DNA-binding transcriptional activator, orchestrates symbiotic root nodule development. Cell Host Microbe *15*, 139-152.

Starker, C.G., Parra-Colmenares, A.L., Smith, L., Mitra, R.M., and Long, S.R. (2006). Nitrogen fixation mutants of *Medicago truncatula* fail to support plant and bacterial symbiotic gene expression. Plant Physiol. *140*, 671-680.

Stracke, S., Kistner, C., Yoshida, S., Mulder, L., Sato, S., Kaneko, T., Tabata, S., Sandal, N., Stougaard, J., Szczyglowski, K., et al. (2002). A plant receptor-like kinase required for both bacterial and fungal symbiosis. Nature *417*, 959-962.

Svistoonoff, S., Hocher, V., and Gherbi, H. (2014). Actinorhizal root nodule symbioses: what is signalling telling on the origins of nodulation? Curr. Opin. Plant Biol. *20*, 11-18.

Tirichine, L., Imaizumi-Anraku, H., Yoshida, S., Murakami, Y., Madsen, L.H., Miwa, H., Nakagawa, T., Sandal, N., Albrektsen, A.S., Kawaguchi, M., et al. (2006). Deregulation of a Ca^{2+}/calmodulin-dependent kinase leads to spontaneous nodule development. Nature *441*, 1153-1156.

Tirichine, N.S., Madsen, L.H., Radutoiu, S., Albrektsen, A.S., Sato, S., Asamizu, E., Tabata, S., Stougaard, J. (2007). A gain-of-function mutation in a cytokinin receptor triggers spontaneous root nodule organogenesis. Science *315*, 104-107.

Tsikou, Z.Y., Holt, D.B., Abel, N.B., Reid, D.E., Madsen, L.H., Bhasin, H., Sexauer, M., Stougaard, J., and Markmann, K. (2018). Systemic control of legume susceptibility to rhizobial infection by a mobile microRNA. Science *362*, 233-236.

van Deynze, A., Zamora, P., Delaux, P.M., Heitmann, C., Jayaraman, D., Rajasekar, S., Graham, D., Maeda, J., Gibson, D., Schwartz, K.D., et al. (2018). Nitrogen fixation in a landrace of maize is supported by a mucilage-associated diazotrophic microbiota. PLoS Biol. *16*, e2006352.

van Velzen, R., Holmer, R., Bu, F., Rutten, L., van Zeijl, A., Liu, W., Santuari, L., Cao, Q., Sharma, T., Shen, D., et al. (2018). Comparative genomics of the nonlegume *Parasponia* reveals insights into evolution of nitrogen-fixing rhizobium symbioses. Proc. Natl. Acad. Sci. USA *115*, E4700-E4709.

Wang, D., Griffitts, J., Starker, C., Fedorova, E., Limpens, E., Ivanov, S., Bisseling, T., and

Long, S. (2010). A nodule-specific protein secretory pathway required for nitrogen-fixing symbiosis. Science *327*, 1126-1129.

Wang, L., Rubio, M.C., Xin, X., Zhang, B., Fang, Q., Wang, Q., Ning, G., Becana, M., and Duanmu, D. (2019). CRISP/Cas9 knockout of leghemoglobin genes in *Lotus japonicus* uncovers their synergistic roles in symbiotic nitrogen fixation. New Phytol. *224*, 818-832.

Weisany, Y.R., and Allahverdipoor, K.H. (2013). Role of some of mineral nutrients in biological nitrogen fixation. Bull. Environ. Pharmacol. Life Sci. *2*, 77-84.

Xie, F., Murray, J.D., Kim, J., Heckmann, A.B., Edwards, A., Oldroyd, G.E.D., and Downie, J.A. (2012). Legume pectate lyase required for root infection by rhizobia. Proc. Natl. Acad. Sci. USA *109*, 633-698.

Yang, J., Xie, X., Yang, M., Dixon R., and Wang, Y. (2017). Modular electron-transport chains from eukaryotic organelles function to support nitrogenase activity. Proc. Natl. Acad. Sci. USA *114*, E2460-E2465.

Yano, K., Shibata, S., Chen, W.L., Sato, S., Kaneko, T., Jurkiewicz, A., Sandal, N., Banba, M., Imaizumi-Anraku, H., Kojima, T., et al. (2009). CERBERUS, a novel U-box protein containing WD-40 repeats, is required for formation of the infection thread and nodule development in the legume-rhizobium symbiosis. Plant J. *60*, 168-180.

Yokota, K., Fukai, E., Madsen, L.H., Jurkiewicz, A., Rueda, P., Radutoiu, S., Held, M., Hossain, M.S., Szczyglowski, K., Morieri, G., et al. (2009). Rearrangement of actin cytoskeleton mediates invasion of *Lotus japonicus* roots by *Mesorhizobium loti*. Plant Cell *21*, 267-284.

Zgadzaj, R., Garrido-Oter, R., Jensen, D.B., Koprivova, A., Schulze-Lefert, P., and Radutoiu, S. (2016). Root nodule symbiosis in *Lotus japonicus* drives the establishment of distinctive rhizosphere, root, and nodule bacterial communities. Proc. Natl. Acad. Sci. USA *113*, E7996-E8005.

病虫害与抗性

何祖华　周俭民　周雪平　李传友　何光存

1 国内外研究进展

作物病虫害是全世界农业的普遍威胁，抗病虫是作物育种的主要目标性状。从 20 世纪初植物病理学家和育种家认识到作物抗病虫性由基因决定以来，世界主要国家的研究人员在抗病虫基因发掘、机制研究和育种上均取得了重大进展。

1.1 我国作物病虫害危害概况

农作物病虫害长期影响我国农业的稳定和绿色生产。同时，随着全球气候变暖，病虫害的流行规律也发生改变，第一是原生性农作物病虫频繁暴发，如稻瘟病、褐飞虱等连年发生；第二是一些次要病虫害逐渐发展成为毁灭性灾害，如稻曲病、黏虫等。近年来，农作物病虫害年发生面积超过 70 亿亩·次（1 亩 $\approx$ 0.067 hm^2），在大量使用农药的情况下，仍然损失粮食 1600 万吨、棉花 50 万吨、油料作物 90 万吨及其他作物 1100 万吨。全国每年防治病虫危害面积达 5.61 亿 hm^2·次，是耕地面积的 4.16 倍，即每年每一块耕地上至少实施防治 4 次，严重影响环境与人类健康。因此，培育抗病虫作物是作物改良的重要目标，也是未来作物绿色生产的主要方向。

1.2 作物抗病虫研究历程

1905 年，英国科学家 Biffen 首次发现植物具有抗病基因。1917 年，美国植物病理学家 Stakman 发现病原真菌"生理小种"，揭示了病原变异和植物抗病的特异性。1947 年，美国植物病理学家 Flor 建立植物抗病性和病原真菌致病性"基因对基因"假说。之后数十年，欧美科学家发现了"激发子"与"植保素"，并建立了"诱导抗性"的概念，先后完成了病原菌无毒基因的克隆（20世纪80年代）、抗病虫基因的克隆（20世纪90年代），以及激发子与受体的研究（21世纪初）。随着国内外对病原菌效应子研究的进展，科学家们建立了植物免疫学的理论框架，认为植物虽然没有动物那样的依赖抗体的抗体适应性免疫（adaptive immunity），但具有与动物类似的先天免疫（innate immunity）（Ausubel，2005）。植物免疫系统的三大支柱包括定位于质膜的模式识别受体（pattern recognition receptor，PRR）、病原效应子和定位于胞质识别病原效应子的由抗病基因编码的免疫受体。植物先天免疫的理论框架包含两个层次：第一层次是通过细胞膜上的模式识别受体PRR识别病原分泌的保守分子结构（病原体相关分子模式pathogen associated molecular pattern，PAMP 或microbe associated molecular pattern，MAMP）激发基础防卫反应，称为病原分子模式诱导的免疫反应（PAMP-triggered immunity，PTI）；第二层次是胞内免疫受体（R蛋白）识别病原效应子激发的专化性防卫反应过程（effector-triggered immunity，ETI）（Chisholm et al.，2006；Dangl et al.，2013；Jones and Dangl，2006；Nelson et al.，2017；Liang et al.，2018）。胞内免疫受体主要为一类NLR（nucleotide-binding domain and leucine-rich repeat receptor）受体蛋白，也是目前作物抗病育种的主要靶标基因（张杰等，2019；Dangl et al.，2013；Deng et al.，2017；Le Roux et al.，2015；Li et al.，2020；Shen et al.，2007）。这一理论框架适用于包括病毒、细菌、真菌、卵菌、线虫、刺吸式害虫及寄生植物等多种病原生物。近10年来，植物天然免疫的基础研究主要集中在两大类受体、病虫效应子、植物免疫信号转导、植物抗性生理与产量的协调、抗病虫基因演化与分子设计育种等方面。围绕这些研究领域，国际上从不同植物中克隆了几百个免疫受体或抗病基因（Dangl et al.，2013；Nelson et al.，2017；钱韦等，2016；张杰等，2019）。

植物激素对抗病虫发挥重要调控作用，同时也调控植物免疫与发育（产量性状）的互作，这是近20年来国际上研究的热点。植物免疫信号下游由防卫激素水杨酸（salicylic acid，SA）、茉莉酸（jasmonic acid，JA）、乙烯（ethylene）等途径组成复杂信号网络（Bürger and Chory，2019；Yang et al.，2013）。其中，水杨酸对抵抗活体营养型和半活体营养型病原菌的免疫反应尤为重要，其受体为NPR1、NPR3和NPR4，分别正向和反向调节植物的免疫反应（Ding et al.，2018；Fu et al.，2012）。与此相反，茉莉酸调节对腐生病原菌和咀嚼式昆虫的防卫反应以及伤害反应（Sun et al., 2011；Yan et al., 2018）。乙烯常与茉莉酸协同参与植物免疫的调控，但乙烯信号单独如何参与植物免疫调控尚不清楚。水杨酸的免疫途径与茉莉酸-乙烯的免疫途径之间在多个层次上发生拮抗，但也存在协同调控。此外，其他生长发育的植物激素，如生长素（auxin）、细胞分裂素（cytokinin）、赤霉素（gibberellin）、脱落酸（abscisic acid）和油菜素甾醇（brassinosteroid）等，也通过交互作用影响植物免疫反应。这些调控作用在作物抗病性与产量等重要农艺性状间起到平衡作用。

1.3 我国抗病虫机制研究进展

我国科学家在植物与微生物互作和抗病虫领域长期耕耘，逐渐在一些领域形成了优势。尤其是近10年来，通过973计划、转基因重大专项和七大作物育种等项目的资助，我国在植物免疫基础理论和作物抗病虫基因克隆、功能解析和分子育种应用上均取得了重大的进展（钱韦等，2016；唐威华等，2017；张杰等，2019）。我国科研人员在国际上率先阐释了宿主免疫受体识别病原分子、免疫激活或抑制的分子机制，提出了植物免疫的"诱饵模型"等；在水稻抗稻瘟病、白叶枯病、条纹叶枯病和褐飞虱，小麦抗白粉病和条锈病，玉米抗丝黑穗病和茎腐病等抗病虫基因的克隆及功能研究上处于国际领先，并将以上研究成果成功应用于抗病虫分子育种；我国在稻曲病的抗性鉴定与资源发掘上也处于国际领先；此外，我国科研人员在植物抗病-发育互作、茉莉酸介导的防卫信号通路和植物病毒等方面的研究也形成了自己独特的优势；特别重要的是，我国在水稻广谱抗病、水稻抗稻飞虱、小麦抗赤霉病、抗性与产量协调、基因编辑创制抗病小麦、植物与卵菌互作和利用RNAi转基因技术抗病等方面的研究已经进入世界前列（Deng et al., 2017；Guo et al., 2018；Li et al., 2017, 2019；Liu et al., 2015；Ma et al., 2017；Mao et al., 2007；Su et al., 2019；Wang et al., 2014, 2018；Zhang et al., 2015, 2016）。根据国家自然科学基金委员会的统计数据，2006～2015年，我国植物病理学研究的论文总篇数和总被引用频次一直稳居世界第二。近5年，我国科研人员在植物与微生物互作领域先后在 Science 和 Cell 等国际顶级学术期刊上发表了12项突破性的研究成果，尤其是最近对植物NLR免疫受体的结构域功能解析（Wang et al.，2019a，2019b），标志着中国在植物与微生物互作领域的研究水平整体上产生了飞跃。

稻瘟病在2012年被列为十大真菌病害之首。目前至少已经有28个抗瘟性基因（包括等位基因）被克隆，这些基因绝大多数编码NLR蛋白（张杰等，2019）。小麦的抗病研究主要集中在锈病、白粉病和赤霉病，目前国际上已经克隆了多个抗锈病和白粉病基因，绝大多数也编码NLR蛋白。小麦抗赤霉病属于多基因控制的数量性状（QTL），且抗源匮乏，至今科研人员只发现少数几个中度抗性基因，而我国科研人员克隆了抗赤霉病的主效QTL基因。我国在其他作物的抗病性研究上也取得了重大的进展，如玉米、大豆和番茄等，克隆了大批主基因和主效QTL抗病基因。马铃薯Y科病毒和双生病毒科病毒组成了世界上种类最多、危害最为严重的病毒，我国科研人员在这方面做了大量病毒基因组和与寄主互作的研究，鉴定出一些抗病相关基因和病毒靶标。另外，我国科研人员在水稻抗条纹叶枯病毒等方面也做出了国际领先的研究成果（Fu et al.，2018；Wu et al.，2017）。植物病毒病基本无药可用，抗病育种是关键，但是植物抗病毒病基因相对较少，主要是通过RNAi转基因或挖掘隐性抗病基因等实现对病毒的靶向抑制，目前国内外已成功商业化木瓜抗环斑病毒转基因品系。

在作物抗虫方面，我国科研人员在发掘作物自身的抗虫基因、解析作物抗虫机制方面取得了一系列重要进展，主要体现在对水稻抗稻飞虱和其他植物抗蚜虫的研究。近10年来，相继成功克隆了十几个抗褐飞虱基因，分别编码不同的蛋白，包括NLR、凝集素受体激酶OsLecRK、DNA结合蛋白等；其他已克隆的抗虫基因有番茄抗蚜虫和线虫基因 *Mi-1.2*，甜瓜抗蚜虫基因 *Vat*，这些基因也都编码NLR受体；此外，小麦和大豆抗蚜虫方面也通过品种鉴定和分子标记分析检测出了多个抗性位点。

我国科研人员在21世纪初率先提出作物抗病与产量协调的研究理念，并通过植物激素之间的交互作用研究植物激素对抗病虫以及其与生长发育性状互作的调控机理。此外，科研人员系统研究了防卫激素水杨酸、茉莉酸和乙烯等信号网络及其与其他发育激素（如赤霉素、生长素、油菜素内酯）的交互作用，发表了多篇重要的代表性论文，对于未来作物的高产高抗平衡育种具有重要的指导意义。

2 针对未来作物分子设计需求的研究内容与突破口

未来作物的特征是优质高产、绿色生产、资源节约、环境友好、广适性和设施农业专用，强化对病虫害的抗性是未来作物育种的重要一环。此外，针对全球气候变暖和新病虫害的出现，需要前瞻性布局抗病虫研究。

2.1 我国抗病虫研究的短板及未来突破口

我国抗病虫研究的短板及未来突破口主要有以下几个方面。

加强重大原创性研究 虽然我国抗病虫研究在过去十几年取得了长足进步，但相比农业学科其他领域，研究团队相对薄弱，在模式识别受体研究领域尤其单薄，新受体的分离鉴定几乎完全是欧美国家的天下。模式识别受体对于广谱抗性和缺乏高抗抗原的病害具有重要意义，因此有必要组织力量通过野生资源分析，结合生化鉴定，获得一批具有自主知识产权的重要受体基因，用以开展跨物种抗病虫育种。此外，胞内免疫受体尤其是广谱抗病受体的作用机制研究有待加强。目前，植物胞内免疫受体蛋白的结构和作用机制解析已有重大突破，这可能为人工合成广谱抗病新基因奠定基础。此外，有必要研究作物中新的免疫调控模式及其信号网络，突破国际上对这一领域固有的模式化研究体系与理念。

解决抗病基因应用与病原变异迅速的矛盾 目前，抗病基因尤其是广谱抗病受体基因的分离鉴定完全依赖前期已知的遗传材料，通过传统的图位克隆方法获得，工作量大且效率低。而且，田间病原微生物常通过群体变异迅速导致品种抗性丧失。如何高效分离鉴定新的广谱抗病基因是急需解决的重大问题。建立抗病基因的全基因组分析方法，高效挖掘新的抗病基因尤其是广谱抗病基因，以及解析作物免疫激发的共性与特异性机制都需要尽快开展。此外，要原创性布局作物抗病基因驯化与广谱抗病重构的前沿领域，创新作物抗病研究体系与分子育种基因资源。

持续性发生的重大病害抗源匮乏 现代农业的规模化种植以及耕作制度和气候的变化，伴随而来的是新旧病虫害暴发。近年来，我国小麦赤霉病发病区域和面积不断扩大，水稻稻曲病和纹枯病危害持续上升，这些病害缺乏主效抗病基因资源。未来急需针对抗病 QTL、感病基因和异源物种中的免疫受体进行广泛的分析,明确相应的分子机制,创立新的分子育种技术路线。此外,由稻瘟病菌进化而来的病原菌已经感染并引发小麦瘟,该病最早在南美洲发生,但近年来在孟加拉国大面积发生。随着春季温度的升高,该病在我国小麦产区大暴发的可能性日益增加。因此,我国有必要未雨绸缪,前瞻性布局未来小麦抗瘟性育种,通过小麦抗瘟基因的发掘,以及采用水稻的广谱抗瘟基因进行小麦转基因或基因编辑育种来解决问题。

作物抗虫基因资源发掘与抗性机制研究仍需加强 目前除了 Bt 转基因抗虫技术外,已经克隆的作物刺吸式害虫的基因在水稻等作物育种中也能够发挥作用,但是类型较少。抗咀嚼式害虫的抗性资源则更少见,尤其是对广食性害虫(如螟虫和菜粉蝶等),进一步挖掘作物抗虫基因,为作物分子设计抗虫品种提供更多的可供选用的基因是下一步研究的重点之一。其他研究重点还包括:理论上深入研究作物与害虫互作分子机制、害虫效应子的作用机制,阐明抗虫生态型的专化性机制;以非粮作物为主,进一步完善 siRNA 介导的抗虫技术体系,争取尽早大面积推广基于 RNAi 的精准抗虫分子育种技术。

新病害的发生和病害复合侵染 随着全球气候变暖及耕作制度的变化,一些新的作物病虫害渐成多发趋势,如水稻穗腐病和稻绿蝽等。我国需要尽早布局对新病虫害的研究,以期为未来作物抗性育种提供新的基因资源。此外,在长期的作物病区存在不同病害的复合侵染,对此进行研究将是一个新的创新领域,同时可为对不同病害的广谱抗病的理论与育种技术研究奠定基础。

适应未来农业工厂化生产的抗病虫育种 随着人工智能的日新月异,未来农业将进入"工厂化农业"新生产模式,相应的育种选择目标也要改变,尤其蔬菜和瓜果类作物。工厂化生产可能产生新的重大病虫害,如霜霉病、青枯病、粉虱和蚜虫等,我国需要前瞻性部署工厂化农业的抗病虫育种。

2.2 未来作物多抗育种的瓶颈问题

未来作物抗病虫育种必须解决三个瓶颈问题。第一,抗病与抗虫的多抗性协调。由于下游 SA 与 JA 激素信号的交互,抗病和抗虫性状在很多情况下存在拮抗作用。水稻稻瘟病-稻飞虱抗性提供了一个理想的抗病虫耦合的多抗理论与育种研究体系,两者的很多抗病基因均是 NLR。此外,目前已有的数据表明这些不同的 NLR 等位基因之间有时会引起杂种劣势的问题，这也为未来作物杂交品种选育提供新的技术指导。第二，抗病虫与高产性状的耦合。作物抗病虫与产量性状有交互影响，二者存在一定程度的相互拮抗，主要由激素途径的交互作用控制。目前，以水稻作为模式作物的抗病虫和与产量性状交互作用的激素调控网络已初步建立，建立与完善这些分子网络可以使未来作物获得多抗与高产的育种性状，关键在于抗病虫的强化如何避免或弥补在某些产量性状上的负效应。例如，抗病虫可能影响穗子、籽粒大小或结实率，找到相应的产量性状基因进行设计育种从而达到高产高抗平衡的育种目标。第三，如何整合高抗的 NLR 受体基因和广谱抗性的 PRR 受体重构作物的广谱抗病性。利用全基因组分析技术，对控制这些免疫性状的基因网络进行系统的耦合效应解析与预测，将是未来作物多抗/高抗分子育种的技术关键。

3 阶段性目标

我国作物抗病虫育种已有近 60 年的历史，主要从 20 世纪的国家科技攻关 "六五" 计划开始，围绕水稻抗稻瘟病、抗白叶枯病、抗稻飞虱和小麦锈病等开展了卓有成效的抗病遗传研究和育种工作，使我国在水稻和麦类抗病育种上趋于国际领先地位。过去 15 年来，我国在作物抗病虫领域不仅产生了一大批高水平原创性成果，并形成了多个优秀研究团队。我国科研人员已经从先前的国际"跟跑者"，逐渐成长为"并跑者"，尤其可喜的是在植物免疫受体结构与功能、植物广谱抗病和抗飞虱等方向上已经成长为"领跑者"，受到国际同行的高度关注。

未来作物的抗病虫育种需要在基础研究上有重大源头创新。一方面需要发掘新的有重要应用价值的抗病虫基因，包括潜在暴发性病虫害的抗性基因；另一方面发展抗病虫育种的核心技术，包括全基因组选择和基因功能的获得性编辑等。2019 年，第十八届 IS-MPMI 大会（The International Society for Molecular Plant-Microbe Interactions XVIII Congress）提出了植物与微生物互作的 10 个重大科学问题：① 植物在与有益微生物互作的同时，如何限制病原微生物；② 环境因素如何影响植物与微生物互作；③ 如何才能将植物免疫基础研究应用到农作物抗病虫害中去；④微生物与微生物互作如何影响植物与微生物互作；⑤非寄主抗性的分子基础是什么；⑥NLR蛋白如何诱导细胞死亡；⑦为什么有的病原微生物需要如此多的效应蛋白，而有的却需要很少；⑧病原菌如何产生新的毒性功能；⑨植物与微生物二元互作研究得到的发现在生态系统中是否能被证实；⑩如何发掘针对广谱腐生菌如纹枯病和完全活体病原菌如稻曲病的抗病基因并有效应用于抗病育种？这10个问题中，问题③、⑥和⑩由我国学者提出或共同提出。这些问题为我国未来作物抗病虫研究提供了很好的提示。我国科学家应力争在植物免疫基础理论与未来作物抗性设计改良上有重大突破，以便更好地服务于我国农作物抗病虫绿色生产这一重大战略需求。

3.1 2035年目标

根据我国作物抗病虫育种发展现状与未来农业生产所面临的问题，至2035年，我国应重点布局以下9个抗病虫研究前沿领域与育种目标，强调理论创新与育种应用的重大产出，在植物免疫受体、全基因组作物抗病虫基因发掘与育种应用等方面引领国际前沿。主要阶段性目标包括：①发现新的植物免疫与广谱抗病新机制，引领国际前沿；②创立作物抗病虫驯化选择与广谱抗性重构的新领域；③普遍开展抗病虫的分子育种和基因编辑精准育种，农业主产区品种推广做到稻瘟病、白叶枯病、稻飞虱、小麦锈病、马铃薯晚疫病、玉米黑穗病和大豆胞囊线虫等高抗性；④在水稻稻曲病、纹枯病和小麦赤霉病抗病育种与应用推广上获得重大突破，育成系列新品种；⑤在咀嚼式害虫的抗性育种上具有新的基因资源与技术突破；⑥建立成熟的以水稻为模式的作物抗病虫多抗育种的体系并推广应用；⑦建立对潜在新病虫害尤其是外来重大危害物种的早期预警和抗性育种对策；⑧普遍推广设施农作物的抗病育种；⑨广泛在抗病虫选育中应用人工智能。

3.2 2050年目标

至2050年的长期目标为全方位引领国际植物抗病虫研究，从理论和核心技术上把控国际发展方向，广泛部署和应用大数据、人工智能与合成生物学等技术于作物抗病虫的育种，包括：①在植物免疫、作物广谱抗病和抗虫领域全方位引领国际前沿；② 主要作物实现重大病虫害的多抗育种，基本做到零农药使用；③ 引领设施农作物的抗病育种研究；④ 作物抗病虫育种进入到合成生物学时代，做到抗病虫基因的精准调控与新基因的创造；⑤ 开发新的可视化分子技术，实现高效的基因整合和改造；⑥ 利用人工智能与互联网技术实现对病虫害的预测与抗性品种自动部署。

致谢：本章在撰写过程中得到了邓一文和尹昕博士的协助，特此致谢！

参考文献

钱韦, 方荣祥, 何祖华 (2016). 植物免疫与作物抗病分子育种的重大理论基础——进展与设想. 中国基础科学. 植物科学专刊 *18*, 38-45.

唐威华, 冷冰, 何祖华 (2017). 植物抗病虫与抗逆. 植物生理学报 *53*, 1333-1336.

张杰, 董莎萌, 王伟, 赵建华, 陈学伟, 郭惠珊, 何光存, 何祖华, 康振生, 李毅, 彭友良, 王国梁, 周雪平, 王源超, 周俭民 (2019). 植物免疫研究与抗病虫绿色防控：进展、机遇与挑战. 中国科学：生命科学 *11*, 1479-1507.

Ausubel, F.M. (2005). Are innate immune signaling pathways in plants and animals conserved? Nat. Immunol. *6*, 973-979.

Bürger, M, and Chory, J. (2019). Stressed out about hormones: how plants orchestrate immunity. Cell Host Microbe *14*, 163-172.

Chisholm, S.T., Coaker, G., Day, B., and Staskawicz, B.J. (2006). Host-microbe interactions: shaping the evolution of the plant immune response. Cell *124*, 803-814.

Dangl, J.L., Horvath, D.M., and Staskawicz, B.J. (2013). Pivoting the plant immune system from dissection to deployment. Science *341*, 746-751.

Deng, Y., Zhai, K., Xie, Z., Yang, D., Zhu, X., Liu, J., Wang, X., Qin, P., Yang, Y., Zhang, G., et al. (2017). Epigenetic regulation of antagonistic receptors confers rice blast resistance with yield balance. Science *355*, 962-965.

Ding, Y., Sun, T., Ao, K., Peng, Y., Zhang, Y., Li, X., and Zhang, Y. (2018). Opposite roles of salicylic acid receptors NPR1 and NPR3/NPR4 in transcriptional regulation of plant immunity. Cell *173*, 1454-1467.

Fu, S., Xu, Y., Li, C., Li, Y., Wu, J., and Zhou, X. (2018). Rice stripe virus interferes with S-acylation of remorin and induces its autophagic degradation to facilitate virus infection. Mol. Plant *11*, 269-287.

Fu, Z.Q., Yan, S., Saleh, A., Wang, W., Ruble, J., Oka, N., Mohan, R., Spoel, S.H., Tada, Y., Zheng, N., et al. (2012). NPR3 and NPR4 are receptors for the immune signal salicylic acid in plants. Nature *486*, 228-232.

Guo, J., Xu, C., Wu, D., Zhao, Y., Qiu, Y., Wang, X., Ouyang, Y., Cai, B., Liu, X., Jing, S., et al. (2018). *Bph6* encodes an exocyst-localized protein and confers broad resistance to planthoppers in rice. Nat. Genet. *50*, 297-306.

Jones, J.D.G., and Dangl, J.L. (2006). The plant immune system. Nature *444*, 323-329.

Le Roux, C., Huet, G., Jauneau, A., Camborde, L., Trémousaygue, D., Kraut, A., Zhou, B., Levaillant, M., Adachi, H., Yoshioka, H., et al. (2015). A receptor pair with an integrated decoy converts pathogen disabling of transcription factors to immunity. Cell *161*, 1074-1088.

Li, G., Zhou, J., Jia, H., Gao, Z., Fan, M., Luo, Y., Zhao, P., Xue, S., Li, N., Yuan, Y., et al. (2019). Mutation of a histidine-rich calcium-binding-protein gene in wheat confers resistance to *Fusarium* head blight. Nat. Genet. *51*, 1106-1112.

Li, W., Deng, Y., Ning, Y., He, Z., and Wang, G.L. (2020). Exploiting broad-spectrum disease resistance in crops: from molecular dissection to breeding. Annu. Rev. Plant Biol. *71*, 575-603.

Li, W., Zhu, Z., Chern, M., Yin, J., Yang, C., Ran, L., Cheng, M., He, M., Wang, K., Wang, J., et al. (2017). A natural allele of a transcription factor in rice confers broad-spectrum blast resistance. Cell *170*, 114-126.

Liang, X.X., and Zhou, J.M. (2018). Receptor-like cytoplasmic kinases: central players in plant receptor kinase-mediated signaling. Annu. Rev. Plant Biol. *69*, 267-299.

Liu, Y., Wu, H., Chen, H., Liu, Y., He, J., Kang, H., Sun, Z., Pan, G., Wang, Q., Hu, J., et al. (2015). A gene cluster encoding lectin receptor kinases confers broad-spectrum and durable insect resistance in rice. Nat. Biotechnol. *33*, 301-305.

Ma, Z., Zhu, L., Song, T., Wang, Y., Zhang, Q., Xia, Y., Qiu, M., Lin, Y., Li, H., Kong, L., et al. (2017). A paralogous decoy protects *Phytophthora sojae* apoplastic effector PsXEG1 from a host inhibitor. Science *355*, 710-714.

Mao, Y.B., Cai, W.J., Wang, J.W., Hong, G.J., Tao, X.Y., Wang, L.J., Huang, Y.P., and Chen, X.Y. (2007). Silencing a cotton bollworm P450 monooxygenase gene by plant-mediated RNAi impairs larval tolerance of gossypol. Nat. Biotechnol. *25*, 1307-1313.

Nelson, R., Wiesner-Hanks, T., Wisser, R., and Balint-Kurti, P. (2017). Navigating complexity to breed disease-resistant crops. Nat. Rev. Genet. *19*, 21-33.

Shen, Q.H., Saijo, Y., Mauch, S., Biskup, C., Bieri, S., Keller, B., Seki, H., Ulker, B., Somssich, I.E., and Schulze-Lefert, P. (2007). Nuclear activity of MLA immune receptors links isolate-specific and basal disease-resistance responses. Science *315*, 1098-1103.

Su, Z., Bernardo, A., Tian, B., Chen, H., Wang, S., Ma, H., Cai, S., Liu, D., Zhang, D., Li, T., et al. (2019). A deletion mutation in *TaHRC* confers *Fhb1* resistance to *Fusarium* head blight in wheat. Nat. Genet. *51*, 1099-1105.

Sun, J.Q., Jiang, H.L., and Li, C.Y. (2011). Systemin/jasmonate-mediated systemic defense signaling in tomato. Mol. Plant *4*, 607-715.

Wang, H., Sun, S., Ge, W., Zhao, L., Hou, B., Wang, K., Lyu, Z., Chen, L., Xu, S., Guo, J., et al. (2020). Horizontal gene transfer of *Fhb7* from fungus underlies *Fusarium* head blight resistance in wheat. Science *368*, eaba5435.

Wang, J., Zhou, L., Shi, H., Chern, M., Yu, H., Yi, H., He, M., Yin, J., Zhu, X., Li, Y. et al. (2018). A single transcription factor promotes both yield and immunity in rice. Science *361*, 126-128.

Wang, J.Z., Hu, M.J., Wang, J., Qi, J.F., Han, Z.F., Wang, G.X., Qi, Y.J., Wang, H.W., Zhou, J.M., and Chai, J.J. (2019a). Reconstitution and structure of a plant NLR resistosome conferring immunity. Science *364*, eaav5870.

Wang, J.Z., Wang, J., Hu, M.J., Wu, S., Qi, J.F., Wang, G.X., Han, Z.F., Qi, Y.J., Gao, N., Wang, H.W., et al. (2019b). Ligand-triggered allosteric ADP release primes a plant NLR complex. Science *364*, eaav5868.

Wang, Y., Cheng, X., Shan, Q., Zhang, Y., Liu, J., Gao, C., and Qiu, J. (2014). Simultaneous editing of three homoeoalleles in hexaploid bread wheat confers heritable resistance to powdery mildew. Nat. Biotechnol. *32*, 947-951.

Wu, L., Yang, R., Yang, Z., Yao, S., Zhao, S., Wang, Y., Li, P., Song, X., Jin, L., Zhou, T., et al. (2017). ROS accumulation and antiviral defence control by microRNA528 in rice. Nat.

Plants *3*, 16203.

Yan, C., Fan, M., Yang, M., Zhao, J., Zhang, W., Su, Y., Xiao, L.T., Deng, H., and Xie, D. (2018). Injury activates Ca^{2+}/calmodulin-dependent phosphorylation of JAV1-JAZ8-WRKY51 complex for jasmonate biosynthesis. Mol. Cell *70*, 136-149.

Yang, D., Yang, Y., and He, Z. (2013). Role of plant hormones and their interplay in rice immunity. Mol. Plant *6*, 675-685.

Zhang, N., Zhang, B., Zuo, W.L., Xing, Y.X., Konlasuk, S., Tan, G., Zhang, Q., Ye, J.R., and Xu, M.L. (2015). A maize wall-associated kinase confers quantitative resistance to head smut. Nat. Genet. *47*, 151-157.

Zhang, T., Zhao, Y.L., Zhao J.H., Wang, S., Jin, Y., Chen, Z.Q, Fang, Y.Y., Hua, C.L., Ding, S.W., and Guo, H.S. (2016). Cotton plants export microRNAs to inhibit virulence gene expression in a fungal pathogen. Nat. Plants *2*, 16153.

盐碱与极端温度适应

种 康　杨淑华　郭 岩　蒋才富

1 国内外研究进展

1.1 植物响应极端温度

农业生产高度依赖气候条件，我国是世界上遭受各种气象灾害最严重的国家之一。随着人类工业活动的增加，大气中二氧化碳浓度不断升高，导致全球气候变暖，局部地区极端温度变化加剧。近年来，我国农业主产区高温热害频发且强度增加，局部地区极端低温灾害有所加重。我国因高低温造成的粮食损失每年高达数亿吨，严重影响着粮食供给，因此应对温度灾害的任务十分紧迫。

低温胁迫（cold stress）是指植物在生长期间遭受低温天气，低温对作物生长发育和产量品质产生不利影响。低温胁迫分为冷胁迫（chilling stress，$0 \sim 15\ ℃$）和冻胁迫（freezing stress，$< 0\ ℃$）（Shi et al.，2018）。为了应对低温胁迫的伤害，植物进化出了多套有效的保护机制，如冷锻炼（Ding et al.，2019）。科研人员在拟南芥、水稻、玉米、小麦、大麦、西红柿和油菜等作物中相继克隆出 *CBF/DREB1*（*C-repeat-binding factor/drought responsive element binding protein 1*）家族基因，将 *CBF* 基因过表达能够显著增强植物的抗冻性，这为培育抗低温胁迫新品种提供了基因资源（Ding et al.，2020；Guo et al.，2018）。

水稻起源于热带和亚热带地区，是世界上最为重要的粮食作物之一，对低温极为敏感。目前，研究者们在水稻的 12 条染色体上定位到 250 多个与水稻萌发、孕穗期等低温相关的数量性状位点（QTL）（Fujino et al.，2008；Kim et al.，2011；Lu et al.，2014；Ma et al.，2015；Mao et al.，2019；Saito et al.，2010；Zhang et al.，2017a；Zhao et al.，2017）。*qLTG3-1*（*quantitative trait locus for low-temperature germinability on chromosome 3-1*）

编码一个富含甘氨酸保守结构域的蛋白质，调控水稻在低温下的萌发（Fujino et al.，2008）。*COLD1*（*chilling-tolerance divergence 1*）基因编码一个 G 蛋白信号调节子，通过与 G 蛋白 α 亚基 RGA1 相互作用调控水稻 Ca^{2+} 内流，进而参与水稻感知低温的过程（Ma et al.，2015；Zhang et al.，2019a）。研究表明，过表达 *qCTS-9* 的水稻转基因植株抗冷害的能力增强（Zhao et al.，2017）。OsGSTZ2 可能通过其谷胱甘肽转移酶活性调控水稻的抗寒性（Kim et al.，2011）。*OsLTG1*（*low temperature growth 1*）编码酪蛋白激酶调节水稻在低温下的营养生长（Lu et al.，2014）。*Ctb1* 编码一个 F-box 蛋白，通过与 E3 泛素连接酶 Skp1 相互作用增强水稻孕穗期对低温的抗性（Saito et al.，2010）。*CTB4a*（*cold tolerance at booting stage*）编码一个受体蛋白激酶，通过增强水稻在低温下的 ATP 酶活性和 ATP 的含量，提高水稻花粉的育性和结实率及产量（Zhang et al.，2017a）。最新克隆的 *HAN1* 基因编码将植物激素 JA 由活性形式变成非活性形式的氧化酶，调控 JA 介导的水稻苗期耐冷性（Mao et al.，2019）。除了以上通过 QTL 方法鉴定的基因，科研人员还发现了一些其他的重要基因参与水稻低温响应。例如，钙离子依赖型蛋白激酶 *CPK24*（*calcium-dependent protein kinase 24*）、*CIPK3*（*CDPK-interacting protein kinase 3*）、*CIPK12* 和 *CIPK15*（Liu et al.，2018a；Xiang et al.，2007）正调控水稻抗寒性。OsMPK3 通过磷酸化 OsICE1 增强其蛋白稳定性和转录活性正调控水稻的抗寒能力（Zhang et al.，2017b）。水稻 MYB 类转录因子 *OsMYB3R-2* 正调控水稻的耐低温能力（Ma et al.，2009），而 OsMYB30 负调控水稻的低温耐受性（Lv et al.，2017a）。bZIP 转录因子 OsbZIP73 是水稻苗期低温耐受同时也是进化中受到强烈选择的粳稻低温基因（Liu et al.，2018b）。玉米作为另外一种重要的经济作物，对其响应低温胁迫的分子机制也有一些研究，但是相对较少。有研究报道，水孔蛋白 PIP2;3 对维持低温条件下玉米叶片细胞的水平衡起着重要的作用（Bilska-Kos et al.，2016）。在拟南芥中过表达玉米 *ZmMKK4* 能显著提高转基因植株的抗冻性（Kong et al.，2011）。和水稻类似，玉米中的钙离子依赖型蛋白激酶 ZmCDPKs 同样在玉米响应低温胁迫中发挥作用（Weckwerth et al.，2015）。Huang 等（2013）通过全基因组关联分析找到 40 个涉及玉米低温萌发和低温生长相关的候选基因。

高温热害是另一种造成农业生产量降低的自然灾害，是指环境温度超过植物适宜温度的上限，从而危害作物的生长发育。根据报道，环境温度每上升 1℃，就能够造成减产 10%。热激转录因子（heat shock transcription factor）在植物响应热胁迫时发挥重要作用（Ding et al.，2020）。研究人员发现小麦 *TaHsf3* 基因和番茄 *LpHsfA1* 基因能够正调控植株耐热能力（Miller and Mittler，2006；Zhang et al.，2013）。HSP（heat shock protein）作为分子伴侣能够协助变性的蛋白重新折叠（Wang et al.，2004）。水稻叶绿体定位的热激蛋白 OsHSP70CP1 是高温下叶绿体发育所必需的。最近科研人员发现，生长在热带的非洲稻已经进化出抵抗高温胁迫的策略。研究者通过图位克隆的方法鉴定到非洲稻抗高温胁迫的主效 QTL—*OgTT1*（*thermo tolerance 1*）。OgTT1 蛋白增强细胞蛋白酶体对泛素化底物的降解速度，从而及时并更加有效的清除高温所造成的失活蛋白，降低毒性蛋白积累，增强水稻抗高温能力（Li et al.，2015）。受体激酶 ERECTA 增强逆境下植物细胞膜的完整性，显著提高转基因作物高温下的结实率和成活率（Shen et al.，2015）。*TOGR1*（*thermo tolerant growth required 1*）编码一个 DEAD-box ENA 解旋酶。TOGR1 作为 Pre-mRNA 分子伴侣保证了高温下细胞分裂所需的 rRNA 有效加工，从而提高植物的耐热性（Wang et al.，2016）。*EG1*（*extra glume 1*）编码一个线粒体定位脂酶，通过介导高温依赖的线粒体脂酶途径保证水稻花器官形成基因正常表达（Zhang et al.，2016）。另外，有些基因调控水稻高温下细胞器的发育，如 *OsFLN1/WLP2* 调控水稻叶绿体在高温下的发育（Lv et al.，2017b）。

1.2 植物响应盐碱胁迫

我国约有 1 亿公顷（1 公顷为 15 亩）盐碱地，其中现代盐碱土约 3700 万公顷，残余盐碱土约 4500 万公顷，潜在盐碱土 1700 万公顷。受灌溉依赖的耕作模式和极端气候（如干热）的影响，耕地盐碱化及其导致的土地地力退化日趋严重，已成为限制我国乃至世界农业可持续发展的主要非生物胁迫因素。培育耐盐碱的优良作物新品种不仅可以提高作物单产，而且可以使盐碱地得以利用，扩大耕地面积。

盐碱胁迫分为盐胁迫（主要为氯化钠）和碱胁迫（主要为碳酸氢盐和碳酸盐）（Zhang et al.，2018）。植物耐受盐胁迫主要包括以下几种途径：维持

K⁺/Na⁺ 平衡（Munns et al.，2012；Yang and Guo，2018）；调控渗透压平衡（Munns et al.，2012）；激活活性氧清除机制（Apel and Hirt，2004）；适应性生长调节（Achard et al.，2006）。

在作物中的研究表明，*HKT1*（*high-affinity potassium transporter 1*）家族基因编码 Na⁺ 通道，在作物抗盐应答中发挥着重要作用。水稻 *SKC1*（*OsHKT1;5*）编码一个在木质部薄壁细胞中表达的 Na⁺ 通道，不同水稻品种中 SKC1 转运 Na⁺ 的活性与其抗盐性密切相关（Ren et al.，2005）。小麦中的研究表明，将一粒小麦（*Tritcum monococcum*）的 *Nax2* 位点导入硬粒小麦（*Triticum durum*）可以显著提高后者在盐胁迫条件下的产量，而 *Nax2* 编码水稻 SKC1 的同源蛋白（Munns et al.，2012）。在玉米中的研究表明，抗盐 QTL 基因 *ZmNC1* 也编码一个 HKT1 家族蛋白，它在调控玉米叶片 Na⁺ 含量和抗盐应答中发挥重要作用（Zhang et al.，2018）。HAK 型离子转运蛋白也参与调控 K⁺/Na⁺ 平衡及作物抗盐应答（Chen et al.，2015；Zhang et al.，2019b）。SOS 途径在模式植物拟南芥中的抗盐作用研究较多（Yang and Guo，2018），但该途径在作物研究相对较少。SOS 途径是所有陆生植物最保守的调控 Na⁺ 离子平衡的机制。已有研究表明，过表达硬粒小麦的 *SOS1* 同源基因可以显著提高拟南芥的抗盐能力（Feki et al.，2014），说明 SOS 途径在作物抗盐应答中也发挥着重要作用。

除了关于 Na⁺、K⁺ 转运蛋白的研究，作物中一些其他耐盐因子也陆续被报道。例如，水稻类受体蛋白激酶 OsSIK1 和 OsSIK2 参与正调控水稻的抗盐过程（Chen et al.，2013；Ouyang et al.，2010）。水稻转录共激活子 DCA1（DST co-activator 1）在水稻耐盐应答中起重要作用（Cui et al.，2015）。小麦 *TaOPR1* 基因受盐诱导表达，过表达该基因可增强小麦的耐盐性（Dong et al.，2013）。小麦蓝光响应因子 TaGBF1 能够负调控小麦抗盐（Sun et al.，2016）。最新的研究报道了一个玉米抗盐碱 QTL 基因 *ZmNSA1*，该基因编码 EF-hand 蛋白，通过解码 Ca²⁺ 信号，增强 SOS1 的活性，从而调控盐碱胁迫下 Na⁺ 稳态（Cao et al.，2020）。大豆转录因子 GmNAC11、GmNAC20 及 GmWRKY27 都参与调控大豆的耐盐性（Hao et al.，2011；Wang et al.，2015）。

相对于抗盐机制的研究，作物碱胁迫应答机制的研究还很薄弱。有研究表明，碱胁迫下，土壤中大部分铁元素以 Fe^{3+} 形式存在，难以被作物利用，从而对作物生长和产量产生负面影响（Gómez-Galera et al.，2012）。水稻铁转运蛋白 OsIRT1/2 能够有效地转运 Fe^{2+} 到细胞内，从而增强水稻对碱性土壤的耐受性（Ishimaru，2006）。MAs 是另外一类植物铁载体，能够整合土壤中 Fe^{3+}，在禾本科植物吸收转运铁的过程中起到重要作用（Inoue et al.，2008）。水稻中 OsYSL15/16 能够将 Fe^{3+}-MAs 复合物从土壤运送到根表皮细胞，同时 OsYSL15/16/18 负责复合物长距离运输，将 Fe^{3+}-MAs 从地下部运输到地上部（Aoyama et al.，2009；Inoue et al.，2009；Koike et al.，2004）。水稻 *OsNAS2* 的过表达植株表现出对铁缺乏耐受性增强，在碱性条件下生长良好（Lee et al.，2012）。过表达铁响应转录因子 OsIDEF1 的植株表现出更强的耐碱性（Nishizawa，2011）。bHLH 型转录因子 OsIRO2 通过增强 Fe^{3+}-DMA 转运载体 *OsYSL1* 的表达，增强水稻的耐碱能力（Ogo et al.，2007，2011）。

2 未来发展趋势与关键突破口

目前我国粮食的单产水平已很高，在现有农业科技水平上依靠传统农业技术与措施已很难再有所作为。利用分子设计育种是解决农作物品种改良的重要手段。现阶段虽然我们对植物抗逆相关的分子和生化反应机制有了初步的了解，然而这些研究结果大多是零散的，距离全面解析植物抗逆机制并最终在作物分子改良中发挥作用为时尚远。因此，研究作物感受和应答盐碱和温度胁迫的遗传和分子生物学基础，深入解析作物响应盐碱/温度胁迫的信号转导和分子调控网络，在农作物/牧草中克隆重要的耐盐碱和高低温基因，将为高抗作物和林草新品种分子设计育种提供依据，是解决农作物品种改良、保障我国粮食安全和可持续发展的迫切需求。

2.1 通过分子设计育种培育高抗作物

为了推进我国绿色生态农业生产、保障我国农业可持续发展的重大需求，利用分子设计培育高抗作物/林草/果蔬新品种应从以下几个方面寻求突破。
① 全面了解我国盐碱和极端温度灾害区域和发生特点，针对不同的地方品

种建立精准的作物抗逆表型鉴定标准，完善作物灾害敏感性评估体系，搭建高效优异基因资源的评价鉴定研究平台。② 利用我国丰富的种质资源，构建广泛自然变异的关联分析群体、特定的遗传连锁群体、人工诱变群体，挖掘优良抗逆品种的优良等位变异并开发抗逆分子标记，利用分子辅助选育创制新的抗逆遗传材料。③ 全面、深入、动态解析植物/作物耐盐碱和温度响应机制，对植物如何"感受"高低温、盐碱等非生物逆境胁迫信号等重大科学问题进行突破，构建作物抗逆调控网络，解析作物抗高低温、高盐胁迫信号途径之间的遗传互作关系。④ 克隆耐盐碱/高温/低温物种中的高抗QTL位点，挖掘主效抗逆基因优良等位变异并开发抗逆分子标记，利用分子辅助选育创制新的抗逆遗传材料发掘优良等位变异。⑤ 建立高通量精准的作物/林草/果蔬转化和基因编辑体系（敲除、置换、碱基编辑等），加速抗逆基因的多基因聚合进程。⑥ 引入大数据和机器学习技术，建立抗逆作物全基因组选择育种智能决策体系，开发作物/林草/果蔬分子育种新技术、新工具，加速育种进程。

2.2 创制未来作物抗逆遗传材料的手段

创制未来作物抗逆遗传材料的手段包括如下几点。① 借鉴模式生物的研究成果，开展作物抗盐碱、耐极端温度的基础和应用研究。例如，拟南芥 SOS 耐盐途径、*CBF* 冷响应信号通路、植物热激蛋白 *HSP* 和 *HSF* 等许多抗逆基因已被克隆，这些基因大多在主要作物中存在直系同源基因。通过反向遗传学手段，利用基因编辑等技术，对这些同源基因进行定向功能解析，挖掘这些基因在作物抗非生物逆境改良中的作用，是未来作物抗盐碱、耐极端温度基础和应用研究的重要组成部分。② 作物关键抗盐碱、耐极端温度 QTL 基因的克隆、优异等位基因的挖掘及其在分子设计育种中的应用。主要作物（如水稻、小麦、玉米等）的不同品系耐盐性存在显著差异，这些差异大多受 QTL 调控。通过 QTL 和 GWAS 方法全面定位和克隆大量的植物抗逆基因，解析这些基因表达的特异性和表达强弱，获得调控这些基因时空表达的启动子，解析相关基因的功能，挖掘其优良等位变异，并开发抗盐碱、耐极端温度分析标记，为未来作物分子基础解析和分子设计育种提供基因资源。③ 设计分子开关，消除抗逆基因对作物生长发育和产量的负面影响。高抗基因的高水平表达往往影响植物的产量和表现型，利用逆境响应元件，设计抗性开关，使之在正常环境下不表达或少量表达，仅在逆境来临时发挥作用，既可以抵抗极端温度对植物造成的伤害，又能够减少逆境基因对作物产量和品质的影响。④ 利用基因编辑、分子标记辅助选育、单倍体育种等技术改良作物抗逆性状，通过全基因组分子选育等手段定向整合，快速高效培育多抗作物品种。⑤ 作物对非生物逆境胁迫应答是一个复杂的生物学过程，每一方面的应答又涉及成百上千的分子水平的改变。可以利用生物信息学对基因组学、表观遗传学、转录组学、蛋白质组学、代谢组学等多组学大数据全面整合，系统全面解析作物抗盐碱、耐极端温度胁迫的分子基础，构建作物的抗逆调控网络，解析作物抗高低温、高盐胁迫信号途径之间的遗传互作关系。⑥ 基于多组学大数据全面解析作物抗逆分子基础，构建基于全基因组选择的作物人工智能育种平台，通过生物信息学及机器学习技术建立抗逆作物全基因组选择育种智能决策体系，是未来高抗作物分子基础解析和"精准高效"育种的发展趋势。

3 阶段性目标

3.1 2035 年目标

为了将关键基因和遗传资源正确、有效、快速地应用于作物抗逆性遗传改良，从现在到 2035 年的主要阶段性目标如下。① 全面了解我国现有盐碱地的土壤特性，包括离子、pH、光照及温度等气候特性，为针对性地培育适合地区种植的耐盐碱品种提供环境和土壤数据。全面了解极端温度对农作物影响的季节和地域特征，进一步完善和建立极端温度灾害的评估体系和模型。② 充分利用我国作物种质资源的丰富度和独特性，建立良好的自然条件筛选体系，创新抗盐碱、极端温度的种质资源，建立种质基因资源库，便于大力挖掘抗盐碱、极端温度新基因。③ 在主要经济作物中完成一批重要抗盐碱、耐极端温度基因的克隆和功能研究，明确这些基因在作物胁迫应答中的分子调控机制，建立它们的遗传互作关系，构建相对完善的作物逆境应答的主要分子调控网络。④ 筛选获得一批优良抗盐碱、极端温度等位基因和开发分子标记，在水稻、小麦、玉米等主要作物中基本实现高抗分子设计育种。⑤ 获得调控抗逆基因时空表达的启动子，设计分子开关，为精准植物抗盐碱、极端温度分子育种设计提供基因资源。⑥广泛应用基因编辑等新兴技术在作物抗盐碱、极端温度种质资源创新中的研究。⑦围绕作物逆境应答的分子基础，水稻、小麦、玉米等主要作物中多组学（包括基因组、转录组、蛋白质组和代谢组）的大数据研究技术体系日趋完善，基于人工智能在抗盐碱、极端温度基因挖掘和育种技术中的迅速发展，主要作物全基因组选择育种的分子理论基础基本成型。⑧为实现快速高效地作物抗盐碱、高低温胁迫的分子育种，需要建立高效基因型依赖性低的作物/林草/果蔬转化体系，精准的作物/林草/果蔬基因编辑体系，以及开发加速分子育种进程的新技术和新工具，培育出种植范围广，适应温度能力强，高产稳产的新型经济作物。

3.2 2050年目标

如果我们能在以上方面取得突破，就能在2050年实现以下目标。①人工智能技术被广泛应用到作物抗逆的分子基础和应用研究领域。②更多新的抗性基因资源被发现，大部分作物实现抗盐碱、温度胁迫的分子设计育种。③作物抗逆的分子基础得到更全面的解析，主要经济作物实现基于全基因组选择的"精准高效"育种。④通过人工智能分子育种平台从全基因组水平对作物的抗逆性状及其他农艺性状进行优势聚合。⑤建立关于植物抗逆基因的"基础研究、标记开发、基因克隆、遗传转化、品种培育、产品推广"的完整产业技术研发体系。

致谢：本章在撰写过程中得到了施怡婷和丁杨林博士的协助，特此感谢！

参考文献

Achard, P., Cheng, H., De Grauwe, L., Decat, J., Schoutteten, H., Moritz, T., Van Der Straeten, D., Peng, J.R., and Harberd, N.P. (2006). Integration of plant responses to environmentally activated phytohormonal signals. Science *311*, 91-94.

Aoyama, T., Kobayashi, T., Takahashi, M., Nagasaka, S., Usuda, K., Kakei, Y., Ishimaru, Y., Nakanishi, H., Mori, S., and Nishizawa, N.K. (2009). OsYSL18 is a rice iron(III)-deoxymugineic acid transporter specifically expressed in reproductive organs and phloem of lamina joints. Plant Mol. Biol. *70*, 681-692.

Apel, K., and Hirt, H. (2004). Reactive oxygen species: metabolism, oxidative stress, and signal transduction. Annu. Rev. Plant Biol. *55*, 373-399.

Bilska-Kos, A., Szczepanik, J., and Sowinski, P. (2016). Cold induced changes in the water balance affect immunocytolocalization pattern of one of the aquaporins in the vascular system in the leaves of maize (*Zea mays* L.). J. Plant Physiol. *205*, 75-79.

Cao, Y., Zhang, M., Liang, X., Li, F., Shi, Y., Yang, X., and Jiang, C. (2020). Natural variation of an EF-hand Ca^{2+}-binding-protein coding gene confers saline-alkaline tolerance in maize. Nat. Commun. *11*, 186.

Chen, G., Hu, Q.D., Luo, L., Yang, T.Y., Zhang, S., Hu, Y.B., Yu, L., and Xu, G.H. (2015). Rice potassium transporter OsHAK1 is essential for maintaining potassium-mediated growth and functions in salt tolerance over low and high potassium concentration ranges. Plant Cell Environ. *38*, 2747-2765.

Chen, L.J., Wuriyanghan, H., Zhang, Y.Q., Duan, K.X., Chen, H.W., Li, Q.T., Lu, X., He, S.J., Ma, B., Zhang, W.K., et al. (2013). An S-domain receptor-like kinase, OsSIK2, confers abiotic stress tolerance and delays dark-induced leaf senescence in rice. Plant Physiol. *163*, 1752-1765.

Chinnusamy, V., Zhu, J., and Zhu, J.K. (2007). Cold stress regulation of gene expression in plants. Trends Plant Sci. *12*, 444-451.

Cui, L.G., Shan, J.X., Shi, M., Gao, J.P., and Lin, H.X. (2015). DCA1 acts as a transcriptional co-activator of DST and contributes to drought and salt tolerance in Rice. PLoS Genet. *11*, e1005617.

Ding, Y., Shi, Y., and Yang, S. (2019). Advances and challenges in uncovering cold tolerance regulatory mechanisms in plants. New Phytol. *222*, 1690-1704.

Ding, Y., Shi, Y., and Yang, S. (2020). Molecular regulation of plant responses to environmental temperatures. Mol. Plant *13*, 544-564.

Dong, W., Wang, M.C., Xu, F., Quan, T.Y., Peng, K.Q., Xiao, L.T., and Xia, G.M. (2013). Wheat oxophytodienoate reductase gene *TaOPR1* confers salinity tolerance via enhancement of abscisic acid signaling and reactive oxygen species scavenging. Plant Physiol. *161*, 1217-1228.

Feki, K., Quintero, F.J., Khoudi, H., Leidi, E.O., Masmoudi, K., Pardo, J.M., and Brini, F. (2014). A constitutively active form of a durum wheat Na^+/H^+ antiporter SOS1 confers high salt tolerance to transgenic *Arabidopsis*. Plant Cell Rep. *33*, 277-288.

Fujino, K., Sekiguchi, H., Matsuda, Y., Sugimoto, K., Ono, K., and Yano, M. (2008). Molecular identification of a major quantitative trait locus, *qLTG3-1*, controlling low-temperature germinability in rice. Proc. Natl. Acad. Sci. USA *105*, 12623-12628.

Gómez-Galera, S., Sudhakar, D., Pelacho, A.M., Capell, T., and Christou, P. (2012). Constitutive expression of a barley Fe phytosiderophore transporter increases alkaline soil tolerance and results in iron partitioning between vegetative and storage tissues under stress. Plant Physiol. Biochem. *53*, 46-53.

Guo, X.Y., Liu, D.F., and Chong, K. (2018). Cold signaling in plants: insights into mechanisms and regulation. J. Integr. Plant Biol. *60*, 745-756.

Hao, Y.J., Wei, W., Song, Q.X., Chen, H.W., Zhang, Y.Q., Wang, F., Zou, H.F., Lei, G., Tian, A.G., Zhang, W.K., et al. (2011). Soybean NAC transcription factors promote abiotic stress tolerance and lateral root formation in transgenic plants. Plant J. *68*, 302-313.

Huang, J., Zhang, J., Li, W., Hu, W., Duan, L., Feng, Y., Qiu, F., and Yue, B. (2013). Genome-wide association analysis of ten chilling tolerance indices at the germination and seedling stages in maize. J. Integr. Plant Biol. *55*, 735-744.

Inoue, H., Kobayashi, T., Nozoye, T., Takahashi, M., Kakei, Y., Suzuki, K., Nakazono, M., Nakanishi, H., Mori, S., and Nishizawa, N.K. (2009). Rice OsYSL15 is an iron-regulated iron(III)-deoxymugineic acid transporter expressed in the roots and is essential for iron uptake in early growth of the seedlings. J. Biol. Chem. *284*, 3470-3479.

Inoue, H., Takahashi, M., Kobayashi, T., Suzuki, M., Nakanishi, H., Mori, S., and Nishizawa, N.K. (2008). Identification and localisation of the rice nicotianamine aminotransferase gene *OsNAAT1* expression suggests the site of phytosiderophore synthesis in rice. Plant Mol. Biol. *66*, 193-203.

Ishimaru, Y. (2006). Rice plants take up iron as an Fe^{3+}-phytosiderophore and as Fe^{2+}. Plant J. *45*, 335-346.

Kim, S.I., Andaya, V.C., and Tai, T.H. (2011). Cold sensitivity in rice (*Oryza sativa* L.) is strongly correlated with a naturally occurring 199V mutation in the multifunctional glutathione transferase isoenzyme GSTZ2. Biochem. J. *435*, 373-380.

Koike, S., Inoue, H., Mizuno, D., Takahashi, M., Nakanishi, H., Mori, S., and Nishizawa, N.K. (2004). OsYSL2 is a rice metal-nicotianamine transporter that is regulated by iron and expressed in the phloem. Plant J. *39*, 415-424.

Kong, X., Pan, J., Zhang, M., Xing, X., Zhou, Y., Liu, Y., Li, D., and Li, D. (2011). ZmMKK4, a novel group C mitogen-activated protein kinase kinase in maize (*Zea mays*), confers salt and cold tolerance in transgenic *Arabidopsis*. Plant Cell Environ. *34*, 1291-1303.

Lee, S., Kim, Y.S., Jeon, U.S., Kim, Y.K., Schjoerring, J.K., and An, G. (2012). Activation of rice nicotianamine synthase 2 (OsNAS2) enhances iron availability for biofortification. Mol. Cells *33*, 269-275.

Li, X.M., Chao, D.Y., Wu, Y., Huang, X., Chen, K., Cui, L.G., Su, L., Ye, W.W., Chen, H., and Chen, H.C. (2015). Natural alleles of a proteasome α2 subunit gene contribute to thermotolerance and adaptation of African rice. Nat. Genet. *47*, 827-833.

Liu, C., Ou, S., Mao, B., Tang, J., Wang, W., Wang, H., Cao, S., Schlappi, M.R., Zhao, B., Xiao, G., et al. (2018b). Early selection of bZIP73 facilitated adaptation of *japonica* rice to cold climates. Nat. Commun. *9*, 3302.

Liu, Y., Xu, C.J., Zhu, Y.F., Zhang, L.N., Chen, T.Y., Zhou, F., Chen, H., and Lin, Y.J. (2018a). The calcium-dependent kinase OsCPK24 functions in cold stress responses in rice. J. Integr. Plant Biol. *60*, 173-188.

Lu, G.W., Wu, F.Q., Wu, W.X., Wang, H.J., Zheng, X.M., Zhang, Y.H., Chen, X.L., Zhou, K.N., Jin, M.N., Cheng, Z.J., et al. (2014). Rice *LTG1* is involved in adaptive growth and fitness under low ambient temperature. Plant J. *78*, 468-480.

Lv, Y., Yang, M., Hu, D., Yang, Z., Ma, S., Li, X., and Xiong, L. (2017a). The OsMYB30 transcription factor suppresses cold tolerance by interacting with a JAZ protein and suppressing beta-amylase expression. Plant Physiol. *173*, 1475-1491.

Lv, Y.S., Shao, G.N., Qiu, J.H., Jiao, G.A., Sheng, Z.H., Xie, L.H., Wu, Y.W., Tang, S.Q., Wei, X.J., and Hu, P.S. (2017b). *White Leaf and Panicle 2*, encoding a PEP-associated protein, is required for chloroplast biogenesis under heat stress in rice. J. Exp. Bot. *68*, 5147-5160.

Ma, Q.B., Dai, X.Y., Xu, Y.Y., Guo, J., Liu, Y.J., Chen, N., Xiao, J., Zhang, D.J., Xu, Z.H., Zhang, X.S., et al. (2009). Enhanced tolerance to chilling stress in *OsMYB3R-2* transgenic rice is mediated by alteration in cell cycle and ectopic expression of sress genes. Plant Physiol. *150*, 244-256.

Ma, Y., Dai, X., Xu, Y., Luo, W., Zheng, X., Zeng, D., Pan, Y., Lin, X., Liu, H., Zhang, D., et al. (2015). *COLD1* confers chilling tolerance in rice. Cell *160*, 1209-1221.

Mao, D.H., Xin, Y.Y., Tan, Y.J., Hu, X.J., Bai, J.J., Liu, Z.Y., Yu, Y.L., Li, L.Y., Peng, C., Fan, T., et al. (2019). Natural variation in the *HAN1* gene confers chilling tolerance in rice and allowed adaptation to a temperate climate. Proc. Natl. Acad. Sci. USA *116*, 3494-3501.

Miller, G., and Mittler, R. (2006). Could heat shock transcription factors function as hydrogen peroxide sensors in plants? Ann. Bot. *98*, 279-288.

Munns, R., James, R.A., Xu, B., Athman, A., Conn, S.J., Jordans, C., Byrt, C.S., Hare, R.A., Tyerman, S.D., Tester, M., et al. (2012). Wheat grain yield on saline soils is improved by an ancestral Na^+ transporter gene. Nat. Biotechnol. *30*, 360-364.

Nishizawa, N. (2011). Enhancing tolerance of rice to alkaline soils using genes involved in Fe acquisition in plants. J. Jap. Soc. Extrem. *7*, 15-19.

Nover, L., Bharti, K., Doring, P., Mishra, S.K., Ganguli, A., and Scharf, K.D. (2001). *Arabidopsis* and the heat stress transcription factor world: how many heat stress transcription factors do we need? Cell Stress Chaperones *6*, 177-189.

Ogo, Y., Itai, R., Nakanishi, H., Kobayashi, T., Takahashi, M., Mori, S., and Nishizawa, N. (2007). The rice bHLH protein OsIRO2 is an essential regulator of the genes involved in Fe uptake under Fe-deficient conditions. Plant Cell Physiol. *48*, 366-377.

Ogo, Y., Itai, R.N., Kobayashi, T., Aung, M.S., Nakanishi, H., and Nishizawa, N.K. (2011). OsIRO2 is responsible for iron utilization in rice and improves growth and yield in calcareous soil. Plant Mol. Biol. *75*, 593-605.

Ouyang, S.Q., Liu, Y.F., Liu, P., Lei, G., He, S.J., Ma, B., Zhang, W.K., Zhang, J.S., and Chen, S.Y. (2010). Receptor-like kinase OsSIK1 improves drought and salt stress tolerance in rice (*Oryza sativa*) plants. Plant J. *62*, 316-329.

Ren, Z.H., Gao, J.P., Li, L.G., Cai, X.L., Huang, W., Chao, D.Y., Zhu, M.Z., Wang, Z.Y., Luan, S., and Lin, H.X. (2005). A rice quantitative trait locus for salt tolerance encodes a sodium transporter. Nat. Genet. *37*, 1141-1146.

Saito, K., Hayano-Saito, Y., Kuroki, M., and Sato, Y. (2010). Map-based cloning of the rice cold tolerance gene *Ctb1*. Plant Sci. *179*, 97-102.

Shen, H., Zhong, X.B., Zhao, F.F., Wang, Y.M., Yan, B.X., Li, Q., Chen, G.Y., Mao, B.Z., Wang, J.J., Li, Y.S., et al. (2015). Overexpression of receptor-like kinase ERECTA improves thermotolerance in rice and tomato. Nat. Biotechnol. *33*, 996-1003.

Shi, Y., Ding, Y., and Yang, S. (2018). Molecular regulation of CBF signaling in cold acclimation. Trends Plant Sci. *23*, 623-637.

Sun, Y., Xu, W., Jia, Y., Wang, M., and Xia, G. (2016). The wheat *TaGBF1* gene is involved in the blue-light response and salt tolerance. Plant J. *84*, 1219-1230.

Thomashow, M.F. (1999). Plant cold acclimation: freezing tolerance genes and regulatory mechanisms. Annu. Rev. Plant Physiol. Plant Mol. Biol. *50*, 571-599.

Wang, D., Qin, B.X., Li, X., Tang, D., Zhang, Y., Cheng, Z.K., and Xue, Y.B. (2016). Nucleolar DEAD-box RNA helicase TOGR1 regulates thermotolerant growth as a pre-rRNA chaperone in rice. PLoS Genet. *12*, e1005844.

Wang, F., Chen, H.W., Li, Q.T., Wei, W., Li, W., Zhang, W.K., Ma, B., Bi, Y.D., Lai, Y.C., Liu, X.L., et al. (2015). GmWRKY27 interacts with GmMYB174 to reduce expression of GmNAC29 for stress tolerance in soybean plants. Plant J. *83*, 224-236.

Wang, W.X., Vinocur, B., Shoseyov, O., and Altman, A. (2004). Role of plant heat-shock proteins and molecular chaperones in the abiotic stress response. Trends Plant Sci. *9*, 244-252.

Weckwerth, P., Ehlert, B., and Romeis, T. (2015). *ZmCPK1*, a calcium-independent kinase member of the *Zea mays CDPK* gene family, functions as a negative regulator in cold stress signalling. Plant Cell Environ. *38*, 544-558.

Xiang, Y., Huang, Y.M., and Xiong, L.Z. (2007). Characterization of stress-responsive *CIPK* genes in rice for stress tolerance improvement. Plant Physiol. *144*, 1416-1428.

Yang, Y.Q., and Guo, Y. (2018). Elucidating the molecular mechanisms mediating plant salt-stress responses. New Phytol. *217*, 523-539.

Zhang, B.Y., Wu, S.H., Zhang, Y.E., Xu, T., Guo, F.F., Tang, H.S., Li, X., Wang, P.F., Qian, W.F., and Xue, Y.B. (2016). A high temperature-dependent mitochondrial lipase EXTRA GLUME1 promotes floral phenotypic robustness against temperature fluctuation in Rice (*Oryza sativa* L.). PLoS Genet. *12*, e1006152.

Zhang, C.Y., Zhang, Z.Y., Li, J.H., Li, F., Liu, H.H., Yang, W.S., Chong, K., and Xu, Y.Y. (2017b). OsMAPK3 phosphorylates OsbHLH002/OsICE1 and inhibits its ubiquitination to activate OsTPP1 and enhances rice chilling tolerance. Dev. Cell *43*, 731-743.

Zhang, J., Li, X. M., Lin, H. X., and Chong, K. (2019a). Crop improvement through temperature resilience. Annu. Rev. Plant Biol. *70*, 753-780.

Zhang, M., Cao, Y.B., Wang, Z.P., Wang, Z.Q., Shi, J.P., Liang, X.Y., Song, W.B., Chen, Q.J., Lai, J.S., and Jiang, C.F. (2018). A retrotransposon in an HKT1 family sodium transporter causes variation of leaf Na^+ exclusion and salt tolerance in maize. New Phytol. *217*, 1161-1176.

Zhang, M., Liang, X., Wang, L., Cao, Y., Song, W., Shi, J., Lai, J., and Jiang, C. (2019b). A HAK family Na^+ transporter confers natural variation of salt tolerance in maize. Nat. Plants *5*, 1297-1308.

Zhang, S.X., Xu, Z.S., Li, P.S., Yang, L., Wei, Y.Q., Chen, M., Li, L.C., Zhang, G.S., and Ma, Y.Z. (2013). Overexpression of *TaHSF3* in transgenic *Arabidopsis* enhances tolerance to extreme temperatures. Plant Mol. Biol. Rep. *31*, 688-697.

Zhang, Z., Li, J., Pan, Y., Li, J., Zhou, L., Shi, H., Zeng, Y., Guo, H., Yang, S., Zheng, W., et al. (2017a). Natural variation in *CTB4a* enhances rice adaptation to cold habitats. Nat. Commun. *8*, 14788.

Zhao, J.L., Zhang, S.H., Dong, J.F., Yang, T.F., Mao, X.X., Liu, Q., Wang, X.F., and Liu, B. (2017). A novel functional gene associated with cold tolerance at the seedling stage in rice. Plant Biotechnol. J. *15*, 1141-1148.

品质与营养

刘巧泉　吴殿星

1 国内外研究进展

品质是农作物最重要的经济性状,品质的优劣决定了农产品的应用价值和市场竞争力。因为作物的最终用途与加工产品的不同,所以人们对其品质需求、评价指标和相关标准也不尽相同。克隆重要农作物品质与营养形成的功能基因,解析品质形成的调控网络,是未来作物品质设计的基础,也是社会健康持续发展的基本保障,这对于提升农产品质量和人民生活水平具有重要意义,使人们在"吃得饱"的基础上"吃得好""吃得健康""吃得放心"。进入21世纪以来,国内外科学家在作物品质基因克隆及其分子生物学研究领域取得了长足进展,获得了一系列原创性研究成果(杨君等,2016;张昌泉等,2016;张勇等,2016;张玉芹等,2016;Bouis and Saltzman,2017;Li et al.,2019;Zhu et al.,2020)。

作物种子形状既与产量有关,又是品质的重要影响因素,尤其对于水稻的籽粒充实和外观品质影响甚大。近年来,有关水稻粒形调控相关基因的克隆与功能研究取得了突出进展,科研人员从不同的水稻种质资源中分离了几十个粒形相关基因(Li et al.,2019;Liu et al.,2018;Zhao et al.,2018)。根据粒形基因决定的表型特征,可将其分为三大类。第一类包括 $D1$、$D2$、$D11$、$D61$ 等,这类基因突变会造成植株矮化从而间接造成小粒表型;第二类基因通常所指粒形控制基因,能够特异性调控籽粒形状;第三类基因是指小圆粒基因,主要发现于粳稻亚种中。在育种实践中,对于外观品质改良具有重要利用价值的主要是第二类基因,包括 $GS3$、$GS2/GL2$、$GL7/GW7$、$GW2$、$GS5$、$GW8$、$GS9$ 等。此外,$GIF1$ 和 $Chalk5$ 等基因可通过对胚乳中储藏物质累积的调控,进而影响籽粒充实和垩白形成。油菜等油料作物的种子大小也与其产量、矿质元素和脂肪酸组成等品质密切关联(Che et al.,

2015；Duan et al, 2015；Hu et al., 2015；Miller et al.，2019）。

按照人体的需求，农作物中的营养成分大体可归纳为碳水化合物（淀粉和糖类）、蛋白质、脂肪、维生素、矿质元素、膳食纤维及水共7类营养成分。其中，碳水化合物、蛋白质和脂肪是三大主成分，维生素和矿质元素是微量营养元素，膳食纤维与抗性淀粉是一类具有润肠通便、调节控制血糖、降血脂等一种或多种生理功能的功能性物质。这些物质的组成、结构及均衡状态决定了作物产品的品质与营养。

淀粉作为禾谷类和块茎类作物的主要成分，其组成与结构与蒸煮食味品质、加工品质、营养功能等密切相关。除食用外，淀粉也是极其重要的工业原料。从分子水平来看，参与淀粉合成与调控的基因都可能对品质形成发挥着重要作用。参与淀粉合成的酶类已较为清晰，主要有ADP-葡萄糖焦磷酸化酶、颗粒结合淀粉合成酶、可溶性淀粉合成酶、淀粉分支酶、淀粉去分支酶、淀粉磷酸化酶和淀粉异构酶等，编码这些酶的基因可统称为SSRG（starch synthesis-related genes）（Tian et al., 2009）。近年来，科研人员对这些基因的等位变异进行了较多研究，挖掘了大量优良等位变异，也利用基因编辑技术创建了很多新的等位变异，为作物品质改良提供了重要的基因资源（Gao et al., 2020；Zhang et al., 2019a；Zhou et al., 2016）。此外，一些参与胚乳淀粉和贮藏蛋白表达调控的转录因子也陆续被克隆（Feng et al., 2018；Li et al., 2018；Wang et al., 2013；Zhang et al., 2016, 2019b）。

相对于以淀粉为主的粮食作物，脂肪酸含量和配比是油料作物品质评价的重要指标。关于油脂合成的生化途径研究已经比较透彻，但其调控机制还不是很清楚。迄今为止，已有大量关于油菜、大豆和玉米脂肪酸合成代谢的数量性状位点（QTL）定位和基因功能研究的报道（Clemente et al., 2009；Li et al., 2013；Lu et al., 2018；Zhang et al., 2018）。与此同时，种子中油的形成受到一些转录因子的调控，如LEC1/2、FUSCA3、WRI1、BCCP2等，这些调控因子的突变均会引起含油量的改变（张玉芹等，2016）。蛋白质和必需氨基酸含量与组成是评价作物营养品质的另一重要指标。从1964年起，一些重要的基因相继被发现，并被用于改良作物中赖氨酸和/或蛋白质的含量（张昌泉等，2016；Galili and Amir, 2013；Peng et al., 2014；Yang et al., 2019）。其中较为成功例子是优质蛋白玉米（quality protein maize,

QPM）（Deng et al.，2017；Li et al.，2020；Liu et al.，2016）和高赖氨酸水稻（Yang et al.，2018a）等的培育及其调控机制解析。麦谷蛋白低分子量亚基和多酚氧化酶活性等基因表达与小麦加工品质密切相关，已发掘了 50 多个育种可用的等位变异和基因特异性标记（张勇等，2016；Hu et al.，2018）。

优异等位变异的挖掘是作物品种改良的重要基础（Fernie et al.，2006）。农作物的主要品质性状存在广泛变异，主要原因是作物品质形成重要基因具有较为丰富的等位变异，如控制稻米蒸煮食味品质的 *Wx* 基因在栽培稻中具有 10 多个等位变异（Zhang et al.，2019a），控制粒形和香味的 *GS3*、*GW5* 和 *BADH* 等基因也有多个复等位变异（Kovach et al.，2009；Song et al.，2007；Weng et al.，2008）。小麦中麦谷蛋白 *Glu-A1*、*Glu-B1* 和 *Glu-D1* 基因等位变异的发现，也将小麦的面筋蛋白分为不同的等级（Nakamura，2000）。

作物品质形成与表达是一个复杂的代谢调控网络。优质蛋白玉米种子中 *opaque2* 基因调控了淀粉合成相关基因表达，在赖氨酸和蛋白质含量提高的同时，淀粉结构相应改变，胚乳变硬或变软，透明度降低（Zhang et al.，2016）。在水稻胚乳中通过对赖氨酸代谢途径调控提高赖氨酸含量，同时诱导了色氨酸代谢途径表达，从而引起了植物胁迫响应（Yang et al.，2018a）。大豆和油菜等油料作物中的油分和蛋白质往往是负相关的，在提高油酸含量的同时，其储藏蛋白含量则下降，反之亦然（Patil et al.，2017）。更有甚者，品质的改良有时会影响作物产量（Zeng et al.，2017）。种子品质性状，如蛋白质、淀粉和油含量等，通常与母本植株发育过程中碳氮平衡（Gutiérrez et al.，2007；Shi et al.，2013）、代谢主干与源库相互作用（Angélovici et al.，2011；Toubiana et al.，2012；Yang et al.，2018b）以及激素调控（Batista et al.，2019）等功能相关。尽管目前人们越来越重视提高种子作物的营养价值和种子质量的经济价值，但是调节代谢过程和碳氮平衡的遗传基础导致种子性状的定量变化尚待探索。

作物品质与营养基因克隆及分子机制的解析为作物改良提供了重要支撑。除了通过杂交、回交等常规育种技术替换或聚合有利基因改良品质外，分子标记辅助选择、转基因、基因组编辑等现代生物技术及合成生物学的发展为改良品质提供了新的途径和方法，并已取得许多卓有成效的研究成。例如，黄金大米（Ye et al.，2000）、虾青素大米（Zhu et al.，2018）、高赖氨酸玉米和大米（Azevedo and Arruda, 2010; Yang et al., 2018a）、高生育酚作物（Wilson and Roberts, 2014; Zhang et al., 2013）、高铁和高锌作物（Gao et al., 2019; Uauy et al., 2006; Zhang et al., 2020）、高抗性淀粉作物（Zhou et al., 2016; Zhu et al., 2012）及淀粉改良作物（Bull et al., 2018; Gao et al., 2020; Regina et al., 2015）等。

2 未来发展趋势与关键突破口

2.1 未来发展趋势

随着生活水平提升及消费需求多样化，作物品质和营养改良呈现多元化目标。作物品质和营养的多元化开发与利用是未来作物改良的一个重要方向。为满足不同蒸煮品质、食味品质、消化品质和加工品质等要求，未来水稻品种对于胚乳淀粉组成与结构呈现多样化需求；甜玉米、黏玉米、油料玉米和膨化玉米的市场需求要求提供不同甜度、淀粉含量和组成、蛋白质溶解度、胚含油量等指标的玉米品种；高、中、低筋小麦蛋白质含量的不同在一定程度上满足了烘焙、面食、糕点、饼干和面筋的加工原料；豆腐、豆浆、豆油等豆制品的市场需求多元化也要求大豆种质资源的多元化。油菜作为主要的油料作物，调控油酸含量和组成可培育不同需求的双低油菜、黄籽油菜。

适应供给侧改革及消费需求高端化，作物品质与产量、不同品质性状间需协调改良。高产又优质的未来作物选育是育种家面临的重大挑战。作物中待改良的营养品质性状往往与其他不良的品质性状紧密相关，比如，高蛋白的玉米和水稻往往与其食味或外观品质呈一定程度的负相关；因此，定向和定量调控作物中营养物质的代谢调控，将是未来作物产量与品质、不同品质性状间协同改良的思路。在挖掘具有协同调控品质相关基因的基础上，合理调控多个性状基因在一定范围内实现高产优质作物的培育是可行的。此外，物质的富足和营养物质的摄入失衡导致人们表现出一系列的"亚健康问题"，在保证主粮食品基本品质和营养含量的前提下，需要不断开发增强人体体质、维护健康的作物新产品。虽然关于作物理化品质、营养品质和专用品质的研究各有成效，但三者往往不可兼顾。因此，协调三者的关系仍需深入研究。

基于各单基因功能解析，基因互作及代谢调控网络仍需深入探明。作物的品质表现是一个复杂的综合性状,这些性状往往是由多个组分决定,其中每个组分是一个典型的数量性状。因此,作物中代谢网络的深入解析是培育高产优质作物的基础,将有助于深化人们对作物基因组和代谢组遗传基础的理解,同时可为利用基因工程进行作物品质改良提供新的信息和方向(Gong et al., 2013; Shachar, 2013)。另外,多条代谢途径之间往往相互关联(Yang et al., 2018a, 2018b; Zhang et al., 2016)。尽管科学家一直在研究各种植物代谢物的生物合成和功能,但我们对代谢物之间的相互作用、代谢物信号传递、代谢物与发育的相互作用及代谢调控在基因型到表型关系中的作用知之甚少。通过代谢调控作物体内物质和能量代谢及其分配,保持其在生长发育和环境适应之间保持最佳的最佳平衡,最终显著影响作物重要的品质性状的品质育种目标尚待探索。转录组学、蛋白质组学和代谢组学等组学发展,为作物品质代谢网络调控的研究将提供有力的技术支持(Basnet et al., 2016; Kusano et al., 2015)。

顺应生产方式变革和生态环境变化,作物品质表现与环境适应性间的关系有待深入研究。农作物品质形成是在大田自然环境下形成的,因此易受生长期尤其是种子形成期温、光、水、肥等的影响。例如,高温胁迫影响稻米垩白和淀粉含量、玉米淀粉品质、小麦籽粒蛋白质含量等,进而影响加工、蒸煮与食味品质等(Madan et al., 2012)。施肥与灌溉等栽培条件的选择和合理运用可以通过影响作物的生长发育和碳氮平衡来调控籽粒品质性状。目前,多学科的交叉应用使得作物中越来越多代谢物质的代谢调控机制正逐步被阐明,但环境影响品质形成的分子机制研究还不够深入。因此,为适应不同环境,作物品质和营养的抗逆机制研究及优质作物培育也迫在眉睫。

2.2 关键突破口

优异种质资源创新与利用 品质性状难于直观鉴定,现有育成品种品质单一。因此需进一步扩大优质、营养和专用类种质资源收集、创新与利用。在此基础上,进一步通过现代遗传学、基因组学和分子生物学等克隆控制品质与营养性状的重要基因,解析其调控机制。

品质与营养相关功能性成分形成遗传学基础 作物中形成品质的功能性成分主要包括淀粉类、蛋白质类和油脂类等,这些成分与消费者健康和幸福紧密相关。针对这些主要的功能性成分形成的品质性状，结合细胞学、分子生物学、系统生物学等研究，在基因组、转录组、蛋白质组和代谢组水平上解析这些主要功能性成分形成的遗传基础和分子基础，明确品质性状形成和演化的特征、规律和趋势，进而推进作物功能性品质改良与微营养生物强化的深度开发。

品质与营养相关重要代谢物质代谢调控网络 作物中淀粉、蛋白质、氨基酸、脂肪酸等重要物质代谢从来都不是一个简单的单线过程，而是一个相互关联的复杂网络，因此，解析植物重要代谢物质的代谢调控网络是品质育种的基础。各种组学和生物信息学等的发展为这些重要代谢物质的调控网络在基因、蛋白质和代谢物水平的研究提供了便利。

作物品质响应环境的调控机制 环境胁迫往往限制优良品质性状的表达。增加作物对非生物和生物胁迫反应的理解可用于在面对环境胁迫和气候变化时保持或提高品质。今后应加强作物品质与环境适应性之间关键机制的研究，从作物整体、器官或组织、分子水平等不同层次上深入研究作物品质对环境胁迫的响应机制及其调控途径，为培育优质、高产、抗逆作物提供理论基础。

3 阶段性目标

3.1 2035 年目标

至 2035 年，阶段性目标如下。

第一，深化分子基础和代谢网络研究。应用遗传学、多重组学、关联分析等技术手段，鉴定主要农作物外观品质、营养品质、加工品质和健康功能品质等相关基因/QTL，明确其对品质性状的遗传贡献，解析其功能及调控网络，阐明品质形成的分子基础；通过检测基因在不同野生和栽培品种中的等位变异，挖掘和鉴定控制品质性状优异等位基因，用于育种研究。

第二，拓宽作物功能性品质研究。在继续加强淀粉、蛋白质、氨基酸、油脂、籽粒形状等品质性状研究同时，重视抗性淀粉、多糖、微量元素、维生素和生物活性成分等研究，改善作物功能性品质，提高其在商业市场上的竞争力。

第三，重视作物品质性状多元化研究。利用现有和新开发的优质资源，通过多基因聚合和多性状整合，培育理化品质、蒸煮食味品质、营养品质、专用品质及产量等方面具有多个优质性状作物新种质，满足市场需求多元化。

第四，强调品质与环境互作机制的研究。针对不同作物在不同环境胁迫下对淀粉、垩白的影响，结合基因组水平、蛋白组水平、代谢物水平和 miRNA 水平对环境胁迫的响应，开展基于作物对不同环境胁迫响应机制的对策研究。

第五，加强育种新技术的开发与应用。应用基因编辑对目标基因进行定向编辑或功能研究；应用基因组学、代谢组学、蛋白质组学研究作物品质形成的遗传效应；密切结合分子标记技术与常规育种技术，推动分子育种实用化，提高品质改良效率；运用生物信息学大数据分析找出共同调控因子，以实现协同改良多个品质性状。

3.2 2050 年目标

至 2050 年，阶段性目标如下。

第一，作物中淀粉、蛋白质、氨基酸、脂肪酸等主要物质的代谢调控及其关联网络的解析，实现定量定向的理想化作物品质改良。

第二，整合作物重要品质性状的表型组、基因组、蛋白质组、代谢组等数据库，根据不同作物品质与营养要求，研制分子设计育种软件，构建分子设计信息系统；建立主要农作物复杂品质性状主效基因选择、全基因组选择等技术，建立高效分子设计育种技术体系。

第三，通过基因聚合、遗传调控等手段，创造有重大育种利用价值的新材料，加强品质亲本创新，改良品质性状及在不同环境下的稳定性，提高优质作物的环境适应性和产量潜力。

致谢：本章在撰写过程中得到了张昌泉和舒小丽的协助，特此致谢！

参考文献

杨君, 马崎英, 王省芬 (2016). 棉花纤维品质改良相关基因研究进展. 中国农业科学 49, 4310-4322.

张昌泉, 赵冬生, 李钱峰, 顾铭洪, 刘巧泉 (2016). 稻米品质性状基因的克隆与功能研究进展. 中国农业科学 *49*, 4267-4283.

张勇, 郝元, 张艳, 何心尧, 夏先春, 何中虎 (2016). 小麦营养和健康品质研究进展. 中国农业科学 *49*, 4284-4298.

张玉芹, 陆翔, 李擎天, 陈受宜, 张劲松 (2016). 大豆品质调控基因克隆和功能研究进展. 中国农业科学 *49*, 4299-4309.

Angelovici, R., Fait, A., Fernie, A.R., and Galili, G. (2011). A seed high lysine trait is negatively associated with the TCA cycle and slows down *Arabidopsis* seed germination. New Phytol. *189*, 148-159.

Azevedo, R.A., and Arruda, P. (2010). High-lysine maize: the key discoveries that have made it possible. Amino Acids *39*, 979-989.

Basnet, R.K., Del, C.D.P., Xiao, D., Bucher, J., Jin, M., Boyle, K., Fobert, P., Visser, R.G., Maliepaard, C., and Bonnema, G. (2016). A systems genetics approach identifies gene regulatory networks associated with fatty acid composition in *Brassica rapa* seed. Plant Physiol. *170*, 568-585.

Batista, R.A., Figueiredo, D.D., Santos-González, J., and Köhler, C. (2019). Auxin regulates endosperm cellularization in *Arabidopsis*. Genes Dev. *33*, 466-476.

Bouis, H.E., and Saltzman, A. (2017). Improving nutrition through biofortification: a review of evidence from HarvestPlus, 2003 through 2016. Global Food Sec. *12*, 49-58.

Bull, S.E., Seung, D., Chanez, C., Mehta, D., Kuon, J.E., Truernit, E., Hochmuth, A., Zurkirchen, I., Zeeman, S.Z., Gruissem, W., et al. (2018). Accelerated *ex situ* breeding of GBSS- and PTST1-edited cassava for modified starch. Sci. Adv. *4*, eaat6086.

Che, R., Tong, H., Shi, B., Liu, Y., Fang, S., Liu, D., Xiao, Y., Hu, B., Liu, L., Wang, H., et al. (2015). Control of grain size and rice yield by GL2-mediated brassinosteroid responses. Nat. Plants *2*, 15195.

Clemente, T.E., and Cahoon, E.B. (2009). Soybean oil: genetic approaches for modification of functionality and total content. Plant Physiol. *151*, 1030-1040.

Deng, M., Li, D., Luo, J., Xiao, Y., Liu, H., Pan, Q., Zhang, X., Zhao, M., and Yan, J. (2017). The genetic architecture of amino acids dissection by association and linkage analysis in maize. Plant Biotechnol. J. *15*, 1250-1263.

Duan, P., Ni, S., Wang, J., Zhang, B., Xu, R., Wang, Y., Chen, H., Zhu, X., and Li, Y. (2015). Regulation of OsGRF4 by OsmiR396 controls grain size and yield in rice. Nat. Plants *2*, 15203.

Fang, Y., Zeng, L., Xu, J., Yu, H., Shi, Z., Pan, J., Zhang, D., Kang, S., Zhu, L., Dong, G., et al. (2015). A rare allele of GS2 enhances grain size and grain yield in rice. Mol. Plant. *8*, 1455-1465.

Feng, F., Qi, W., Lv, Y., Yan, S., Xu, L., Yang, W., Yuan, Y., Chen, Y., Zhao, H., and Song, R. (2018). OPAQUE11 is a central hub of the regulatory network for maize endosperm development and nutrient metabolism. Plant Cell *30*, 375-396.

Fernie, A.R., Tadmor, Y., and Zamir, D. (2006). Natural genetic variation for improving crop quality. Curr. Opin. Plant Biol. *9*, 196-202.

Galili G., and Amir R. (2013). Fortifying plants with the essential amino acids lysine and methionine to improve nutritional quality. Plant Biotechnol. J. *11*, 211-222.

Gao, H.R., Gadlage, M.J., Lafitte, H.R., Lenderts, B., Yang, M.Z., Schroder, M., Farrell, J., Snopek, K., Peterson, D., Feigenbutz, L., et al. (2020). Superior field performance of waxy corn engineered using CRISPR-Cas9. Nat. Biotechnol. *38*, 579-581.

Gao, S., Xiao, Y., Xu, F., Gao, X., Cao, S., Zhang, F., Wang, G., Sanders, D., and Chu, C. (2019). Cytokinin-dependent regulatory module underlies the maintenance of zinc nutrition in rice. New Phytol. *224*, 202-215.

Gibbon, B.C., Wang, X., and Larkins, B.A. (2003). Altered starch structure is associated with endosperm modification in quality protein maize. Proc. Natl. Acad. Sci. USA *100*, 15329-15334.

Gong, L., Chen, W., Gao, Y., Liu, X., Zhang, H., Xu, C.G., Yu, S., Zhang, Q., and Luo, J. (2013). Genetic analysis of the metabolome exemplified using a rice population. Proc. Natl. Acad. Sci. USA *110*, 20320-20325.

Gutiérrez, R.A., Lejay, L.V., Dean, A., Chiaromonte, F., Shasha, D.E., and Coruzzi, G.M. (2007). Qualitative network models and genome-wide expression data define carbon/nitrogen-responsive molecular machines in *Arabidopsis*. Genome Biol. *8*, R7.

Hu, X., Peng, Y., Ren, X., Peng, J., Nevo, E., Ma, W., and Sun, D. (2018). Allelic variation of low molecular weight glutenin subunits composition and the revealed genetic diversity in durum wheat (*Triticum turgidum* L. ssp. *durum* (Desf)). Breed. Sci. *68*, 524-535.

Kovach, M., Calingacion, M., Fitzgerald, M., and Mccouch, S. (2009). The origin and evolution of fragrance in rice (*Oryza sativa* L.). Proc. Natl. Acad. Sci. USA *106*, 14444-14449.

Kusano, M., Yang, Z., Okazaki, Y., Nakabayashi, R., Fukushima, A., and Saito, K. (2015). Using metabolomic approaches to explore chemical diversity in rice. Mol. Plant *8*, 58-67.

Li, C.B., Yue, Y.H., Chen, H.J., Qi, W.W., and Song, R.T. (2018). The ZmbZIP22 transcription factor regulates 27-kD γ-zein gene transcription during maize endosperm development. Plant Cell *30*, 2402-2424.

Li, C.S., Xiang, X.L., Huang, Y.C., Zhou, Y., An, D., Dong, J.Q., Zhao, C.X., Liu, H.J., Li, Y.B., Wang, Q., Du, C.G., Messing, J.M., Larkins, B.A., Wu, Y.R., and Wang, W.Q. (2020). Long-read sequencing reveals genomic structural variations that underlie creation of quality protein maize. Nat. Commun. *11*, 17.

Li, H., Peng, Z., Yang, X., Wang, W., Fu, J., Wang, J., Han, Y., Chai, Y., Guo, T., Yang, N., et al. (2013). Genome-wide association study dissects the genetic architecture of oil biosynthesis in maize kernels. Nat. Genet. *45*, 43-72.

Li, N., Xu, R., and Li, Y.H. (2019). Molecular networks of seed size control in plants. Ann. Rev. Plant Biol. *70*, 435-463.

Liu, H., Shi, J., Sun, C., Gong, H., Fan, X., Qiu, F., Huang, X., Feng, Q., Zheng, X., Yuan, N., et al. (2016). Gene duplication confers enhanced expression of 27-kDa γ-zein for endosperm modification in quality protein maize. Proc. Natl. Acad. Sci. USA *113*, 4964-4969.

Liu, Q., Han, R., Wu, K., Zhang, J., Ye, Y., Wang, S., Chen, J., Pan, Y., Li, Q., Xu, X., et al. (2018). G-protein βγ subunits determine grain size through interaction with MADS-domain transcription factors in rice. Nat. Commun. *9*, 852.

Lu, S., Sturtevant, D., Aziz, M., Jin, C., Chapman, K.D., and Guo, L. (2018). Spatial analysis of lipid metabolites and expressed genes reveals tissue-specific heterogeneity of lipid metabolism in high- and low-oil *Brassica napus* L. seeds. Plant J. *94*, 915-932.

Madan, P., Jagadish, S.V.K., Craufurd, P.Q., Fitzgerald, M., Lafarge, T., and Wheeler, T.R. (2012). Effect of elevated CO_2 and high temperature on seed-set and grain quality of rice. J. Exp. Bot. *63*, 3843-3852.

Miller, C., Wells, R., McKenzie, N., Trick, M., Ball, J., Fatihi, A., Dubreucq, B., Chardot, T.,

Lepiniec, L., and Bevan, M.W. (2019). Variation in expression of the HECT E3 ligase UPL3 modulates LEC2 levels, seed size, and crop yields in *Brassica napus*. Plant Cell *31*, 2370-2385.

Nakamura, H. (2000). Allelic variation at high-molecular-weight glutenin subunit Loci, *Glu-A1*, *Glu-B1* and *Glu-D1*, in Japanese and Chinese hexaploid wheats. Euphytica *112*, 187.

Patil, G., Mian, R., Vuong, T., Pantalone, V., Song, Q., Chen, P., Shannon, G.J., Carter, T.C., and Nguyen, H.T. (2017). Molecular mapping and genomics of soybean seed protein: a review and perspective for the future. Theor. Appl. Genet. *130*, 1975-1991.

Peng, B., Kong, H., Li, Y., Wang, L., Zhong, M., Sun, L., Gao, G.J., Zhang, Q., Luo, L., Wang, G., et al. (2014). OsAAP6 functions as an important regulator of grain protein content and nutritional quality in rice. Nat. Commun. *5*, 4847.

Regina, A., Berbezy, P., Kosar-Hashemi, B., Li, S., Cmiel, M., Larroque, O., Bird, A.R., Swain, S.M., Cavanagh, C., Jobling, S.A., et al. (2015). A genetic strategy generating wheat with very high amylose content. Plant Biotechnol. J. *13*, 1276-1286.

Shachar, H.Y. (2013). Metabolic network flux analysis for engineering plant systems. Curr. Opin. Biotech. *24*, 247-255.

Shi, W., Muthurajan, R., Rahman, H., Selvam, J., Peng, S., Zou, Y., and Jagadish, K.S. (2013). Source-sink dynamics and proteomic reprogramming under elevated night temperature and their impact on rice yield and grain quality. New Phytol. *197*, 825-837.

Song, X.J., Huang, W., Shi, M.N., Zhu, M.Z., Lin, H.X. (2007). A QTL for rice grain width and weight encodes a previously unknown RING-type E3 ubiquitin ligase. Nat. Genet. *39*, 623-630.

Tian, Z., Qian, Q., Liu, Q., Yan, M., Liu, X., Yan, C., Liu, G., Gao, Z., Tang, S., Zeng, D., et al. (2009). Allelic diversities in rice starch biosynthesis lead to a diverse array of rice eating and cooking qualities. Proc. Natl. Acad. Sci. USA *106*, 21760-21765.

Toubiana, D., Semel, Y., Tohge, T., Beleggia, R., Cattivelli, L., Rosental, L., Nikoloski, Z., Dani, Z., Fernie, A.R., and Fait, A. (2012). Metabolic profiling of a mapping population exposes new insights in the regulation of seed metabolism and seed, fruit, and plant relations. PLoS Genet. *8*, e1002612.

Uauy, C., Distelfeld, A., Fahima, T., Blechl, A., and Dubcovsky, J. (2006). A *NAC* gene regulating senescence improves grain protein, zinc, and iron content in wheat. Science *314*, 1298-1301.

Wang, J.C., Xu, H., Zhu, Y., Liu, Q.Q., and Cai, X.L. (2013). OsbZIP58, a basic leucine zipper transcription factor, regulates starch biosynthesis in rice endosperm. J. Exp. Bot. *64*, 3453-3466.

Weng, J., Gu, S., Wan, X., Gao, H., Guo, T., Su, N., Lei, C., Zhang, X., Cheng, Z., Guo, X., et al. (2008). Isolation and initial characterization of *GW5*, a major QTL associated with rice grain width and weight. Cell Res. *18*, 1199-1209.

Wilson, S.A., and Roberts, S.C. (2014). Metabolic engineering approaches for production of biochemicals in food and medicinal plants. Curr. Opin. Biotechnol. *26*, 174-182.

Yang, J., Fu, M., Ji, C., Huang, Y., and Wu, Y. (2018b). Maize Oxalyl-CoA decarboxylase1 degrades oxalate and affects the seed metabolome and nutritional quality. Plant Cell *30*, 2447-2462.

Yang, Q.Q., Zhao, D.S., Zhang, C.Q., Wu, H.Y., Li, Q.F., Gu, M.H., Sun, S.S.M., and Liu, Q.Q. (2018a). A connection between lysine and serotonin metabolism in rice endosperm. Plant Physiol. *176*, 1965-1980.

Yang, Y., Guo, M., Sun, S., Zou, Y., Yin, S., Liu, Y., Tang, S., Gu, M., Yang, Z., and Yan, C. (2019). Natural variation of *OsGluA2* is involved in grain protein content regulation in rice. Nat. Commun. *10*, 1949.

Ye, X.D., Al-Babili, S., Klöti, A., Zhang, J., Lucca, P., Beyer, P., and Potrykus, I. (2000). Engineering the provitamin A (β-carotene) biosynthetic pathway into (carotenoid-free) rice endosperm. Science *287*, 303-305.

Zeng, D., Tian, Z., Rao, Y., Dong, G., Yang, Y., Huang, L., Leng, Y., Xu, J., Sun, C., Zhang, G., et al. (2017). Rational design of high-yield and superior-quality rice. Nat. Plants *3*, 17031.

Zhang, C., Zhu, J., Chen, S., Fan, X., Li, Q., Lu, Y., Wang, M., Yu, H., Yi, C., Tang, S., et al. (2019a). Wx^{lv}, the ancestral allele of rice *Waxy* gene. Mol. Plant *12*, 1157-1166.

Zhang, G.Y., Liu, R.R., Zhang, P., Li, Y., Tang, K.X., Liang, G.H., and Liu, Q.Q. (2013). Increased alpha-tocotrienol content in seeds of transgenic rice overexpressing *Arabidopsis* γ-tocopherol methyltransferase. Transgenic Res. *22*, 88-99.

Zhang, J., Wang, X., Lu, Y., Bhusal, S., Song, Q., Cregan, P.B., Yen, Y., Brown, M., and Jiang,

G.L. (2018). Genome-wide scan for seed composition provides insights into soybean quality improvement and the impacts of domestication and breeding. Mol. Plant *11*, 460-472.

Zhang, Z., Dong, J. Ji, C., Wu, Y., and Messing, J. (2019b). NAC-type transcription factors regulate accumulation of starch and protein in maize seeds. Proc. Natl. Acad. Sci. USA *116*, 11223-11228.

Zhang, Z., Gao, S., and Chu, C. (2020). Improvement of nutrient use efficiency in rice: current toolbox and future perspectives. Theor. Appl. Genet. *133*, 1365-1384.

Zhang, Z., Zheng, X., Yang, J., Messing, J., and Wu, Y. (2016). Maize endosperm-specific transcription factors O_2 and PBF network the regulation of protein and starch synthesis. Proc. Natl. Acad. Sci. USA *113*, 10842-10847.

Zhao, D.S., Li, Q.F., Zhang, C.Q., Zhang, C., Yang, Q.Q., Pan, L.X., Ren, X.Y., Lu, J., Gu, M.H., and Liu, Q.Q. (2018). *GS9* acts as a transcriptional activator to regulate rice grain shape and appearance quality. Nat. Commun. *9*, 1240.

Zhou, H., Wang, L., Liu, G., Meng, X., Jing, Y., Shu, X., Kong, X., Sun, J., Yu, H., Smith, S.M., et al. (2016). Critical roles of soluble starch synthase SSIIIa and granule-bound starch synthase Waxy in synthesizing resistant starch in rice. Proc. Natl. Acad. Sci. USA *113*, 12844-12849.

Zhu, L.J., Gu, M.H., Meng, X.L., Cheung, S.C.K., Yu, H.X., Huang J., Sun, Y., Shi, Y.C., and Liu, Q.Q. (2012). High-amylose rice improves indices of animal health in normal and diabetic rats. Plant Biotechnol. J. *10*, 353-362.

Zhu, Q., Wang, B., Tan, J., Liu, T., Li, L., and Liu, Y.G. (2020). Plant synthetic metabolic engineering for enhancing crop nutritional quality. Plant Commun. *1*, 100017

Zhu, Q., Zeng, D., Yu, S., Cui, C., Li, J., Li, H., Chen, J., Zhang, R., Zhao, X., Chen, L., et al. (2018). From golden rice to aSTARice: bioengineering astaxanthin biosynthesis in rice endosperm. Mol. Plant *11*, 1440-1448.

特殊功用作物改良

罗 杰　王国栋　漆小泉　陈晓亚

1 国内外研究进展

植物通过次生代谢途径合成上百万种化合物,这些物质不但是植物生长、发育及防御各种逆境胁迫的基础,也是人类所需食品、药品、化工产品等的重要来源。植物次生代谢研究的突破对国计民生能够产生重大的推动作用。例如,在萜类代谢产物中,对赤霉素的研究直接促成了第一次绿色革命,对青蒿素的研究使之成为迄今最有效的抗疟药物。然而,迄今我们对植物次生代谢途径及调控机制的认识非常少,成为利用植物次生代谢产物的瓶颈问题。

目前,我国城乡居民温饱问题基本解决,正朝着全面建成小康社会的目标迈进。因此,我国粮食安全的内涵正处于由"量"到"质"的转变时期,需要农业生产供应充足而富有营养的食物。人们对作物营养的需求越来越"个性化",近来面向特殊人群需求的功能性作物新品种培育取得了显著的进展。例如,高血压患者专用的富含 γ-氨基丁酸的水稻品种,糖尿病患者专用的富含抗性淀粉的小麦和水稻品种已进入市场。总体来说,特殊营养品质改良为未来农业发展开辟广阔的发展空间,也是今后作物育种的一个重要特征。近年来,国际知名杂志 *Science* 和 *Plant Cell* 连续发表专刊和多篇综述,阐述了植物代谢研究对解决人类营养与健康问题的重要意义及面临的各种挑战(Fitzpatrick et al., 2012; Martin et al., 2011; Traka and Mithen, 2011)。营养品质是评价农产品品质性状的关键,除淀粉、蛋白质和脂肪等大量营养素外,营养品质还受到很多次生代谢物的影响,包括特殊营养成分(也包括一些对人体健康有害的抗营养因子)。作物特殊营养品质性状不但受到代谢途径本身影响,而且受到复杂的网络调控,属于多基因控制的复杂性状。

值得强调的是,并不是所有的植物次生代谢产物都对人体健康有益,作物中还存在一些小分子化合物，它们破坏或阻碍营养物质的消化与吸收，从而影响人体健康，称之为"抗营养因子"。已报道的植物抗营养因子有数百种之多，主要有蛋白酶抑制剂、植酸、单宁、棉酚、芥子酸、甾醇类生物碱等（Cardoso et al., 2017; Gupta et al., 2015; Romanucci et al., 2016; Tian et al., 2018; Vagadia et al., 2017）。随着人们对营养品质与食物安全的要求不断提高，培育富含特殊营养成分，同时降低（或去除）抗营养因子作物新品种是保障食物安全、提高人们生活水平的重要手段。

不管是作物的产量性状还是品质性状都决定于代谢产物在特定器官中的积累。据估计，一种植物能产生5000至25 000个化合物，其中大部分（>70%）迄今仍不清楚。近年来，基因组学、转录组学、蛋白质组学及代谢组学的快速发展，为人们从整体水平上定性、定量和动态分析代谢过程和代谢产物创造了良好条件。目前的质谱技术已经能鉴定代谢组中30%的化合物，代谢组学在基因克隆与功能分析，以及作物重要农艺性状的改良中取得了明显的成效（Saito and Matsuda, 2010）。水稻、玉米、大豆等主要作物基因组测序和自然群体的重测序相继完成，为作物代谢途径的解析取得显著进展带来重大机遇（Chen et al., 2014; Fang et al., 2017; Tieman et al., 2017; Zhou et al., 2015; Zhu et al., 2018a）。以代谢组和基因组为基础建立起来的技术体系不但为深入解析复杂的代谢途径及其调控机制等基础理论研究提供了新方法，还可应用于高通量定位控制代谢产物的关键基因/位点（mQTL），为分子设计培育富含多种特殊营养成分的作物新品种提供基础（Christ et al., 2018; Fang et al., 2019）。目前有针对性地提高作物特殊营养成分的分子操作思路和手段还比较初步，通常是利用自主特异性启动子将已知的代谢途径基因导入底盘植物中。2000年，科研人员将4个控制类胡萝卜素合成的关键基因转入水稻，使得本身并不合成胡萝卜素的水稻胚乳中的胡萝卜素含量显著升高，这也就是著名的"黄金大米"的来源（Ye et al., 2000）。利用类似的手段，华南农业大学等团队通过遗传转化将4个虾青素（astaxanthin）合成的关键基因导入水稻胚乳中，使其特异表达，获得虾青素积累到十几微克/克干重的含量的水稻（Zhu et al., 2018b）。随着基因编辑技术的发展和日趋成熟，该技术将广泛应用于植物基因功能研究和作物育种改良（Li et al., 2013; Nekrasov et al., 2013; Shan et al., 2013）。例如，中国水稻研究所研究人员采用 CRISPR/Cas9 技术对控制水稻香味的 *BADH2* 基因进行敲除，使得突变体材料中香味物质显著增加，加快了优质香稻的育种进程（Shen et al.，2017）；近期，中国科学院遗传与发育生物学研究所的研究人员利用 CRISPR/Cas 基因编辑技术通过编辑维生素 C 生物合成途径的基因 *SLGGP1* 编码框上游的序列，增加了 *SLGGP1* 转录本的翻译效率，进而增加维生素 C 的含量（Li et al.，2018；Zhang et al.，2018）。

代谢作为生物体最基本的生命活动方式，本质上是由代谢物通过酶连接而成的一个复杂网络。作物代谢研究的主要目的之一就是为了建立代谢模型，为进一步的作物改良提供指导。目前，建立植物代谢模型有两种主要途径：一种途径是建立代谢过程的基于限制的动力学模型，该方法主要依赖于公共数据库中代谢网络结构和酶学生化信息，简单但准确性较差（没有考虑作物的特异性），无法为特定作物分子育种提供精准的理论指导（Blank and Kuepfer，2010；Fischer and Sauer，2005）；另外一种途径是建立代谢过程的动力学模型，这需要大量代谢途径中所涉及的化合物变化信息和关键酶的动力学特性（K_m、K_{cat} 等常数）（Szecowka et al.，2013）。虽然工作量大，但这种模型的建立及其与其他高通量数据的结合，将极大地促进作物代谢工程研究，快速提升我国作物品质精准改良的能力。

总体来说，通过克隆控制特殊营养成分等生物合成的关键基因与调控因子，开发紧密连锁的分子标记，利用分子育种途径培育富含特殊营养成分的作物品种是今后作物育种的新方向。

2 未来发展趋势与关键突破口

2.1 未来发展趋势

2.1.1 营养健康的未来作物

均衡营养是人体健康的重要前提，作物产品作为人类食物的主要来源之一，是保障人体营养健康的源头。随着生活水平不断提高，我国居民对食物需求发生了新的变化，主要表现为由以前"吃得饱"逐渐转变为"吃得好"，更要"吃得安全""吃出健康"。与此相对应，联合国《2030 年可持续发展议程》(Transforming Our World: the 2030 Agenda for Sustainable Development)也将"终止饥饿、保障粮食安全,提高营养,促进农业可持续发展"作为其主要目标之一。为实现该目标,越来越多的国际机构开始探索从食物源头解决营养问题的途径,呼吁调整以往以产量为导向的农业生产方式,将"营养健康"作为农业发展的重要方向。因此,有必要大力促进未来作物朝着营养和健康驱动的方向转型,以应对当前全球面临的重大挑战。

2.1.2 特殊功能化的未来作物

伴随经济社会的发展和国民的膳食结构的变化,人们生活节奏加快、工作压力加大,心血管疾病等富贵病快速增长,老年病增多,亚健康人群增加,对于具有保健功能的作物产品的需求非常迫切,让消费者从主要作物的功能化产品中得到各种营养素和功能因子的补充将成为未来作物研究的重要方向。

在以满足温饱为主要目标时,各类人群对于作物产品的需求基本一致。随着温饱问题的解决,不同消费群体对于个性化作物产品需求得到极大释放。事实上,受年龄、性别、身体内在状态、生活习惯、环境情况等多种内外因素影响,不同类型人群对于营养健康作物产品都有着特殊需求。在消费日益细分,消费群体日渐变窄的情况下,针对不同人群、同一人群不同阶段等个性化"订制"作物也是未来作物发展的重要方向。

2.1.3 多营养强化的未来作物

人体所需营养素是非常多的,但是传统的作物产品营养成分相对单一,很难通过一种或少数几种作物产品的摄入满足人体均衡营养的需求。因此,我国古代就提出了"五谷为养、五果为助、五畜为益、五菜为充"的说法。另外,目前作物产品中营养成分的含量相对较低,需要较为大量摄入才能满足人体的需要。但是,人类生活和工作方式的改变,节奏的加快使得在日常生活中大量食用多种作物产品有较大困难。因此,开发多营养成分、高营养密度的作物(如同时富含多种必需氨基酸、维生素及类黄酮等营养健康成分的作物)也是未来作物发展的重要方向。多营养成分、高营养密度作物产品的开发,对于减少耕地需求、降低运输、贮存成本、推动可持续发展也有重要意义。

2.2 关键突破口

2.2.1 营养健康牵引型未来作物的培育开发

针对农业发展与营养健康所面临的新挑战，牢牢把握国际上"从农业源头解决营养问题"的总体发展趋势，加强营养和健康相关代谢生物学研究，适时调整未来作物研究方向和重点，设立主要作物营养健康牵引型农业重大研发专项，开展高通量营养和健康成分代谢组分析及功能鉴定，营养和健康性状为标志的新型作物产品创制，高营养和健康农产品人群效应评价及监测，高营养和健康作物产品国际化推广研究等全链条基础及应用研究，助力农业供给侧结构性改革的实施和深化。

2.2.2 制定作物产品营养健康品质标准

食物是人类营养的主要来源，作物在我国居民食物结构中占有主导地位，居民整体营养健康水平的提高依赖源头高品质作物产品的供给。在缺乏系统性基础研究和原始创新的情况下，科学、合理、系统的作物产品营养品质标准体系尚未建立。缺乏有效指导科研和生产，满足农业发展需求和居民营养需求的作物产品营养质量标准，已经成为制约我国农业提质增效的"卡脖子"因素。作物产品供给质量提升的前提是确定合理准确的评价标准，构建作物营养质量调控、干预、改良甚至从头设计的科学标准体系。在科学、合理、统筹的前提下，制定作物产品营养质量控制标准、作物产品营养价值评价、加工营养控制、营养检测和监控等系列标准，引导形成"优质优价"的农产品及其加工产品的交易市场，促进农业生产适应我国农产品供给的区域化、复杂性和差异化特征，满足我国农业转型发展和居民膳食结构优化的需要。

2.2.3 代谢物与作物产品的品质与产量直接相关，需要多学科合作

未来农业的发展一定是多学科交叉的系统工程，关于作物营养品质强化改良方向（除"恶"扬"善"，恶为抗营养因子，善为对人体健康有益的代

谢物），需要在大规模挖掘高营养品质遗传资源与代谢调控网络的基础上，结合基因组编辑、合成生物学等新兴技术，加快培育可推广、易应用的营养优质作物新品种。

3 阶段性目标

3.1 2035年目标

至 2035 年的阶段性目标如下。

第一，建立系统分析植物代谢谱方法与技术，完成水稻、小麦、大豆、番茄、薯类、谷子等主要作物代谢谱分析，建立代谢谱数据库；了解特殊营养成分的组织特异性分布特征，发掘一批富含特殊营养成分的种质资源（同时考虑降低和去除抗营养因子的含量）；培育开发系列功能化、个性化、多营养成分和高营养密度的作物新品种。

第二，筛选营养健康品质优良（营养均衡、营养密度高）的植物；培育营养均衡、高营养密度的新作物。

第三，建立主要作物营养和健康农产品检测、评价中心；以作物科学、食品科学与营养科学相结合的系统性研究成果为科学依据，建立和编制适宜我国农业发展和居民营养需求的作物产品营养标准体系；制定主要作物营养和健康作物产品营养标签体系；制定主要作物营养和健康作物品种审定和认定标准，建立绿色通道；制定营养和健康作物产品市场准入标准及政策；制定营养和健康作物产品摄入标准。

3.2 2050年目标

至 2050 年的阶段性目标如下。

第一，建立完善、高效的作物营养健康品质育种体系，培育适应营养健康需求的作物新品种。

第二，培育一批特殊营养功能化、多营养强化的品质优良新作物。

第三，建立完善的营养健康牵引型作物认证、评价、推广体系，推动实现"精准营养"。

参考文献

Blank, L.M., and Kuepfer, L. (2010). Metabolic flux distributions: genetic information, computational predictions, and experimental validation. Appl. Microbiol. Biotechnol. *86*, 1243-1255.

Cardoso, L.D., Pinheiro, S.S., Martino, H.S.D., and Pinheiro-Sant'Ana, H.M. (2017). *Sorghum* (*Sorghum bicolor* L.): nutrients, bioactive compounds, and potential impact on human health. Crit. Rev. Food Sci. *57*, 372-390.

Chen, W., Gao, Y.Q., Xie, W.B., Gong, L., Lu, K., Wang, W.S., Li, Y., Liu, X.Q., Zhang, H.Y., Dong, H.X., et al. (2014). Genome-wide association analyses provide genetic and biochemical insights into natural variation in rice metabolism. Nat. Genet. *46*, 714-721.

Christ, B., Pluskal, T., Aubry, S., and Weng, J.K. (2018). Contribution of untargeted metabolomics for future assessment of biotech crops. Trends Plant Sci. *23*, 1047-1056.

Fang, C., Ma, Y.M., Wu, S.W., Liu, Z., Wang, Z., Yang, R., Hu, G.H., Zhou, Z.K., Yu, H., Zhang, M., et al. (2017). Genome-wide association studies dissect the genetic networks underlying agronomical traits in soybean. Genome Biol. *18*, 161.

Fang, C.Y., Fernie, A.R., and Luo, J. (2019). Exploring the diversity of plant metabolism. Trends Plant Sci *24*, 83-98.

Fischer, E., and Sauer, U. (2005). Large-scale in vivo flux analysis shows rigidity and suboptimal performance of *Bacillus subtilis* metabolism. Nat. Genet. *37*, 636-640.

Fitzpatrick, T.B., Basset, G.J.C., Borel, P., Carrari, F., DellaPenna, D., Fraser, P.D., Hellmann, H., Osorio, S., Rothan, C., Valpuesta, V., et al. (2012). Vitamin deficiencies in humans: can plant science help? Plant Cell *24*, 395-414.

Gupta, R.K., Gangoliya, S.S., and Singh, N.K. (2015). Reduction of phytic acid and enhancement of bioavailable micronutrients in food grains. J. Food Sci. Tech. *52*, 676-684.

Li, J.F., Norville, J.E., Aach, J., McCormack, M., Zhang, D.D., Bush, J., Church, G.M., and Sheen, J. (2013). Multiplex and homologous recombination-mediated genome editing in *Arabidopsis* and *Nicotiana benthamiana* using guide RNA and Cas9. Nat. Biotechnol. *31*, 688-691.

Li, T.D., Yang, X.P., Yu, Y., Si, X.M., Zhai, X.W., Zhang, H.W., Dong, W.X., Gao, C.X., and Xu,

C. (2018). Domestication of wild tomato is accelerated by genome editing. Nat. Biotechnol. *36*, 1160-1163.

Martin, C., Butelli, E., Petroni, K., and Tonelli, C. (2011). How can research on plants contribute to promoting human health? Plant Cell *23*, 1685-1699.

Nekrasov, V., Staskawicz, B., Weigel, D., Jones, J.D.G., and Kamoun, S. (2013). Targeted mutagenesis in the model plant *Nicotiana benthamiana* using Cas9 RNA-guided endonuclease. Nat. Biotechnol. *31*, 691-693.

Romanucci, V., Pisanti, A., Di Fabio, G., Davinelli, S., Scapagnini, G., Guaragna, A., and Zarrelli, A. (2016). Toxin levels in different variety of potatoes: alarming contents of alpha-chaconine. Phytochem. Lett. *16*, 103-107.

Saito, K., and Matsuda, F. (2010). Metabolomics for functional genomics, systems biology, and biotechnology. Ann. Rev. Plant Biol. *61*, 463-489.

Shan, Q.W., Wang, Y.P., Li, J., Zhang, Y., Chen, K.L., Liang, Z., Zhang, K., Liu, J.X., Xi, J.J., Qiu, J.L., et al. (2013). Targeted genome modification of crop plants using a CRISPR-Cas system. Nat. Biotechnol. *31*, 686-688.

Shen, L., Hua, Y., Fu, Y., Li, J., Liu, Q., Jiao, X., Xin, G., Wang, J., Wang, X., Yan, C., et al. (2017). Rapid generation of genetic diversity by multiplex CRISPR/Cas9 genome editing in rice. Sci. China Life Sci. *60*, 506-515.

Szecowka, M., Heise, R., Tohge, T., Nunes-Nesi, A., Vosloh, D., Huege, J., Feil, R., Lunn, J., Nikoloski, Z., Stitt, M., et al. (2013). Metabolic fluxes in an illuminated *Arabidopsis* rosette. Plant Cell *25*, 694-714.

Tian, X., Ruan, J.X., Huang, J.Q., Yang, C.Q., Fang, X., Chen, Z.W., Hong, H., Wang, L.J., Mao, Y.B., Lu, S., et al. (2018). Characterization of gossypol biosynthetic pathway. Proc. Natl. Acad. Sci. USA *115*, E5410-E5418.

Tieman, D., Zhu, G.T., Resende, M.F.R., Lin, T., Taylor, M., Zhang, B., Ikeda, H., Liu, Z.Y., Fisher, J., Zemach, I., et al. (2017). A chemical genetic roadmap to improved tomato flavor. Science *355*, 391-394.

Traka, M.H., and Mithen, R.F. (2011). Plant science and human nutrition: challenges in assessing health-promoting properties of phytochemicals. Plant Cell *23*, 2483-2497.

Vagadia, B.H., Vanga, S.K., and Raghavan, V. (2017). Inactivation methods of soybean trypsin inhibitor—a review. Trends Food Sci. Technol. *64*, 115-125.

Ye, X.D., Al-Babili, S., Kloti, A., Zhang, J., Lucca, P., Beyer, P., and Potrykus, I. (2000). Engineering the provitamin A (beta-carotene) biosynthetic pathway into (carotenoid-free) rice endosperm. Science *287*, 303-305.

Zhang, H.W., Si, X.M., Ji, X., Fan, R., Liu, J.X., Chen, K.L., Wang, D.W., and Gao, C.X. (2018). Genome editing of upstream open reading frames enables translational control in plants. Nat. Biotechnol. *36*, 894-898.

Zhou, Z.K., Jiang, Y., Wang, Z., Gou, Z.H., Lyu, J., Li, W.Y., Yu, Y.J., Shu, L.P., Zhao, Y.J., Ma, Y.M., et al. (2015). Resequencing 302 wild and cultivated accessions identifies genes related to domestication and improvement in soybean. Nat. Biotechnol. *33*, 408-414.

Zhu, G.T., Wang, S.C., Huang, Z.J., Zhang, S.B., Liao, Q.G., Zhang, C.Z., Lin, T., Qin, M., Peng, M., Yang, C.K., et al. (2018a). Rewiring of the fruit metabolome in tomato breeding. Cell *172*, 249-261.

Zhu, Q.L., Zeng, D.C., Yu, S.Z., Cui, C.J., Li, J.M., Li, H.Y., Chen, J.Y., Zhang, R.Z., Zhao, X.C., Chen, L.T., et al. (2018b). From golden rice to aSTARice: bioengineering astaxanthin biosynthesis in rice endosperm. Mol. Plant *11*, 1440-1448.

基因表达调控

丁 勇　曹晓风

1 国内外研究进展

1.1 总述

真核生物基因表达调控是一个复杂的过程，主要由转录调控和转录后调控组成。在转录调控过程中，转录启动子的顺式元件和增强子、转录因子、染色体的结构状态，以及 DNA 的甲基化修饰状态参与转录调控；RNA 多种的修饰和非编码 RNA（包括 miRNA 和 siRNA 在内的小 RNA 和长非编码 RNA）参与转录后调控。

1.1.1 顺式元件和转录因子

转录起始时的转录复合体结合于启动子区（如 TATA box 类）的通用顺式元件，激活基因的转录。除这些最通用的顺式元件外，还存在特异的顺式元件，决定转录的特异性。这些特异的顺式元件可以位于基因的启动子区或转录起始的远端（可达 10 kb 以上的距离），也可位于基因的内部。增强子调节转录的强度，同样可以位于基因转录起始的远端。在动物中，这些区域通常受组蛋白 H3 第 4 位——甲基化（H3K4me1）修饰。转录因子通过与特异的顺式元件和增强子结合，决定基因的转录状态。

1.1.2 染色体状态

真核生物的 DNA 缠绕在组蛋白上，组蛋白的 N 端及核心区易受到多种共价修饰，这些修饰改变了染色体的微环境，被不同蛋白质识别，调控基因的转录；在转录的过程中，RNA 聚合酶状态也在改变着染色体修饰的状态。组蛋白上多种共修饰和不同的修饰位点已经陆续被发现。此外，组蛋白还存在着多种不同的变体，这些组蛋白变体影响着核小体的稳定性，调节 RNA

聚合酶的在相应基因上的富集和移动速率。

转录因子与染色体修饰因子形成复合体，与顺式元件协同作用完成基因组蛋白修饰的特异性，促进或抑制基因的转录。此外，同一条染色体内部和不同的染色体间形成拓扑结构域（topologically associating domains，TADs），通过高级结构调节基因的转录。

1.1.3 DNA修饰

遗传信息的主要载体DNA易受到甲基化的共价修饰，这类共价修饰可以发生在胞嘧啶C和腺嘌呤A上。无论在动物还是植物中的胞嘧啶C受甲基化修饰的程度很高时，通常会抑制基因的转录；而腺嘌呤A的甲基化（N^6-adenine，6mA）频率则很低，与个体的发育进程相关，且与基因的转录呈正相关。

1.1.4 RNA修饰

随着RNA研究的深入，多种RNA修饰被发现。与DNA相似，mRNA的A（N^6-methyladenosine，m^6A）和C（5-methylcytosine，m^5C）都可以被甲基化，m^6A修饰可以加速pre-mRNA的加工以及mRNA的转运和稳定性。除此以外，A上还存在N^1-methyladenosine（m^1A）和N^6,2′-O-dimethyladenosine（m^6A_m）等修饰，这些新的修饰的生物学功能还有待进一步阐述。

1.1.5 非编码RNA

非编码RNA是一类不编码蛋白质的分子，以RNA作为主体参与基因表达调控。根据长度主要分为small RNA（主要包括miRNA和siRNA）和long non-coding RNA（lncRNA）。small RNA对靶基因的调控方式包括：靶转录本的切割、靶基因的DNA甲基化和靶基因的翻译抑制。与小分子RNA相比较，植物中关于lncRNA的研究相对较少，还有很多非编码的调控机制尚不清楚。

1.2 国内研究进展

1.2.1 小RNA研究进展

中国科学院遗传与发育生物学研究所曹晓风研究员课题组系统研究了水

稻中各类小 RNA 的生成机制（Liu et al.，2005，2007；Song et al.，2012a，2012b；Wei et al.，2014），揭示了转座子来源的 siRNA 调控旁邻基因表达，精细调控农艺性状的分子机制；北京生命科学研究所何新建课题组对小 RNA 介导的 DNA 甲基化机制做出了多项重要发现（Tan et al.，2018；Zhang et al.，2012，2013）；清华大学戚益军课题组揭示了小 RNA 进入植物 AGO 蛋白的分子机制（Cui et al.，2016；Fang et al.，2015，2019；Li et al.，2018；Liu et al.，2018；Mi et al.，2008；Ye et al.，2012，2016），发现了一类参与 DNA 损伤修复的新型小 RNA（Gao et al.，2014；Liu et al.，2017；Wei et al.，2012），以及长链非编码 RNA 促进转录的功能（Zhao et al.，2018）；中国科学院微生物研究所郭惠珊、方荣祥，北京大学李毅，浙江大学周雪平，福建农林大学吴建国等课题组在 RNA 干扰抗植物病毒的机制研究方面也取得一系列重要进展（Cui et al.，2020；Li et al.，2017；Mei et al.，2018，Wu et al.，2017；Yao et al.，2019；Zhang et al.，2016，2020）。在非编码 RNA 的功能和应用研究方面，中国科学院植物生理生态研究所陈晓亚课题组发现在植物中表达棉铃虫基因的双链 RNA（dsRNA）可抑制棉铃虫生长（Mao et al.，2007），发现 miR156 参与对植物防御响应和抗虫能力的动态调控（Mao et al.，2017）；中国科学院遗传与发育生物学研究所李家洋课题组发现 miR156 可以通过调控水稻理想株型形成关键基因 *IPA1* 的水平进而调控产量（Jiao et al.，2010）；中山大学陈月琴课题组发现 miR397 通过影响 BR 信号通路促进水稻的籽粒增大和产量提高，而 miR408 可以通过下调靶基因 UCL8 控制产量（Zhang et al.，2013）；中国科学院遗传与发育生物学研究所储成才和李云海、武汉大学李绍清以及中国水稻研究所的钱前课题组，共同发现 miR396 通过调控 GRF4 和 GRF6 控制水稻籽粒大小控制产量（Che et al.，2015；Duan et al.，2015；Gao et al.，2015；Hu et al.，2015）；中国科学院植物生理生态研究所王佳伟研究植物 miRNA 的生物学功能取得多项研究成果（Gou et al.，2011；Wang et al.，2011；Yu et al.，2012），特别是该课题组发现 miR156 可参与多年生植物的开花时间调控（Zhou et al.，2013）。华中农业大学张启发课题组和华南农大庄楚雄课题组发现一个 siRNA/长非编码 RNA 可调控水稻的光敏雄性不育（Ding et al.，2012；Zhou et al.，2014）；张启发课题组还鉴定到一个非编码位点 PMS1T 通过产生 phasiRNA 参与水稻的光敏不育（Fan et al., 2016）；中国农业科学院王海洋课题组发现光敏色素通过小分子 RNA 参与植物庇荫反应（Xie et al., 2017）；中国科学院植物生理生态研究所何祖华课题组发现抗病基因 *Pigm* 位点附近的转座子来源的 siRNA 介导了水稻广谱抗稻瘟病及产量间的平衡（Deng et al., 2017）；中国农业大学秦峰课题组和华中农业大学王石平课题组分别发现转座子来源的 siRNA 调控玉米抗旱及水稻抗病（Mao et al., 2015；Zhang et al., 2016）。

1.2.2 组蛋白修饰研究进展

中国科学院植物生理生态研究所何跃辉课题组发现组蛋白甲基化修饰的建立（Li et al., 2016；Luo et al., 2019；Tao et al., 2017，2019；Wang et al., 2014a；Yuan et al., 2016）；中国科学技术大学丁勇课题组发现组蛋白磷酸化新的修饰位点，以及转录对组蛋白甲基化修饰的反馈调控（Jiang et al., 2018a，2018b；Lu et al., 2017；Su et al., 2017；Tian et al., 2019；Zheng et al., 2018）；北京大学邓兴旺课题组揭示了杂种优势的表观修饰差异（Shen et al., 2012；Wang et al., 2014b；Zhu et al., 2016）；中国科学院遗传与发育生物学研究所傅向东课题组发现组蛋白修饰参与营养高效利用（Wu et al., 2020）；华南农业大学陈乐天课题组发现顺式元件招募组蛋白修饰因子（Xie et al., 2018）；复旦大学董爱武课题组发现组蛋白修饰和分子伴侣的功能（Bu et al., 2014；Jin et al., 2015；Liu et al., 2016；Zhang et al., 2015a；Zhu et al., 2017）。中国科学院植物生理生态研究所何祖华课题组发现组蛋白甲基化参与水稻花器官生成（Yan et al., 2015）。

华中农业大学周道绣课题组发现组蛋白甲基化和乙酰化参与植物生长发育调控（Liu et al., 2015；Zhou et al., 2016，2017），鉴定了水稻组蛋白丁酰化和巴豆酰化修饰与转录之间的关系（Lu et al., 2018）。中国科学院植物生理生态研究所朱健康课题组发现组蛋白修饰与 DNA 甲基化之间的关系（Duan et al., 2017a；Qian et al., 2014；Tang et al., 2016；Xiao et al., 2019；Yang et al., 2018a）；中国农业大学巩志忠课题组发现组蛋白乙酰化和甲基化调控 DNA 甲基化和基因组稳定性（Zhang et al., 2016；Zhao et al., 2014）；西北农林科技大学孙其信课题组发现组蛋白乙酰化参与热胁迫（Hu et al.，2015）；华中农业大学王学路课题组发现组蛋白乙酰化与油菜素内酯之间的关系（Hao et al.，2016）；中国农业大学姚颖垠课题组发现组蛋白乙酰化参与玉米种子萌发过程（Yang et al.，2016）；华南植物园侯兴亮课题组发现组蛋白乙酰化参与光信号（Tang et al.，2017）；中国科学院植物研究所刘永秀课题组发现组蛋白乙酰化和组蛋白泛素化的功能（Cao et al.，2015；Wang et al.，2016）；此外，福建农林大学秦源课题组比较了组蛋白变体H2.A 与多个组蛋白修饰之间的关系（Dai et al.，2017）。

一系列组蛋白去甲基化酶的功能被相继发现。中国科学院遗传与发育生物学研究所曹晓风课题组系统阐述了多个组蛋白位点去甲基化酶的功能（Cui et al.，2013，2016；Lu et al.，2010，2011；Zhang et al.，2015b），并与南方科技大学杜嘉木课题和中国科学院植物研究所金京波课题组合作，解析了组蛋白第 4 位去甲基化酶的结构和功能（Liu et al.，2019a；Yang et al.，2018c），与中国科学院植物生理生态研究所何祖华和南方科技大学杜嘉木合作，解析了组蛋白第 27 位去甲基化酶的结构和功能（Liu et al.，2019b；Zheng et al.，2019）。多个新功能的识别蛋白相继被报道，中国科学院植物生理生态研究所何跃辉研究组发现可识别第 4 位和第 27 位甲基的蛋白（Li et al.，2018），南方科技大学的杜嘉木课题组与威斯康星大学麦迪逊分校钟雪花课题组合作也发现类似的识别蛋白（Qian et al.，2018；Yang et al.，2018b）。清华大学的李海涛课题组、孙前文课题组和李不龙课题组合作证实植物 HP1 蛋白的功能（Zhao et al.，2019）。

北京大学的贾桂芳课题组与美国芝加哥大学的何川课题组发现 RNA 的 m^6A 去甲基化和甲基化识别蛋白（Duan et al.，2017b；Wei et al.，2018）。中国农业科学院的谷晓峰课题组与新加坡国立大学的俞皓课题组合作发现拟南芥 RNA 的 m^5C 修饰（Cui et al.，2017）和 DNA 的 m^6A 修饰（Liang et al.，2018）。谷晓峰课题组还鉴定出水稻'日本晴'和'93-11'中 DNA 的 6mA 修饰水平差异（Zhang et al.，2018）。

1.3 国外研究进展

国外多个课题组在表观修饰及其机制阐述的研究中处于领先位置。例如，美国加州大学洛杉矶分校 Steven Jacobson 系统阐述了 DNA 甲基化的形成机制，并阐述了组蛋白甲基化与 DNA 甲基化之间的关系（Gallego-Bartolome et al.，2019；Johnson et al.，2014）；美国加州大学河滨分校 Xuemei Chen 和德州农工大学 Xiuren Zhang 等在小 RNA 形成领域开展了大量系统的研究（Song et al.，2019；Wang et al.，2018；Yu et al.，2019）；英国 John Innes Center 的 Caroline Dean 阐述了春化过程中组蛋白修饰参与 *FLC* 的沉默，以及长非编码 RNA *COOLAIR* 参与这一调控过程（Whittaker and Dean，2017）；奥地利 Austrian Academy of Sciences, Gregor Mendel Institute of Molecular Plant Biology 的 Frédéric Berger 解答了新合成的组蛋白如何完成 H3K27me3 的科学难题（Jiang and Berger，2017），Doris Wangner 发现 PRC2 复合体如何识别基因的 PRE 位点（Xiao et al.，2017）；新加坡国立大学的 Hao Yu 阐述了多种组蛋白修饰和 RNA 甲基化在植物生长发育中的作用（Cui et al.，2017；Liang et al.，2018）。

2 针对未来作物分子设计需求的研究内容与突破口

未来作物的特点包括：适应集约化生产、优质高产、资源节约、环境友好、广适性、优质专用。植物体在生长发育过程中受多种环境的影响，植物对个体生长发育和对环境的适应是基因表达的精细调控过程。因此，解析基因转录调控，是理解作物高产、优质、抗逆的基础。近年来的研究表明，转录和转录后水平的调控在植物的生长、发育和环境适应性中发挥重要作用。其中非编码 RNA、组蛋白修饰、DNA 修饰和 RNA 修饰领域的研究，是近年来生物学研究的热点和新方向之一。

这其中的关键科学问题包括：① 决定农作物复杂性状的关键非编码 RNA、组蛋白修饰、DNA 修饰和 RNA 修饰因子；② 非编码 RNA、组蛋白修饰、DNA 修饰和 RNA 修饰建立和维持的分子机制；③非编码 RNA、组蛋白修饰、DNA 修饰和 RNA 修饰调控转录的分子机制；④ 非编码 RNA、组蛋白修饰、DNA 修饰和 RNA 修饰参与非生物胁迫的分子机制；⑤ 非编码 RNA、组蛋白修饰、DNA 修饰和 RNA 修饰适应环境调节生长进程的分子机制；⑥ 非编码 RNA、组蛋白修饰、DNA 修饰和 RNA 修饰如何参与代谢通路以及作物品质的调控。

主要的研究内容包括：① 主要农作物中非编码 RNA 的种类，组蛋白修饰、DNA 修饰和 RNA 修饰的种类；② 非编码 RNA、组蛋白修饰、DNA 修饰和 RNA 修饰在作物产量性状中的作用；③ 非编码 RNA、组蛋白修饰、DNA 修饰和 RNA 修饰在植物非生物胁迫应答中的作用；④ 非编码 RNA、组蛋白修饰、DNA 修饰和 RNA 修饰如何感知外界环境信号；⑤ 非编码 RNA、组蛋白修饰、DNA 修饰和 RNA 修饰调控基因转录过程中的作用；⑥ 非编码 RNA、组蛋白修饰、DNA 修饰和 RNA 修饰如何参与在代谢合成通路及作物品质中的作用。

围绕上述的研究方向，以作物为材料，可以开展以下几个方面的研究。① 新的非编码 RNA、组蛋白修饰、DNA 修饰和 RNA 修饰的鉴定。主要包括重要农作物中非编码 RNA、组蛋白修饰、DNA 修饰和 RNA 修饰的系统发现，阐述其在作物生长发育过程中的功能，研究其在产量品质等重要农艺性状中的作用。② 研究非编码 RNA、组蛋白修饰、DNA 修饰和 RNA 修饰与环境之间的交流。主要包括在外界环境改变情况下，非编码 RNA、组蛋白修饰、DNA 修饰和 RNA 修饰的变化。研究非编码 RNA 以及 DNA、RNA 和组蛋白修饰感知外界环境变化，以如何参与环境应答和适应调节。③ 非编码 RNA、组蛋白修饰、DNA 修饰和 RNA 修饰的维持和信息的传递。主要包括非编码 RNA 以及 DNA、RNA 和组蛋白修饰在细胞重新编程过程的建立和维持，如何决定细胞命运。相应的修饰一旦建立，其信息如何传递；不同器官组织的表观遗传学差异是什么。④ 非编码 RNA、组蛋白修饰、DNA 修饰和 RNA 修饰如何参与植物生长发育进程。主要包括非编码 RNA、组蛋白修饰、DNA 修饰和 RNA 修饰如何协同内源激素和生长因子，调节生长发育的可塑性，决定生长发育的进程。⑤ 组蛋白修饰、DNA 修饰和 RNA 修饰与植物合成代谢及作物品质之间的关系。

3 阶段性目标

五年发展目标：① 在植物非编码 RNA、组蛋白修饰、DNA 修饰和 RNA 修饰的机理研究方面，取得多项突破性的研究进展；② 获得若干个决定重要农艺性状的非编码 RNA、组蛋白修饰因子、DNA 修饰因子及 RNA

修饰因子等。

3.1 2035 年目标

2035 年目标包括阐述多个生命过程中的重大表观调控的科学问题，取得一批具有国际影响和"从 0 到 1"原始创新的重要研究成果；培养和形成一支在植物染色质与表观遗传学前沿领域做出国际一流成果的研究队伍和领军人才。

3.2 2050 年目标

至 2050 年，建立世界一流研究队伍，在植物染色质与表观遗传学前沿领域中发挥引领作用。

参考文献

Bu, Z., Yu, Y., Li, Z., Liu, Y., Jiang, W., Huang, Y., and Dong, A.W. (2014). Regulation of *Arabidopsis* flowering by the histone mark readers MRG1/2 via interaction with CONSTANS to modulate FT expression. PLoS Genet. *10*, e1004617.

Cao, H., Li, X., Wang, Z., Ding, M., Sun, Y., Dong, F., Chen, F., Liu, L.A., Doughty, J., Li, Y., and Liu, Y.X. (2015). Histone H2B monoubiquitination mediated by HISTONE MONOUBIQUITINATION1 and HISTONE MONOUBIQUITINATION2 is involved in anther development by regulating tapetum degradation-related genes in rice. Plant Physiol. *168*, 1389-1405.

Che, R., Tong, H., Shi, B., Liu, Y., Fang, S., Liu, D., Xiao, Y., Hu, B., Liu, L., and Wang, H. (2015). Control of grain size and rice yield by GL2-mediated brassinosteroid responses. Nat. Plants *2*, 1-8.

Cui, C., Wang, J.J., Zhao, J.H., Fang, Y.Y., He, X.F., Guo, H.S., and Duan, C.G. (2020). A *Brassica* miRNA regulates plant growth and immunity through distinct modes of action. Mol. Plant *13*, 231-245.

Cui, X., Jin, P., Cui, X., Gu, L., Lu, Z., Xue, Y., Wei, L., Qi, J., Song, X., and Luo, M. (2013). Control of transposon activity by a histone H3K4 demethylase in rice. Proc. Natl. Acad. Sci.

USA *110*, 1953-1958.

Cui, X., Liang, Z., Shen, L., Zhang, Q., Bao, S., Geng, Y., Zhang, B., Leo, V., Vardy, L.A., and Lu, T. (2017). 5-Methylcytosine RNA methylation in *Arabidopsis thaliana*. Mol. Plant *10*, 1387-1399.

Cui, X., Lu, F., Qiu, Q., Zhou, B., Gu, L., Zhang, S., Kang, Y., Cui, X., Ma, X., and Yao, Q. (2016). REF6 recognizes a specific DNA sequence to demethylate H3K27me3 and regulate organ boundary formation in *Arabidopsis*. Nat. Genet. *48*, 694.

Dai, X., Bai, Y., Zhao, L., Dou, X., Liu, Y., Wang, L., Li, Y., Li, W., Hui, Y., and Huang, X. (2017). H2A.Z represses gene expression by modulating promoter nucleosome structure and enhancer histone modifications in *Arabidopsis*. Mol. Plant *10*, 1274-1292.

Deng, Y., Zhai, K., Xie, Z., Yang, D., Zhu, X., Liu, J., Wang, X., Qin, P., Yang, Y., and Zhang, G. (2017). Epigenetic regulation of antagonistic receptors confers rice blast resistance with yield balance. Science *355*, 962-965.

Ding, J., Lu, Q., Ouyang, Y., Mao, H., Zhang, P., Yao, J., Xu, C., Li, X., Xiao, J., and Zhang, Q. (2012). A long noncoding RNA regulates photoperiod-sensitive male sterility, an essential component of hybrid rice. Proc. Natl. Acad. Sci. USA *109*, 2654-2659.

Duan, C.G., Wang, X., Xie, S., Pan, L., Miki, D., Tang, K., Hsu, C.C., Lei, M., Zhong, Y., and Hou, Y.J. (2017a). A pair of transposon-derived proteins function in a histone acetyltransferase complex for active DNA demethylation. Cell Res. *27*, 226.

Duan, H.C., Wei, L.H., Zhang, C., Wang, Y., Chen, L., Lu, Z., Chen, P.R., He, C., and Jia, G. (2017b). ALKBH10B is an RNA N^6-methyladenosine demethylase affecting *Arabidopsis* floral transition. Plant Cell *29*, 2995-3011.

Duan, P., Ni, S., Wang, J., Zhang, B., Xu, R., Wang, Y., Chen, H., Zhu, X., and Li, Y. (2015). Regulation of OsGRF4 by OsmiR396 controls grain size and yield in rice. Nat. Plants *2*, 1-5.

Fan, Y., Yang, J., Mathioni, S.M., Yu, J., Shen, J., Yang, X., Wang, L., Zhang, Q., Cai, Z., and Xu, C. (2016). PMS1T, producing phased small-interfering RNAs, regulates photoperiod-sensitive male sterility in rice. Proc. Natl. Acad. Sci. USA *113*, 15144-15149.

Fang, X., Shi, Y., Lu, X., Chen, Z., and Qi, Y. (2015). CMA33/XCT regulates small RNA production through modulating the transcription of dicer-like genes in *Arabidopsis*. Mol. Plant *8*, 1227-1236.

Fang, X., Zhao, G., Zhang, S., Li, Y., Gu, H., Li, Y., Zhao, Q., and Qi, Y. (2019). Chloroplast-to-nucleus signaling regulates microRNA biogenesis in *Arabidopsis*. Dev. Cell *48*, 371-382. e374.

Gallego-Bartolome, J., Liu, W., Kuo, P.H., Feng, S., Ghoshal, B., Gardiner, J., Zhao, J.M.C., Park, S.Y., Chory, J., and Jacobsen, S.E. (2019). Co-targeting RNA polymerases IV and V promotes efficient *de novo* DNA methylation in *Arabidopsis*. Cell *176*, 1068-1082. e1019.

Gao, F., Wang, K., Liu, Y., Chen, Y., Chen, P., Shi, Z., Luo, J., Jiang, D., Fan, F., and Zhu, Y. (2015). Blocking miR396 increases rice yield by shaping inflorescence architecture. Nat. Plants *2*, 1-9.

Gao, M., Wei, W., Li, M.M., Wu, Y.S., Ba, Z., Jin, K.X., Li, M.M., Liao, Y.Q., Adhikari, S., and Chong, Z. (2014). Ago2 facilitates Rad51 recruitment and DNA double-strand break repair by homologous recombination. Cell Res. *24*, 532-541.

Gou, J.Y., Felippes, F.F., Liu, C.J., Weigel, D., and Wang, J.W. (2011). Negative regulation of anthocyanin biosynthesis in *Arabidopsis* by a miR156-targeted SPL transcription factor. Plant Cell *23*, 1512-1522.

Hao, Y., Wang, H., Qiao, S., Leng, L., and Wang, X. (2016). Histone deacetylase HDA6 enhances brassinosteroid signaling by inhibiting the BIN2 kinase. Proc. Natl. Acad. Sci. *113*, 10418-10423.

Hu, J., Wang, Y., Fang, Y., Zeng, L., Xu, J., Yu, H., Shi, Z., Pan, J., Zhang, D., and Kang, S. (2015). A rare allele of *GS2* enhances grain size and grain yield in rice. Mol. Plant *8*, 1455-1465.

Hu, Z., Song, N., Zheng, M., Liu, X., Liu, Z., Xing, J., Ma, J., Guo, W., Yao, Y., and Peng, H. (2015). Histone acetyltransferase GCN 5 is essential for heat stress-responsive gene activation and thermotolerance in *Arabidopsis*. Plant J. *84*, 1178-1191.

Jiang, D., and Berger, F. (2017). DNA replication-coupled histone modification maintains Polycomb gene silencing in plants. Science *357*, 1146-1149.

Jiang, P., Wang, S., Jiang, H., Cheng, B., Wu, K., and Ding, Y. (2018a). The COMPASS-like complex promotes flowering and panicle branching in rice. Plant Physiol. *176*, 2761-2771.

Jiang, P., Wang, S., Zheng, H., Li, H., Zhang, F., Su, Y., Xu, Z., Lin, H., Qian, Q., and Ding, Y. (2018b). SIP 1 participates in regulation of flowering time in rice by recruiting OsTrx1 to

Ehd1. New Phytol. *219*, 422-435.

Jiao, Y., Wang, Y., Xue, D., Wang, J., Yan, M., Liu, G., Dong, G., Zeng, D., Lu, Z., and Zhu, X. (2010). Regulation of OsSPL14 by OsmiR156 defines ideal plant architecture in rice. Nat. Genet. *42*, 541.

Jin, J., Shi, J., Liu, B., Liu, Y., Huang, Y., Yu, Y., and Dong, A. (2015). MRG702, a reader protein of H3K4me3 and H3K36me3, is involved in brassinosteroid-regulated growth and flowering time control in rice. Plant Physiol. *168*, 1257-1285.

Johnson, L.M., Du, J., Hale, C.J., Bischof, S., Feng, S., Chodavarapu, R.K., Zhong, X., Marson, G., Pellegrini, M., and Segal, D.J. (2014). SRA-and SET-domain-containing proteins link RNA polymerase V occupancy to DNA methylation. Nature *507*, 124.

Li, F., Zhao, N., Li, Z., Xu, X., Wang, Y., Yang, X., Liu, S. S., Wang, A., and Zhou, X. (2017). A calmodulin-like protein suppresses RNA silencing and promotes geminivirus infection by degrading SGS3 via the autophagy pathway in *Nicotiana benthamiana*. PLoS Pathogens *13*, e1006213.

Li, Z., Fu, X., Wang, Y., Liu, R., and He, Y. (2018). Polycomb-mediated gene silencing by the BAH-EMF1 complex in plants. Nat. Genet. *50*, 1254.

Li, Z., Jiang, D., Fu, X., Luo, X., Liu, R., and He, Y. (2016). Coupling of histone methylation and RNA processing by the nuclear mRNA cap-binding complex. Nat. Plants *2*, 16015.

Liang, Z., Shen, L., Cui, X., Bao, S., Geng, Y., Yu, G., Liang, F., Xie, S., Lu, T., and Gu, X. (2018). DNA N^6-adenine methylation in *Arabidopsis thaliana*. Dev. Cell *45*, 406-416.

Liu, B., Chen, Z., Song, X., Liu, C., Cui, X., Zhao, X., Fang, J., Xu, W., Zhang, H., and Wang, X. (2007). *Oryza sativa* dicer-like4 reveals a key role for small interfering RNA silencing in plant development. Plant Cell *19*, 2705-2718.

Liu, B., Li, P., Li, X., Liu, C., Cao, S., Chu, C., and Cao, X. (2005). Loss of function of OsDCL1 affects microRNA accumulation and causes developmental defects in rice. Plant Physiol. *139*, 296-305.

Liu, B., Wei, G., Shi, J., Jin, J., Shen, T., Ni, T., Shen, W.H., Yu, Y., and Dong, A. (2016). SET DOMAIN GROUP 708, a histone H3 lysine 36—specific methyltransferase, controls flowering time in rice (*Oryza sativa*). New Phytol. *210*, 577-588.

Liu, C., Xin, Y., Xu, L., Cai, Z., Xue, Y., Liu, Y., Xie, D., Liu, Y., and Qi, Y. (2018). *Arabidopsis*

ARGONAUTE 1 binds chromatin to promote gene transcription in response to hormones and stresses. Dev. Cell *44*, 348-361. e347.

Liu, J., Feng, L., Gu, X., Deng, X., Qiu, Q., Li, Q., Zhang, Y., Wang, M., Deng, Y., and Wang, E. (2019a). An H3K27me3 demethylase-HSFA2 regulatory loop orchestrates transgenerational thermomemory in *Arabidopsis*. Cell Res. *29*, 379-390.

Liu, M., Ba, Z., Costa-Nunes, P., Wei, W., Li, L., Kong, F., Li, Y., Chai, J., Pontes, O., and Qi, Y. (2017). IDN2 interacts with RPA and facilitates DNA double-strand break repair by homologous recombination in *Arabidopsis*. Plant Cell *29*, 589-599.

Liu, P., Zhang, S., Zhou, B., Luo, X., Zhou, X., Cai, B., Jin, Y.H., Niu, D., Lin, J., and Cao, X. (2019b). The histone H3K4 demethylase JMJ16 represses leaf senescence in *Arabidopsis*. Plant Cell *31*, 430-443.

Liu, X., Zhou, S., Wang, W., Ye, Y., Zhao, Y., Xu, Q., Zhou, C., Tan, F., Cheng, S., and Zhou, D.X. (2015). Regulation of histone methylation and reprogramming of gene expression in the rice inflorescence meristem. Plant Cell *27*, 1428-1444.

Lu, C., Tian, Y., Wang, S., Su, Y., Mao, T., Huang, T., Chen, Q., Xu, Z., and Ding, Y. (2017). Phosphorylation of SPT5 by CDKD；2 is required for VIP5 recruitment and normal flowering in *Arabidopsis thaliana*. Plant Cell *29*, 277-291.

Lu, F., Cui, X., Zhang, S., Jenuwein, T., and Cao, X. (2011). *Arabidopsis* REF6 is a histone H3 lysine 27 demethylase. Nat. Genet. *43*, 715.

Lu, F., Cui, X., Zhang, S., Liu, C., and Cao, X. (2010). JMJ14 is an H3K4 demethylase regulating flowering time in *Arabidopsis*. Cell Res. *20*, 387.

Lu, Y., Xu, Q., Liu, Y., Yu, Y., Cheng, Z.Y., Zhao, Y., and Zhou, D.X. (2018). Dynamics and functional interplay of histone lysine butyrylation, crotonylation, and acetylation in rice under starvation and submergence. Genome Biol. *19*, 144.

Luo, X., Chen, T., Zeng, X., He, D., and He, Y. (2019). Feedback regulation of FLC by FLOWERING LOCUS T (FT) and FD through a 5′FLC promoter region in *Arabidopsis*. Mol. Plant *12*, 285-288.

Mao, H., Wang, H., Liu, S., Li, Z., Yang, X., Yan, J., Li, J., Tran, L.S.P., and Qin, F. (2015). A transposable element in a *NAC* gene is associated with drought tolerance in maize seedlings. Nat. Commun. *6*, 1-13.

Mao, Y.B., Cai, W.J., Wang, J.W., Hong, G.J., Tao, X.Y., Wang, L.J., Huang, Y.P., and Chen, X.Y. (2007). Silencing a cotton bollworm P450 monooxygenase gene by plant-mediated RNAi impairs larval tolerance of gossypol. Nat. Biotechnol. *25*, 1307-1313.

Mao, Y.B., Liu, Y.Q., Chen, D.Y., Chen, F.Y., Fang, X., Hong, G.J., Wang, L.J., Wang, J.W., and Chen, X.Y. (2017). Jasmonate response decay and defense metabolite accumulation contributes to age-regulated dynamics of plant insect resistance. Nat. Commun. *8*, 13925.

Mei, Y., Wang, Y., Hu, T., Yang, X., Lozano-Duran, R., Sunter, G., and Zhou, X. (2018). Nucleocytoplasmic shuttling of geminivirus C4 protein mediated by phosphorylation and myristoylation is critical for viral pathogenicity. Mol. Plant *11*, 1466-1481.

Mi, S., Cai, T., Hu, Y., Chen, Y., Hodges, E., Ni, F., Wu, L., Li, S., Zhou, H., and Long, C. (2008). Sorting of small RNAs into *Arabidopsis* argonaute complexes is directed by the 5′ terminal nucleotide. Cell *133*, 116-127.

Qian, S., Lv, X., Scheid, R. N., Lu, L., Yang, Z., Chen, W., Liu, R., Boersma, M.D., Denu, J.M., and Zhong, X. (2018). Dual recognition of H3K4me3 and H3K27me3 by a plant histone reader SHL. Nat. Commun. *9*, 2425.

Qian, W., Miki, D., Lei, M., Zhu, X., Zhang, H., Liu, Y., Li, Y., Lang, Z., Wang, J., and Tang, K. (2014). Regulation of active DNA demethylation by an α-crystallin domain protein in *Arabidopsis*. Mol. Cell *55*, 361-371.

Shen, H., He, H., Li, J., Chen, W., Wang, X., Guo, L., Peng, Z., He, G., Zhong, S., and Qi, Y. (2012). Genome-wide analysis of DNA methylation and gene expression changes in two *Arabidopsis* ecotypes and their reciprocal hybrids. Plant Cell *24*, 875-892.

Song, J., Wang, X., Song, B., Gao, L., Mo, X., Yue, L., Yang, H., Lu, J., Ren, G., and Mo, B. (2019). Prevalent cytidylation and uridylation of precursor miRNAs in *Arabidopsis*. Nat. Plants *5*, 1260-1272.

Song, X., Li, P., Zhai, J., Zhou, M., Ma, L., Liu, B., Jeong, D.H., Nakano, M., Cao, S., and Liu, C. (2012a). Roles of DCL4 and DCL3b in rice phased small RNA biogenesis. Plant J. *69*, 462-474.

Song, X., Wang, D., Ma, L., Chen, Z., Li, P., Cui, X., Liu, C., Cao, S., Chu, C., and Tao, Y. (2012b). Rice RNA-dependent RNA polymerase 6 acts in small RNA biogenesis and spikelet development. Plant J. *71*, 378-389.

Su, Y., Wang, S., Zhang, F., Zheng, H., Liu, Y., Huang, T., and Ding, Y. (2017). Phosphorylation of histone H2A at serine 95: a plant-specific mark involved in flowering time regulation and H2A.Z deposition. Plant Cell *29*, 2197-2213.

Tan, L.M., Zhang, C.J., Hou, X.M., Shao, C.R., Lu, Y.J., Zhou, J.X., Li, Y.Q., Li, L., Chen, S., and He, X.J. (2018). The PEAT protein complexes are required for histone deacetylation and heterochromatin silencing. EMBO J. *37*, e98770.

Tang, K., Lang, Z., Zhang, H., and Zhu, J.K. (2016). The DNA demethylase ROS1 targets genomic regions with distinct chromatin modifications. Nat. Plants *2*, 16169.

Tang, Y., Liu, X., Liu, X., Li, Y., Wu, K., and Hou, X. (2017). *Arabidopsis* NF-YCs mediate the light-controlled hypocotyl elongation via modulating histone acetylation. Mol. Plant *10*, 260-273.

Tao, Z., Hu, H., Luo, X., Jia, B., Du, J., and He, Y. (2019). Embryonic resetting of the parental vernalized state by two B3 domain transcription factors in *Arabidopsis*. Nat. Plants *5*, 424.

Tao, Z., Shen, L., Gu, X., Wang, Y., Yu, H., and He, Y. (2017). Embryonic epigenetic reprogramming by a pioneer transcription factor in plants. Nature *551*, 124.

Tian, Y., Zheng, H., Zhang, F., Wang, S., Ji, X., Xu, C., He, Y., and Ding, Y. (2019). PRC2 recruitment and H3K27me3 deposition at FLC require FCA binding of COOLAIR. Sci. Adv. *5*, eaau7246.

Wang, J.W., Park, M.Y., Wang, L.J., Koo, Y., Chen, X.Y., Weigel, D., and Poethig, R.S. (2011). miRNA control of vegetative phase change in trees. PLoS Genet *.7*, e1002012.

Wang, Y., Gu, X., Yuan, W., Schmitz, R.J., and He, Y. (2014a). Photoperiodic control of the floral transition through a distinct polycomb repressive complex. Dev. Cell *28*, 727-736.

Wang, Y., Wang, X., Deng, W., Fan, X., Liu, T.T., He, G., Chen, R., Terzaghi, W., Zhu, D., and Deng, X.W. (2014b). Genomic features and regulatory roles of intermediate-sized non-coding RNAs in *Arabidopsis*. Mol. Plant *7*, 514-527.

Wang, Z., Chen, F., Li, X., Cao, H., Ding, M., Zhang, C., Zuo, J., Xu, C., Xu, J., and Deng, X. (2016). *Arabidopsis* seed germination speed is controlled by SNL histone deacetylase-binding factor-mediated regulation of AUX1. Nat. Commun. *7*, 13412.

Wang, Z., Ma, Z., Castillo-González, C., Sun, D., Li, Y., Yu, B., Zhao, B., Li, P., and Zhang, X. (2018). SWI2/SNF2 ATPase CHR2 remodels pri-miRNAs via Serrate to impede miRNA production. Nature *557*, 516-521.

Wei, L., Gu, L., Song, X., Cui, X., Lu, Z., Zhou, M., Wang, L., Hu, F., Zhai, J., and Meyers, B.C. (2014). Dicer-like 3 produces transposable element-associated 24-nt siRNAs that control agricultural traits in rice. Proc. Natl. Acad. Sci. USA *111*, 3877-3882.

Wei, L.H., Song, P., Wang, Y., Lu, Z., Tang, Q., Yu, Q., Xiao, Y., Zhang, X., Duan, H.C., and Jia, G. (2018). The m^6A reader ECT2 controls trichome morphology by affecting mRNA stability in *Arabidopsis*. Plant Cell *30*, 968-985.

Wei, W., Ba, Z., Gao, M., Wu, Y., Ma, Y., Amiard, S., White, C.I., Danielsen, J.M.R., Yang, Y.G., and Qi, Y. (2012). A role for small RNAs in DNA double-strand break repair. Cell *149*, 101-112.

Whittaker, C., and Dean, C. (2017). The FLC locus: a platform for discoveries in epigenetics and adaptation. Annu. Rev. Cell Biol. *33*, 555-575.

Wu, J., Yang, R., Yang, Z., Yao, S., Zhao, S., Wang, Y., Li, P., Song, X., Jin, L., and Zhou, T. (2017). ROS accumulation and antiviral defence control by microRNA528 in rice. Nat. Plants *3*, 1-7.

Wu, K., Wang, S., Song, W., Zhang, J., Wang, Y., Liu, Q., Yu, J., Ye, Y., Li, S., and Chen, J. (2020). Enhanced sustainable green revolution yield via nitrogen-responsive chromatin modulation in rice. Science *367*, eaaz2046.

Xiao, J., Jin, R., Yu, X., Shen, M., Wagner, J.D., Pai, A., Song, C., Zhuang, M., Klasfeld, S., and He, C. (2017). *Cis* and *trans* determinants of epigenetic silencing by Polycomb repressive complex 2 in *Arabidopsis*. Nat. Genet. *49*, 1546.

Xiao, X., Zhang, J., Li, T., Fu, X., Satheesh, V., Niu, Q., Lang, Z., Zhu, J.K., and Lei, M. (2019). A group of SUVH methyl DNA binding proteins regulate expression of the DNA demethylase ROS1 in *Arabidopsis*. J. Integr. Plant Biol. *61*, 110-119.

Xie, Y., Liu, Y., Wang, H., Ma, X., Wang, B., Wu, G., and Wang, H. (2017). Phytochrome-interacting factors directly suppress MIR156 expression to enhance shade-avoidance syndrome in *Arabidopsis*. Nat. Commun. *8*, 1-11.

Xie, Y., Zhang, Y., Han, J., Luo, J., Li, G., Huang, J., Wu, H., Tian, Q., Zhu, Q., and Chen, Y. (2018). The intronic *cis* element SE1 recruits *trans*-acting repressor complexes to repress the expression of *ELONGATED UPPERMOST INTERNODE1* in rice. Mol. Plant *11*, 720-735.

Yan, D., Zhang, X., Zhang, L., Ye, S., Zeng, L., Liu, J., Li, Q., and He, Z. (2015). *CURVED CHIMERIC PALEA 1* encoding an EMF1 like protein maintains epigenetic repression of *OsMADS58* in rice palea development. Plant J. *82*, 12-24.

Yang, D.L., Zhang, G., Wang, L., Li, J., Xu, D., Di, C., Tang, K., Yang, L., Zeng, L., and Miki, D. (2018a). Four putative SWI2/SNF2 chromatin remodelers have dual roles in regulating DNA methylation in *Arabidopsis*. Cell Discov. *4*, 55.

Yang, H., Liu, X., Xin, M., Du, J., Hu, Z., Peng, H., Rossi, V., Sun, Q., Ni, Z., and Yao, Y. (2016). Genome-wide mapping of targets of maize histone deacetylase HDA101 reveals its function and regulatory mechanism during seed development. Plant Cell *28*, 629-645.

Yang, Z., Huang, Y., Yang, J., Yao, S., Zhao, K., Wang, D., Qin, Q., Bian, Z., Li, Y., and Lan, Y. (2020). Jasmonate signaling enhances RNA silencing and antiviral defense in rice. Cell Host Microbe *1*, 89-103.

Yang, Z., Qian, S., Scheid, R.N., Lu, L., Chen, X., Liu, R., Du, X., Lv, X., Boersma, M.D., and Scalf, M. (2018b). EBS is a bivalent histone reader that regulates floral phase transition in *Arabidopsis*. Nat. Genet. *50*, 1247.

Yang, Z., Qiu, Q., Chen, W., Jia, B., Chen, X., Hu, H., He, K., Deng, X., Li, S., and Tao, W.A. (2018c). Structure of the *Arabidopsis* JMJ14-H3K4me3 complex provides insight into the substrate specificity of KDM5 subfamily histone demethylases. Plant Cell *30*, 167-177.

Yao, S., Yang, Z., Yang, R., Huang, Y., Guo, G., Kong, X., Lan, Y., Zhou, T., Wang, H., and Wang, W. (2019). Transcriptional regulation of miR528 by OsSPL9 orchestrates antiviral response in rice. Mol. Plant *12*, 1114-1122.

Ye, R., Chen, Z., Lian, B., Rowley, M.J., Xia, N., Chai, J., Li, Y., He, X.J., Wierzbicki, A.T., and Qi, Y. (2016). A Dicer-independent route for biogenesis of siRNAs that direct DNA methylation in *Arabidopsis*. Mol. Cell *61*, 222-235.

Ye, R., Wang, W., Iki, T., Liu, C., Wu, Y., Ishikawa, M., Zhou, X., and Qi, Y. (2012). Cytoplasmic assembly and selective nuclear import of *Arabidopsis* Argonaute4/siRNA complexes. Mol. Cell *46*, 859-870.

Yu, S., Galvão, V.C., Zhang, Y.C., Horrer, D., Zhang, T.Q., Hao, Y.H., Feng, Y.Q., Wang, S., Schmid, M., and Wang, J.W. (2012). Gibberellin regulates the *Arabidopsis* floral transition through miR156-targeted SQUAMOSA PROMOTER BINDING–LIKE transcription factors. Plant Cell *24*, 3320-3332.

Yu, Y., Zhang, Y., Chen, X., and Chen, Y. (2019). Plant noncoding RNAs: hidden players in development and stress responses. Annu. Rev. Cell Biol. *35*, 407-431.

Yuan, W., Luo, X., Li, Z., Yang, W., Wang, Y., Liu, R., Du, J., and He, Y. (2016). A *cis* cold memory element and a *trans* epigenome reader mediate Polycomb silencing of *FLC* by vernalization in *Arabidopsis*. Nat. Genet. *48*, 1527.

Zhang, C., Cao, L., Rong, L., An, Z., Zhou, W., Ma, J., Shen, W.H., Zhu, Y., and Dong, A. (2015a). The chromatin-remodeling factor AtINO80 plays crucial roles in genome stability maintenance and in plant development. Plant J. *82*, 655-668.

Zhang, C., Ding, Z., Wu, K., Yang, L., Li, Y., Yang, Z., Shi, S., Liu, X., Zhao, S., and Yang, Z. (2016). Suppression of jasmonic acid-mediated defense by viral-inducible microRNA319 facilitates virus infection in rice. Mol. Plant *9*, 1302-1314.

Zhang, C.J., Ning, Y.Q., Zhang, S.W., Chen, Q., Shao, C.R., Guo, Y.W., Zhou, J.X., Li, L., Chen, S., and He, X.J. (2012). IDN2 and its paralogs form a complex required for RNA-directed DNA methylation. PLoS Genet. *8*, e1002693.

Zhang, C.J., Zhou, J.X., Liu, J., Ma, Z.Y., Zhang, S.W., Dou, K., Huang, H.W., Cai, T., Liu, R., and Zhu, J.K. (2013). The splicing machinery promotes RNA-directed DNA methylation and transcriptional silencing in *Arabidopsis*. EMBO J. *32*, 1128-1140.

Zhang, H., Li, L., He, Y., Qin, Q., Chen, C., Wei, Z., Tan, X., Xie, K., Zhang, R., and Hong, G. (2020). Distinct modes of manipulation of rice auxin response factor OsARF17 by different plant RNA viruses for infection. Proc. Natl. Acad. Sci. USA *117*, 9112-9121.

Zhang, J., Xie, S., Zhu, J. K., and Gong, Z. (2016). Requirement for flap endonuclease 1 (FEN 1) to maintain genomic stability and transcriptional gene silencing in *Arabidopsis*. Plant J. *87*, 629-640.

Zhang, Q., Liang, Z., Cui, X., Ji, C., Li, Y., Zhang, P., Liu, J., Riaz, A., Yao, P., and Liu, M. (2018). N^6-methyladenine DNA methylation in *japonica* and *indica* rice genomes and its association with gene expression, plant development, and stress responses. Mol. Plant *11*, 1492-1508.

Zhang, S., Zhou, B., Kang, Y., Cui, X., Liu, A., Deleris, A., Greenberg, M.V., Cui, X., Qiu, Q., and Lu, F. (2015b). C-terminal domains of histone demethylase JMJ14 interact with a pair of

NAC transcription factors to mediate specific chromatin association. Cell Discov. *1*, 15003.

Zhao, S., Cheng, L., Gao, Y., Zhang, B., Zheng, X., Wang, L., Li, P., Sun, Q., and Li, H. (2019). Plant HP1 protein ADCP1 links multivalent H3K9 methylation readout to heterochromatin formation. Cell Res. *29*, 54.

Zhao, X., Li, J., Lian, B., Gu, H., Li, Y., and Qi, Y. (2018). Global identification of *Arabidopsis* lncRNAs reveals the regulation of MAF4 by a natural antisense RNA. Nat. Commun. *9*, 1-12.

Zhao, Y., Xie, S., Li, X., Wang, C., Chen, Z., Lai, J., and Gong, Z. (2014). *REPRESSOR OF SILENCING5* encodes a member of the small heat shock protein family and is required for DNA demethylation in *Arabidopsis*. Plant Cell *26*, 2660-2675.

Zheng, H., Zhang, F., Wang, S., Su, Y., Ji, X., Jiang, P., Chen, R., Hou, S., and Ding, Y. (2018). MLK1 and MLK2 coordinate RGA and CCA1 activity to regulate hypocotyl elongation in *Arabidopsis thaliana*. Plant Cell *30*, 67-82.

Zheng, S., Hu, H., Ren, H., Yang, Z., Qiu, Q., Qi, W., Liu, X., Chen, X., Cui, X., and Li, S. (2019). The *Arabidopsis* H3K27me3 demethylase JUMONJI 13 is a temperature and photoperiod dependent flowering repressor. Nat. Commun. *10*, 1303.

Zhou, C.M., Zhang, T.Q., Wang, X., Yu, S., Lian, H., Tang, H., Feng, Z.Y., Zozomova-Lihová, J., and Wang, J.W. (2013). Molecular basis of age-dependent vernalization in *Cardamine flexuosa*. Science *340*, 1097-1100.

Zhou, H., Zhou, M., Yang, Y., Li, J., Zhu, L., Jiang, D., Dong, J., Liu, Q., Gu, L., and Zhou, L. (2014). RNase Z S1 processes Ub L40 mRNAs and controls thermosensitive genic male sterility in rice. Nat. Commun. *5*, 1-9.

Zhou, S., Jiang, W., Long, F., Cheng, S., Yang, W., Zhao, Y., and Zhou, D.X. (2017). Rice homeodomain protein WOX11 recruits a histone acetyltransferase complex to establish programs of cell proliferation of crown root meristem. Plant Cell *29*, 1088-1104.

Zhou, S., Liu, X., Zhou, C., Zhou, Q., Zhao, Y., Li, G., and Zhou, D.X. (2016). Cooperation between the H3K27me3 chromatin mark and non-CG methylation in epigenetic regulation. Plant Physiol. *172*, 1131-1141.

Zhu, P., Wang, Y., Qin, N., Wang, F., Wang, J., Deng, X. W., and Zhu, D. (2016). *Arabidopsis* small nucleolar RNA monitors the efficient pre-rRNA processing during ribosome biogenesis. Proc. Natl. Acad. Sci. USA *113*, 11967-11972.

Zhu, Y., Rong, L., Luo, Q., Wang, B., Zhou, N., Yang, Y., Zhang, C., Feng, H., Zheng, L., and Shen, W.H. (2017). The histone chaperone NRP1 interacts with WEREWOLF to activate GLABRA2 in *Arabidopsis* root hair development. Plant Cell *29*, 260-276.

高通量表型技术

郭庆华

表现型（phenotype），简称表型，是指在一定环境条件下，具有特定基因型的个体所表现出来的所有性状特征的总和（Johannsen，1911）。作为研究"基因型-表型-环境"作用机制的重要桥梁，表型的精准监测和分析对作物育种和农业管理具有重要指导和评估作用，能够加速整个育种进程，并为精准农业监测中的资源调控和管理策略制定提供重要的数据支撑。

近十年来，高通量测序技术蓬勃发展，极大地促进了基因组学的研究，推进了精准育种的步伐。然而，受限于低水平表型识别与监测能力，基因组学研究获得的基因信息往往无法完全被利用，低通量的表型数据处理同样限制了受环境影响的复杂性状的筛选与评估，严重阻碍了作物育种进程（Araus and Cairns，2014）。表型组学的发展对于推动基因组学与表型、环境的交互研究至关重要。

1 国内外研究进展

植物表型组学研究涉及生物体的多个层次，是植物在亚细胞、细胞、组织、器官、单株和群体水平上的所有表型的组合（Dhondt et al.，2013）（图1）。在微观水平即亚细胞、细胞与组织水平上，传统的表型组学技术手段主要是利用显微镜进行人工观察、成像对比和分析；在宏观水平即器官、单株和群体水平上，主要依赖于手工测量和二维影像采集结合影像数据处理分析，但这些方法均费时费力、主观性强，且准确性低，难以满足大规模遗传育种筛选的需求。高通量表型监测技术的出现推动了表型组学走向了高通量、多维度、精准化的新时代。在微观水平上，主要通过以细胞高通量成像技术（high-throughput imaging，HTI）为代表的表型监测技术进行高通量表型监测（Pegoraro and Misteli，2017）；在宏观水平上，主要通过各类近端遥感传感

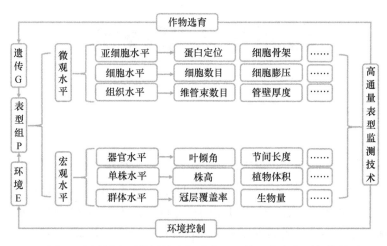

图 1　高通量表型监测技术与遗传、环境因素和表型组的关系

器进行高通量表型监测，如 X-CT 成像扫描仪、探地雷达、RGB 相机、激光雷达扫描仪、多光谱和高光谱传感器等，以及各类传感器整合形成的高通量表型监测平台等（White and Conley，2013）。

1.1　微观水平高通量表型监测技术

微观水平（亚细胞、细胞与组织水平）的表型监测对分辨率具有较高要求，且需要深入植株内部，这就往往造成了处理通量的牺牲，因此微观水平上实现高通量表型监测的步伐较为缓慢。当前阶段，在研究中主流的微观水平表型监测方法仍为较低通量的方法，其主要原理为利用人工或激光等裁切样品剖面，进而使用光学或电子成像，获取细胞数量、几何、颜色、纹理等信息。

近些年来，将自动高通量显微镜与自动图像分析有机结合的细胞高通量成像技术正逐渐兴起，成为当前研究的热门，并且是微观水平上高通量表型监测技术的主要发展方向之一（Pegoraro and Misteli，2017）。尤其是随机光学重建法的发明使得细胞高通量成像技术实现了细胞表型观测的三维化，并突破了衍射极限，可在微观水平同时对多种活细胞表型进行高通量、多时段的无偏监测，将 HTI 技术推向了一个新的高潮（Rust et al.，2006）。

通过分离感兴趣的细胞表型特征，或者 RNA 干扰、转基因、CRISPR/Cas9 等技术相关联，HTI 技术可以应用于细胞的表型筛查、表型分析与精准成像三个方面（Boutros et al.，2015；Pegoraro and Misteli，2017）。在细胞表型筛查方面，研究人员可以对特定表型特征进行筛选，进行囊泡等细胞结构或蛋白质等化合物的定位（Collinet et al.，2010；Glory and Murphy，2007）、确定特定基因的功能（Liberali et al.，2015）或者批量筛选影响某一细胞途径的化合物等（Desbordes and Studer，2013）。在细胞表型分析方面，研究人员可以同时监测多种表型并应用统计方法对其进行分析，依据表型相似性对基因、蛋白质、小分子等影响途径进行聚类（Young et al.，2008），或者按照时间序列分析生物过程的空间与动力学信息（Perez-Rangel et al.，2015）。在细胞精准成像方面，研究人员可以在细胞群体中精准识别具有特定表型的细胞，进行小概率生物学事件研究（Roukos et al., 2013）。尽管HTI技术具有高通量、高精度、三维成像的优点，但是它同样具有微观水平下表型监测技术普遍存在的问题，即前期样品准备过程过于烦琐且具有破坏性，无法在真实环境下进行样品的观测，无法反映在真实生存环境中亚细胞、细胞、组织的真实变化情况。因此，HTI技术并不是高通量表型监测技术的终点，优化样品制备过程，探索更加先进的成像技术，对于加速微观水平下高通量表型检测研究至关重要，仍是未来的首要任务之一。

目前，HTI技术产品已经逐步走向了商业化，被广泛应用于研究与应用领域，如PerkinElmer公司的Opera Phenix与Operetta，GE Healthcare公司的IN Cell Analyzer 6000，Molecular Devices公司的ImageXpress Ultra与ImageXpress Micro Confocal等（Li et al.，2016）。在软件方面，如Fiji/ImageJ、CellProfiler、BioConductor、Icy、BioImageXD和ADAPT plugin等HTI图像处理软件已经出现，能够兼容多种计算机系统，且大部分为开源软件（Smith et al.，2018）。我国HTI技术的研究起步较晚，基础较为薄弱，但是相比之前也有了很大的进步。国内研发较为成功的MoticBA600Mot全自动显微镜，已经做到了区域扫描、高精度自动化控制、自动调焦和自动回溯等，正逐步赶超国际领先水平。

1.2 宏观水平高通量表型监测技术

宏观水平（器官、单株和群体水平）的高通量植物表型监测主要是对植物形态参数（如株高、茎直径、叶面积和根长等），以及生理参数（如叶绿素含量、光合速率和氮含量等），进行快速、无损、高效的监测（Araus and Cairns，2014）。目前已经有多类成熟的传感器被广泛应用于科学研究及生产生活中，可应对不同的表型监测需求。在宏观水平上的表型监测基于土壤阻隔可划分为地下部与地上部两部分。对地下部根的监测主要为 X-CT 成像技术、磁共振成像（magnetic resonance imaging，MRI）技术、探地雷达等，对地上部茎、叶、花、果实与种子的表型监测中较常应用的传感器有 RGB 相机、激光雷达传感器、深度相机、多光谱、高光谱成像仪、热红外相机，以及叶绿素荧光成像仪等（表 1）。

表 1 宏观水平高通量表型监测技术常见传感器优缺点对比

监测部位	传感器	获取信息	优势	不足
地下部	X-CT 成像技术	X 射线衰减系数	可获取根部三维结构信息，可监测植物细根表型	成本较高，信噪比较低，需使用特制容器
	磁共振成像技术	磁共振信号	可获取根部三维结构信息，成像速度快，分辨率高	成本较高，数据处理复杂，无法监测较大植株
	探地雷达	电磁波信息	可获得根部三维结构信息，效率高，可实地无损测量	数据处理复杂，无法监测较细的根部
地上部	RGB 相机	几何特征信息及可见光光谱信息	价格低廉，直观便捷，在几何特征信息及较简单的植被参数测量中具有较高精确度，可以通过飞行时间技术生成点云	数据量大，数据采集易受环境干扰，不适用于复杂环境
	激光雷达传感器	三维结构信息及回波信息	可获取空间三维结构信息，抗干扰能力强，不受自然光影响，尺度覆盖范围广	价格较高，易受风的影响
	深度相机	三维结构信息	价格低廉，可提供空间三维结构信息	数据处理困难，分辨率较低，误差较大，易受环境影响，尺度覆盖较低
	多光谱、高光谱成像仪	光谱信息	可获取光谱数据，具有成熟的商业化系统	数据处理复杂，易受环境干扰
	热红外相机	红外波段光谱信息	可检测作物生长状态，尺度覆盖较高	易受环境影响，难以进行多时段表型监测
	叶绿素荧光成像仪	叶绿素荧光信息	可对叶绿素荧光 II，荧光蛋白进行检测	荧光激发困难，尺度覆盖较低

1.3 地下部高通量表型监测技术

地下部植物表型监测由于土壤不易被穿透，因此发展较为缓慢。目前使用最为广泛的根系高通量表型监测方法是使用凝胶或者营养液代替土壤栽培植物后，利用 RGB 相机或平板扫描仪对根系进行监测（Kuijken et al., 2015），但其与真实的土壤往往相差较大，无法反映植物根系在真实环境下的实际表型。在真实土壤中，利用一些穿透能力较强的射线，如 X-CT 技术使用的 X 射线，MRI 技术使用的射频电磁波，探地雷达使用的高频电磁波，均可以深入地下检测根部特征信息，原位获取根部三维结构信息，因此已经成为根系结构研究中的主流。尽管这些技术成功使得根部实现了原位、三维、高通量地下部表型检测，但由于土壤阻隔，监测结果易受土壤环境影响，同样也面临着难以进行真值验证，难以多时段、多样点比较的问题。开发新的根系监测方法、合理应用现有数据仍将是未来根系表型监测的重点。

1.4 地上部高通量表型监测技术

地上部植物表型受益于近些年来光学技术的进步，以及生产生活中巨大的需求，技术革新速度飞速。基于上文中提及的各类传感器，地上部高通量表型监测技术的发展目前主要有两个方向，即降低使用成本和传感器集成（彩图1）。

降低使用成本方向意即经济实惠，其主要目的是降低高通量表型监测技术使用的门槛，满足在实际应用过程中的经济、便携要求。其中的典型代表为 RGB 相机，RGB 相机可以获取植株反射的可见光信息，随着近些年来科学技术的发展，RGB 相机成本不断降低，且其可与智能手机相结合进行表型数据采集软件的开发（Han et al., 2017），对于所监测植株的几何表型特征，如叶长、叶宽（Wang and Xu, 2018）、冠幅（Makanza et al., 2018）等均具有较高的精确度。

传感器集成方向即将多种传感器集成至高通量表型监测平台上，从而满足植物高通量表型检测技术研究深入后提出的多时空、多尺度需求。高通量表型监测平台的概念非常宽泛，从简单的三脚架平台，到复杂的无人机平台，均可搭载传感器进行表型监测（Araus and Cairns, 2014）。按照平台使用位置的不同，可划分为室内与室外高通量表型平台。

室内表型监测平台由于环境单一，可移动植株，因此研究较为成熟。例如，比利时 CropDesign 公司的 TraitMill（Reuzeau, 2007）、法国农业科学研究院的 Phenoscope（Tisne et al., 2013）和 PHENOPSIS 表型平台（Granier et al.,

2006)、华中农业大学研发的断层扫描仪和数字化考种机（Yang et al.，2014）等。但是室内表型平台整体效率较低，且无法反映植物在真实群体下的性状，具有较大的局限。

室外表型监测平台由于可以反映植物的真实状况，因此受到了研究者的关注。室外表型平台中，应用最为广泛，且最受关注的平台包括：车载移动监测平台、田间固定监测平台及无人机监测平台。其中，车载监测平台可依据实际需求搭载多种传感器，具有迁移性强、灵活性高、价格较低的优点，适用于中等面积的大田，为发展较为成熟的一类室外可移动表型监测平台。例如，德国的 BreedVision 平台（Busemeyer et al.，2013）、澳大利亚科技企研所植物产业组研发的车载平台（Deery et al.，2014）等。田间固定表型监测平台通过在田间搭建固定的支架，或者在既有的观测塔平台上安装相应的传感器进行表型监测，可进行编程控制其运动轨迹，自动化采集数据，适用于需要长期监测的样地，但同时也具有无法移动，建设与维护费用较高的缺点。例如，我国具有自主知识产权，由中国科学院植物研究所研发的 Crop3D 平台（Guo et al.，2016）及英国的 Field Scanalyzer 平台（Virlet et al.，2017）等。无人机载平台较地面平台更加机动灵活，可以在同等时间内覆盖更大的地理区域，且能够避免因为时间变化而造成的误差影响，适用于大型的大田作物表型监测，可以进行多尺度多空间格局的研究。例如，NASA 研发的 Pathfinder-Plus 平台（Herwitz et al.，2004）及澳大利亚研发的无人机表型平台（Merz and Chapman，2011）等。

2 未来发展趋势与关键突破口

中国作为人口大国，以占世界不足 7% 的耕地养育着世界上 1/5 的人口，面临着数倍于其他国家的粮食安全压力，迫切需要充分挖掘种质资源，选育并推广高产、稳产的优质种源以填补巨大粮食缺口。调查数据显示，优质种源对我国作物增产的贡献率约为 40%，低于国际上 50% 左右的贡献率水平。造成这一情况的主要原因之一是缺少对种质资源的系统鉴定和深入研究，尤其是种质资源中优良表型性状及其相关遗传信息的挖掘。作为基因型和环境变量互作的结果，表型监测对于研究者剖析与产量和胁迫耐受性相关的遗传学数量性状，以及在精准农业背景下监测田间变量以实时调控资源分配具有重要意义。通过作物表型高通量监测平台，将作物表型监测贯穿至育种室内亲本选择、田间试验设计、综合观测筛选及最终的考种和品质评估等全环节，可以帮助育种学家进一步深挖作物种质资源，也可以协助农学家实现高效的生长动态监测，从而合理布控资源，提供资源有效利用率。新兴科技的真正投产存在一定周期，尤其是应用到农业上，高通量表型技术真正服务于育种和智慧农业仍存在以下需要突破的技术关口。

2.1 从室内到田间的大型高通量表型监测平台设施

表型监测平台是获取表型数据的重要手段，已受到越来越多的重视，技术推广力度也越来越大。目前，国际上已出现了一些较为成熟、商业化的表型监测系统，而我国作物表型监测技术起步略晚，相关技术研究从整体上而言进展较慢，许多科研机构仍处于引进、推广国外先进平台的层面。迫切需要研发处于技术领先水平的、自主知识产权的表型检测平台。

同时，当前表型平台的应用中也存在一些问题，如表型数据不能很好地与基因型数据匹配，高通量表型监测带来的数据冗余、表型参数尺度较为单一等。因此，需要结合 HTI 技术、激光雷达技术、光学成像技术、多光谱/高光谱成像技术、热红外成像技术、CT 技术等，搭建多尺度的作物表型高通量监测平台，实现从微观到室内到田间到大区域上的作物表型全生育期监测，辅以环境变量实时监控，建立起"基因型 - 表型 - 环境"的国家级数据库。并同步搭建配套软件平台，通过对多源表型数据的处理获取不同尺度的作物表型参数，并将这些参数用于环境监测、决策诊断等。最后通过对应用示范效果的评估等，对该技术的应用构建规范标准，为国家基本政策制定提供技术支撑。

2.2 多源表型数据融合

多源数据融合涉及两个层面：一方面，研究者需将同一传感器在不同时期获取的数据进行融合，从而实现全生育期性状的动态监测和分析；另一方面，作物精确三维结构的获取依赖于 HTI、激光雷达、多光谱/高光谱相机等技术获取的高维数据，如何进行不同维度、不同尺度的数据融合，尤其是在实现从微观到室内到室外多源数据在缺乏坐标信息的情况下实现跨尺度融合，仍是研究者进行从形态到生理乃至机理的综合型研究面临的难题之一。此外，如何从高维数据中计算结构特征，对高维数据进行特征提取、目标探测、分类分割也仍待进一步探讨。

2.3 大数据时代的表型数据的整合、管理、分析

目前，表型组学已逐渐进入"大数据"时代，体现出多领域（表型、基因组学等）、多层次（传统的小到中等扩展到大规模组学）和多尺度（从细胞到整个植物的作物形态，结构和生理数据）特征，针对个体的表型信息管理和分析方法已经不能满足新时期表型组学关联分析和系统构建的需求，如何精确有效地评估、理解和解释这些多源数据所涵盖的特征，挖掘出功能基因组的有价值的数量特征是植物表型开发和应用中的关键问题（Zhao et al., 2019）。大样本统计和与传统农艺性状的关联分析对表型组学的数据管理和相关算法研究提出了新的挑战。

为了应对这些挑战，作物科学界需要将国家和国际层面的人工智能技术与合作研究相结合，研究分析作物表型信息的新理论，构建有效的技术体系，并创建一系列工具，综合整合从多模态、多尺度、不同表型环境和基因型条件下获得的表型大数据，同时引入基于人工智能在深度学习、数据融合、混合智能和群体智能方面的最新成果，推进基于三维点云和传统遥感影像的表型分析。并制定统一的表型数据发布、共享和应用标准，构建具有高度集成化、互操作性、可共享性的国家级表型组学数据管理和分析平台。

2.4 学科交叉，表型 - 基因型 - 环境的综合分析

综合分析表型 - 基因型 - 环境之间的相互作用，需要对各种影响因素进行精确建模，然而目前仍需克服一些科学和技术挑战。例如，模型建模的有效性、适用性，以及模型中各模块之间相互作用机制的准确性都需要进一步验证；同时，还需解决模型尺度推绎问题，验证多尺度下各模块之间的反馈和调节机制。为了从技术层面加速这一过程，需要真正地将工程学科与理论学科相结合，寻求新的解决方案，揭示控制植物生长和发育的生物过程，识

别目标环境感兴趣的关键特征，以促进气候变化背景下的高产作物育种技术发展。同时，实现国家和国际层面的多学科交叉与协同，来应对植物表型组学在大数据背景下的巨大挑战。

3 阶段性目标

随着全球环境的持续恶化及人口的持续增长，粮食安全压力日益增大，实现粮食增产以弥补巨大的粮食缺口刻不容缓。在此背景下，大力发展作物育种，响应国家"藏粮于技"的号召，以科技创新和增长带动农业信息化和智能化进程，缩短育种周期，最终提升粮食产能，是当前以及未来一段时间内表型研究的重中之重。

3.1 2035年目标——大数据时代下的表型设施物联网构建，完成种源挖掘

全球范围内的高通量表型技术研发已经全面展开，美国、澳大利亚及欧洲一些发达国家（德国、法国、荷兰等）已于十余年前投入此项研究，并先后建成了作物表型组学研究机构。我国《国家重大科技基础设施建设"十三五"规划》将作物表型组学研究设施建设列为规划之一，目标是在未来20年内在全国范围内以主要经济粮食作物重点研发基地为依托，研发并建成"空中-田间-室内"多尺度的高通量精准表型监测设施基地，促进整个表型组学的研究进程。

随着大数据时代的到来，物联网（internet of things，IoT）概念不断延伸，作为整合组学研究中的重要一环，多尺度表型设施平台应该完成基于第五代移动互联网技术的物联网建设（彩图2）。预期到2035年，以表型设施物联网为依托，建立起"基因型-表型-环境"的多源、多层次、多尺度的国家级表型数据库，制定统一的表型数据发布、共享和应用协议。搭建起高度集成化、互操作性、可共享性的数据管理和分析平台，然后结合海量基因库，解析并揭示表型-基因型-环境互作的内在机制，并构建具有预测和评估作用的模型，最终为优质种源筛选提供一套高效、精确、切实可行的方法和体系，提升我国种质资源利用广度和深度。

3.2 2050年目标——人工智能与多组学研究结合，切实解决粮食危机

不同于工业自动化程度，我国农业自动化、现代化及规模化程度都尚有很大提升空间。人工智能与农业领域的应用探究始于21世纪之初，对提高农业产能有极大的挖掘潜力。基于2035年预期的表型物联网设施，未来一个时期内，全国范围内将完成海量农业相关基础数据的收集和积累，如何实现智能数据挖掘分析并给出及时决策指导将成为下一个技术突破口。

预计到2050年，通过人工智能的产业化应用，依托于大数据表型数据、基因型数据和环境相关数据等平台，将复杂的表型筛选、分析工作简单化，支撑农业技术人员在播种早期根据土地资源特征筛选合适的种源，在农作物发育早期通过自动化智能监测给出合理的，具有预见性的补种、水肥施加等调控指导。在育种领域，做到目标表型快速筛选，将种质筛选的周期压缩至原先的数倍之下。最终，在全球范围内，带动农业生产的标准化和精准化，推动大国农业转型，提高粮食供给体系质量、竞争力和适应力，从源头上确保食品安全，也提高生产效率，最终解决粮食危机。

致谢：本章在撰写过程中得到了中国科学院战略性先导科技专项(A类)资助(项目编号：XDA24020202)，并得到了吴芳芳、宋师琳和苏艳军的协助，特此感谢！

参考文献

Araus, J.L., and Cairns, J.E. (2014). Field high-throughput phenotyping: the new crop breeding frontier. Trends Plant Sci. *19*, 52-61.

Boutros, M., Heigwer, F., and Laufer, C. (2015). Microscopy-based high-content screening. Cell *163*, 1314-1325.

Busemeyer, L., Mentrup, D., Moeller, K., Wunder, E., Alheit, K., Hahn, V., Maurer, H.P., Reif, J.C., Wuerschum, T., Mueller, J., et al. (2013). BreedVision—a multi-sensor platform for non-destructive field-based phenotyping in plant breeding. Sensors *13*, 2830-2847.

Chen, D., Neumann, K., Friedel, S., Kilian, B., Chen, M., Altmann, T., Klukas, C. (2014). Dissecting the phenotypic components of crop plant growth and drought responses based on high-throughput image analysis. Plant Cell *26*, 4636-4655.

Collinet, C., Stoeter, M., Bradshaw, C.R., Samusik, N., Rink, J.C., Kenski, D., Habermann, B., Buchholz, F., Henschel, R., Mueller, M.S., et al. (2010). Systems survey of endocytosis by multiparametric image analysis. Nature *464*, 243-249.

Deery, D., Jimenez-Berni, J., Jones, H., Sirault, X., and Furbank, R. (2014). Proximal remote sensing buggies and potential applications for field-based phenotyping. Agronomy *4*, 349-379.

Desbordes, S.C., and Studer, L. (2013). Adapting human pluripotent stem cells to high-throughput and high-content screening. Nat. Protoc. *8*, 111-130.

Dhondt, S., Wuyts, N., and Inze, D. (2013). Cell to whole-plant phenotyping: the best is yet to come. Trends Plant Sci. *18*, 433-444.

Glory, E., and Murphy, R.F. (2007). Automated subcellular location determination and high-throughput microscopy. Dev. Cell *12*, 7-16.

Granier, C., Aguirrezabal, L., Chenu, K., Cookson, S.J., Dauzat, M., Hamard, P., Thioux, J.J., Rolland, G., Bouchier-Combaud, S., Lebaudy, A., et al. (2006). PHENOPSIS, an automated platform for reproducible phenotyping of plant responses to soil water deficit in *Arabidopsis thaliana* permitted the identification of an accession with low sensitivity to soil water deficit. New Phytol. *169*, 623-635.

Guo, Q., Wu, F., Pang, S., Zhao, X., Chen, L., Liu, J., Xue, B., Xu, G., Li, L., Jing, H., et al. (2016). Crop 3D: a platform based on LiDAR for 3D high-throughput crop phenotyping. Sci. Sin. Vitae *46*, 1210-1221.

Han, Y.Y., Wang, K.Y., Liu, Z.Q., Zhang, Q., Pan, S.H., Zhao, X.Y., and Wang, S.F. (2017). A crop trait information acquisition system with multitag-based identification technologies for breeding precision management. Comput. Electron. Agr. *135*, 71-80.

Herwitz, S.R., Johnson, L.F., Dunagan, S.E., Higgins, R.G., Sullivan, D.V., Zheng, J., Lobitz, B.M., Leung, J.G., Gallmeyer, B.A., Aoyagi, M., et al. (2004). Imaging from an unmanned aerial vehicle: agricultural surveillance and decision support. Comput. Electron. Agr. *44*, 49-61.

Johannsen W. (1911). The genotype conception of heredity. Int. J. Epidemiol. *43*, 989-1000.

Kuijken, R.C.P., van Eeuwijk, F.A., Marcelis, L.F.M., and Bouwmeester, H.J. (2015). Root phenotyping: from component trait in the lab to breeding. J. Exp. Bot. *66*, 5389-5401.

Li, L., Zhou, Q., Voss, T.C., Quick, K.L., and LaBarbera, D.V. (2016). High-throughput imaging: focusing in on drug discovery in 3D. Methods *96*, 97-102.

Liberali, P., Snijder, B., and Pelkmans, L. (2015). Single-cell and multivariate approaches in genetic perturbation screens. Nat. Rev. Genet. *16*, 18-32.

Makanza, R., Zaman-Allah, M., Cairns, J.E., Magorokosho, C., Tarekegne, A., Olsen, M., and Prasanna, B.M. (2018). High-throughput phenotyping of canopy cover and senescence in maize field trials using aerial digital canopy imaging. Remote Sens. *10*, 330.

Merz, T., and Chapman, S. (2011). Autonomous unmanned helicopter system for remote sensing missions in unknown environments//H. Eisenbeiss, M. Kunz, and H. Ingensand. International Conference on Unmanned Aerial Vehicle in Geomatics, 143-148.

Pegoraro, G., and Misteli, T. (2017). High-throughput imaging for the discovery of cellular mechanisms of disease. Trends Genet. *33*, 604-615.

Perez-Rangel, M., Quiroz-Figueroa, F.R., Gonzalez-Castaneda, J., and Valdez-Vazquez, I. (2015). Microscopic analysis of wheat straw cell wall degradation by microbial consortia for hydrogen production. Int. J. Hydrogen Energ. *40*, 151-160.

Reuzeau, C. (2007). TraitMill (TM): a high throughput functional genomics platform for the phenotypic analysis of cereals. In Vitro Cell. Dev-An. *43*, S4-S4.

Roukos, V., Voss, T.C., Schmidt, C.K., Lee, S., Wangsa, D., and Misteli, T. (2013). Spatial dynamics of chromosome translocations in living cells. Science *341*, 660-664.

Rust, M.J., Bates, M., and Zhuang, X. (2006). Sub-diffraction-limit imaging by stochastic optical reconstruction microscopy (STORM). Nat. Methods *3*, 793-795.

Smith, K., Piccinini, F., Balassa, T., Koos, K., Danka, T., Azizpour, H., and Horvath, P. (2018). Phenotypic image analysis software tools for exploring and understanding big image data from cell-based assays. Cell Sys. *6*, 636-653.

Tisne, S., Serrand, Y., Bach, L., Gilbault, E., Ben Ameur, R., Balasse, H., Voisin, R., Bouchez, D., Durand-Tardif, M., Guerche, P., et al. (2013). Phenoscope: an automated large-scale phenotyping platform offering high spatial homogeneity. Plant J. *74*, 534-544.

Virlet, N., Sabermanesh, K., Sadeghi-Tehran, P., and Hawkesford, M.J. (2017). Field scanalyzer: an automated robotic field phenotyping platform for detailed crop monitoring. Function. Plant Biol. *44*, 143-153.

Wang, Y., and Xu, L. (2018). Unsupervised segmentation of greenhouse plant images based on modified Latent Dirichlet Allocation. PeerJ *6*, e5036.

White, J.W., and Conley, M.M. (2013). A flexible, low-cost cart for proximal sensing. Crop Science *53*, 1646-1649.

Yang, W., Guo, Z., Huang, C., Duan, L., Chen, G., Jiang, N., Fang, W., Feng, H., Xie, W., Lian, X., et al. (2014). Combining high-throughput phenotyping and genome-wide association studies to reveal natural genetic variation in rice. Nat. Commun. *5*, 5087.

Young, D.W., Bender, A., Hoyt, J., McWhinnie, E., Chirn, G.W., Tao, C.Y., Tallarico, J.A., Labow, M., Jenkins, J.L., Mitchison, T.J., et al. (2008). Integrating high-content screening and ligand-target prediction to identify mechanism of action. Nat. Chem. Biol. *4*, 59-68.

Zhao, C., Zhang, Y., Du, J., Guo, X., Gu, S., Wang, J., Fan, J. (2019). Crop phenomics: current status and perspectives. Front. Plant Sci. *10*, 714.

功能基因高效解析

赖锦盛　严建兵　黄学辉

作物功能基因组研究的长远目标是弄清楚作物基因组中所有基因的生物学功能，尤其是解析出控制重要农艺和经济性状（如产量、品质和抗逆）的功能基因，并将相关研究成果应用于作物品种改良，带动植物遗传学的快速发展和分子育种体系建立。最近十多年里，在水稻、玉米等重要农作物的高质量参考基因组完成的基础上，国内外科研人员系统构建了功能基因组组学研究平台和材料体系，开发了用于作物功能基因高通量遗传解析的分析方法，大大加速了重要农艺性状的功能基因组学研究。未来，分析方法和材料体系上的创新有望为作物功能基因的高效解析提供更有力的支撑，并直接服务于作物分子设计育种。

1 国内外研究进展

随着二代测序的普及和三代技术的发展，许多粮食作物的全基因组序列图谱被成功构建。2002年，科学家们成功破译水稻基因组之后，又陆续完成了对玉米、大豆、马铃薯、小麦、大麦、棉花、番茄等重要粮食作物的基因组测序和品种资源的重测序（International Barley Genome Sequencing Consortium，2011，2012；Jiao et al.，2017；Ling et al.，2013；Sun et al.，2018；Tomato Genome Consortium，2012；Wang et al.，2012），这些基础组学数据信息和材料方法体系的建立极大地推动了农艺性状基因的挖掘与功能基因组的研究。

1.1 突变体基因的高效克隆

突变体创制是研究基因功能的重要途径之一。在长期的遗传学基础研究和育种实践中，水稻、玉米等重要农作物已经积累了大量的突变体，包括来自田间的偶发自然突变、离子束突变群体、快中子突变群体、甲基磺酸乙酯突变群体，T-DNA 插入突变群体等。伴随着配套的高效表型筛选方法，很多突变体已经被成功用于株型、花期、抗逆等性状的功能基因的挖掘与功能分析。在作物遗传学研究中，借助个体间的杂交创建突变体定位群体，进而通过连锁分析定位到控制突变表型的关键基因是常见的正向遗传学方法。在基因定位过程中，对于突变体群体的基因型鉴定大多是依赖于插入缺失标记（InDel）、简单重复序列标记等传统的分子标记进行遗传作图，这些方法需要花费大量的时间，而且作图分辨率不高。

随着高通量测序技术的快速发展和日益普及（Goodwin et al.，2016），基因组测序技术也被广泛用于各种作物突变体的遗传定位分析中（Schneeberger et al.，2009）。在水稻、玉米等作物遗传学研究中，科研人员针对突变体等质量性状的图位克隆采取了一些较简便的方法。例如，在水稻中，Abe 等（2012）开发出了一套 MutMap 方法体系，将突变体与对应的野生型材料杂交后产生 F_2 群体，之后将分离群体中两种表型的个体混合后测序，检测等位基因频率在全基因组中的分布，出现极端等位基因频率的区域即为候选位点；通过突变体与对应的野生型材料的基因组序列比较，还有望直接鉴定到候选位点中发生突变的基因，从而大大加速和简化了突变体基因的遗传定位。

1.2 数量性状位点的高精度解析

在作物遗传育种中，数量性状位点（QTL）定位一直是常用的遗传学研究方法（Fan et al.，2006；Shomura et al.，2008）。QTL 的精细定位依赖于较大的连锁群体和精细的基因型图谱。随着测序技术的发展，通过高通量测序对每个个体进行全基因组测序或酶切位点旁邻区域测序（Goodwin et al.，2016），可在短时间内从测序实验结果中获得大量个体基因组内的大量单核苷酸多态（SNP）位点，再运用一定的算法解析，便可获得高密度的基因型图谱，这有利于将 QTL 锁定到一个相对较小的范围。

在水稻遗传学研究中，中国科学院国家基因研究中心的人员开发了一套高通量的基因分型方法及相关软件包 SEG-Map 软件，通过将 150 份重组自交系的 DNA 样本加上不同的接头后进行测序，针对产生的低覆盖率测序数据通过 SEG-Map 软件可以获得高精度的基因型图谱（Huang et al.，2009；

Zhao et al.，2010）。华中农业大学的研究人员在此基础上进行了进一步的改进，实现了在没有已知亲本的情况下对水稻进行高精度基因分型（Xie et al.，2010）。目前，这些基于重测序方法的基因分型已经大大加速了水稻等农作物的遗传解析，同时有效提高了这些农作物基因组中 QTL 定位的精度。

BSR-seq 是玉米中快速克隆突变基因的一个有效方法，在番茄中也得以应用。同时，华中农业大学开发出一套 QTG-Seq 方法用于重要基因的快速克隆（Zhang et al.，2019）。在这个方法中，科研人员首先通过分离群体定位到 4 个株高 QTL，在 BC_1F_1 群体中筛选出只在 $qPH7$ 杂合而其他 3 个 QTL 都纯合的单株自交，获得大量 BC_1F_2 家系，进一步获得极端株高表型的单株混池，通过全基因组混池测序，最后利用新开发的基于最大似然法的似然比平滑统计量 smoothLOD 直接鉴定出目标候选基因。

马铃薯晚疫病会导致马铃薯茎叶死亡、块茎腐烂，严重损害马铃薯和番茄的产量。在 19 世纪 40 年代，由于晚疫病的暴发，爱尔兰马铃薯的产量急剧减少，引起大范围的饥荒。英国 Jonathan D. G. Jones 课题组通过对马铃薯的野生近缘种进行测序分析，利用抗性基因富集测序技术（RenSeq）和单分子实时测序技术（SMRT RenSeq）分离到了一个新的抗晚疫病的基因 *Rpi-amr3*（Witek et al.，2016）。此外，黄三文课题组对马铃薯无性繁殖自交衰退机制进行了研究，构建了 3 个马铃薯自交群体，并开发了一种不依赖于亲本基因分型新型算法，利用此群体和算法鉴定了 5 个纯合致死位点和 4 个影响长势的位点，进一步成功克隆到一个致死突变基因 *AR1*，发现其能调控胚的发育（Zhang et al.，2019）。

以上这些方法的开发和改进为重要农艺性状的遗传解析提供了有力的帮助。

1.3 作物全基因组关联分析

全基因组关联研究（genome-wide association study，GWAS）是一种用于寻找 DNA 序列变异与表型差异之间关系的数量遗传学研究方法，这种方法的分辨度依赖于所研究物种的遗传多样性和连锁不平衡的衰退距离，通常高于连锁分析方法，对应的基因型图谱也更为密集。随着高通量测序成本的降低，使得通过测序对大量样品进行高精度基因分型成为可能。

最近十多年里，GWAS在医学、遗传学领域中发展迅速，被广泛应用（Wellcome Trust Case Control Consortium，2007）。在人类遗传疾病的GWAS研究中，研究者一般会收集大量患者的血液标本及正常人群的血液标本，然后提取样品的基因组DNA进行全基因组范围的基因分型，最终从近百万个分子标记中找出与该表型相关的标记及旁邻的候选基因。到目前为止，世界各地的研究人员已经对数百种遗传疾病（如各类肿瘤、心血管疾病、糖尿病、肥胖症、精神疾病等）进行了GWAS分析，鉴定到了大批疾病易感基因区域和相关基因。这些研究成果对相对应遗传疾病的分子诊断和药物设计起到了一定的推动作用。

以自然群体为主要研究对象的全基因组关联研究方法也在作物遗传学研究中实现并逐渐发挥重要作用（Atwell et al.，2010；Tian et al.，2011；Xiao et al.，2017；Yang et al.，2014；Zhou et al.，2015）。研究人员利用高通量测序技术快速鉴定了水稻、玉米等作物的不同品种间的序列差异（Yang et al.，2011，2014），并构建了高密度的单倍体型图（Bukowski et al.，2018；Chia et al.，2012；Gore et al.，2009；Huang et al.，2010；Lai et al.，2010；Yano et al.，2016）。2010年，中国科学院国家基因研究中心利用第二代测序技术对近千份水稻品种进行重测序，鉴定了几百万个SNP位点并构建了一张高密度的水稻单倍型图谱，还成功应用于14个农艺性状的全基因组关联分析（Huang et al.，2010，2011）。2011年，我国科学家首次对野生稻和栽培稻的基因组进行大规模的遗传多样性分析，为挖掘野生稻优良基因，加快高产、优质水稻品种培育奠定了理论基础（Xu et al.，2011）。近期，中国农业科学院作物科学研究所等团队对从89个国家收集的3000份水稻品种进行了深度重测序，命名为3000份水稻重测序项目，得到了平均14倍覆盖深度的基因组测序数据，鉴定发现了约1890万个SNP和InDel信息（Wang et al.，2018），并利用这套材料资源开展了农艺性状的解析。这些基因序列变异的发掘为研究水稻基因功能，指导基因组育种具有重大的科学意义和实用价值（Zhou and Huang，2019；McCouch et al.，2016）。与此同时，通过结合基因注释信息及基因表达谱信息，在水稻基因组研究中通过GWAS获得的一些重要候选基因也陆续得以确定并通过功能基因组学实验验证，如控制粒

型的 *GLW7* 基因（Si et al.，2016），控制芒长的 *GAD1/RAE2* 基因（Jin et al.，2016），控制耐冷的 *bZIP73* 基因（Liu et al.，2018）。

在玉米的功能基因组学研究中，华中农业大学等多家机构科研人员首先发展了一套具有广泛代表性的多样性群体（Yang et al.，2011），利用基因分型芯片、RNA-Seq 测序技术和 GWAS 方法相结合，对 368 份玉米自交系群体进行了全基因组关联分析，鉴定得到 74 个与玉米油含量及脂肪酸组成显著相关的遗传位点，其中大量候选基因编码参与油分代谢通路的酶（Li et al.，2013）；这项工作为玉米籽粒中油分合成的遗传机制提供了新的视角，并有助于高油玉米的分子设计育种。该群体被广泛应用到玉米品质、产量、抗病、抗旱等多达几十个性状的全基因组关联分析中，一大批关键的功能基因被鉴定，Xiao 等（2017）的综述文章对此进行了全面的总结，其中包括，中国科学院植物研究所科研人员利用该群体开展玉米苗期抗旱性评估，通过 GWAS 方法在第 9 号染色体上鉴定到一个抗旱主效基因 *ZmVPP1*，并经过功能实验确定了基因功能；*ZmVPP1* 基因编码一个定位于液泡膜上的质子泵-焦磷酸水解酶；CIMBL55 基因组的 *ZmVPP1* 的启动子中含有一个长度为 366bp 的插入缺失多态，该缺失片段中含有 3 个干旱应答的 MYB 结合顺式作用元件，能够影响 *ZmVPP1* 在干旱胁迫下的基因表达量。该研究对玉米抗旱性的遗传改良具有重要意义，也为玉米抗旱品种的培育提供了线索（Wang et al.，2016）。

番茄是世界第一大蔬菜作物，它的口感与风味是消费者最为关注的两个方面。Zhu 等（2018）结合 610 个番茄近缘品种的重测序数据、399 个品种的转录组数据和 442 个品种的代谢组数据，将代谢全基因组关联研究（mGWAS）和数量性状位点（QTL）两种技术进行了整合，阐述了全球范围内培育的果实代谢物含量改变的原因，并挖掘到调控果实重量的基因。

我国在棉花的基因组学研究方面也取得了良好进展。Ma 等（2018）对我国 419 份陆地棉核心种质的重测序，并对 12 个环境中 13 个纤维相关性状进行统计，通过 GWAS 发掘了一大批调控重要性状的位点和基因。Du 等（2018）收集了不同棉区 230 份亚洲棉和 13 份草棉材料，成功构建了棉花 A 基因组的高质量变异图谱，并从驯化角度揭示中国的亚洲棉可能最早发源于华南地区，并且还鉴定到了一批与抗病、油分相关的候选基因。Fang 等（2017）对 318 份棉花品种的全基因组进行重测序，进一步挖掘到了 25 个与品种改良，119 个与产量和纤维品质等相关的关联位点。

2 针对未来作物分子设计需求的研究内容与突破口

2.1 测序技术与基因分型的发展

作物功能基因的高效解析依赖于基因组学技术的革命性发展与持续进步。在过去的十多年里，第二代测序技术在多个领域，如转录组（transcriptome）研究、全基因组变异分析（variation detection）和大规模基因分型（genotyping）中发挥了重要作用，也大大推动了作物遗传学的研究。如今，第三代测序——单分子测序技术（PacBio 技术和 Nanopore 技术）逐渐成熟。例如，英国牛津纳米孔公司设计的小型、便携的 MinION 等产品，可以用 U 盘大小的设备全自动化且快速地完成生物学样本长片段的测序。在不远的将来，基因型分析技术将继续改进，基因测序也会更为方便、快速和低成本。因此，将来我们关注的不再是基因分型本身，而是如何通过更精妙的实验设计、更合理的分析手段去解决作物功能基因组学中关于重要农艺性状基因及其基因调控网络等复杂的生物学问题。

2.2 遗传群体与定位分析方法的创新

在利用自然群体进行 GWAS 分析的过程中，科研人员发现群体结构等因素的影响使得包括花期、株高、产量在内等的重要农艺性状在用全基因组关联分析进行定位时容易出现大量假阳性。为此，科研人员在群体的设计方面进行了改进。美国康奈尔大学 Edward Buckler 研究团队在玉米遗传学研究中，成功地开发出了一套被称作巢式联合作图（nested association mapping，NAM）的新方法（Buckler et al.，2009），使用普通的标准基因型杂交得到了 5000 种重组自交系（recombination inbred line，RIL）群体。对最初的亲本玉米进行测序，并对重组自交系进行基因型检测，最终得到足以覆盖每一个重组自交系的单倍体型图。利用该群体进行全基因组关联分析，检测到大量与农艺性状，如叶片形态结构、小斑病抗性（Kump et al.，2011）、开花期和籽粒营养成分等关联的显著位点,并能够检测到位点上多个等位基因。研究结果还表明,存在大量微效 QTL 控制许多重要性状,如开花期等。华中农业大学科研人员进一步提出 ROAM（random-open-parent association mapping）的遗传设计群体（Xiao et al.,2016,2017）,该群体不依赖于特定的共有亲本,而可以是任意群体的组合,降低了组建群体的难度。基于这个群体,科研人员比较系统的探讨了玉米产量、株型和籽粒等的遗传结构,鉴定了大量和这些性状有关的 QTL,发现稀有等位基因在玉米数量性状遗传变异中起重要作用（Liu et al.,2017；Pan et al.,2017；Xiao et al.,2016）。

此外,科研人员也尝试在水稻和小麦等作物中构建和使用多亲本高世代互交（multi-parent advanced generation inter-cross,MAGIC）（Bandillo et al.,2013；Dell'Acqua et al.,2015；Huang et al.,2015）等多亲本杂交衍生的遗传群体,进行复杂性状的解析,这也是近年来发展起来的新型遗传作图及资源创新群体。MAGIC 群体的一个重要特点是应用多亲本间相互杂交及后代的多轮杂交。一般而言,多亲本首先两两杂交,获得的杂交种 F_1 再相互杂交,产生新的杂交种（F_{1s}）。然后新的 F_1 间再交叉杂交,产生下一代 F_1。该代 F_1 也可继续成对杂交,也可开始自交,并连续数代自交产生高度纯合的近交系。由于采用多个亲本的相互交配设计,可以提高群体中基因的重组频率,打破有利与不利位点的遗传连锁累赘,增加后代群体的基因多样性和基因型多样性。

这些群体设计上的改进及对应分析方法上的突破,有望在将来进一步提升作物功能基因解析的效果。

2.3 QTL 遗传互作与表型预测

一些主要农作物中产量、籽粒品质和抗病性的 QTL 基因解析正越来越快速,功能研究也越来越细致和清楚。这些进展为作物数量遗传学研究及分子设计育种提供了重要机遇——我们可以创建更理想的等位基因型,并更加自由地设计出育种路线图。随之而来还有很多挑战需要面对,包括如何有效地解析 QTL 与环境的互作、QTL 之间的遗传互作（Forsberg et al.,2017）及 QTL 与杂种优势的关系。这些关系都不会是简单的线性关系,多种因素交织在一起,使得问题非常复杂。更理想的遗传群体、高通量的组学数据、基因功能数据和新的数据分析方法（如深度学习算法）可能都会为解决这些科学问题提供帮助。

作物复杂性状的数量遗传学研究将会步入一个新的时代——从"表型到基因"过渡到"基因-性状"阶段。在前一阶段，主要关注的焦点是复杂性状的遗传解析和基因功能的逐一鉴定。当大多数功能基因已得到验证，将通过数学建模等方法，把这些QTL基因重新组合在一起，结合对基因-环境互作和基因调控网络分子机制的了解，将能够从系统的角度找出每个基因型的农艺性状表现（Zuo and Li，2014），从而为作物分子设计育种提供强有力的支持。

3 阶段性目标

3.1 2035 年目标

至2035年，应该实现以下目标：主要农作物（水稻、小麦、玉米、大豆、棉花、马铃薯、番茄等）的基因组信息，骨干种质群体的基因组SNP变异和结构变异图谱清晰；基于全基因组关联分析的自然群体中重要性状位点基因克隆的方法和基于基因组混池测序技术的突变体中基因快速克隆的方法体系更加成熟；主要农作物全基因组的调控元件（增强子等）、基因组三维染色体构想信息图谱、全基因组表观修改图谱基本完成；主要农作物的重要农艺性状的主效QTL位点和微效QTL位点信息取得重要进展，控制性状的QTL位点间的遗传关系及相互调控网络基本清晰；重要性状的大部分表型自动化获取快速准确；主要性状的表型和环境相互作用的关系有了根本性的认识。

3.2 2050 年目标

至2050年，应能实现以下目标：主要农作物（水稻、小麦、玉米、大豆、棉花、马铃薯、番茄等）基因组大部分（超过70%）基因和调控元件的功能基本清晰；人工设计的染色体片段对主要农作物的性状有突破性的补充完善；主要农作物的综合农艺性状大幅度改良，农业生产效率和抵御极端气候变化的能力大幅度提高。

参考文献

Abe, A., Kosugi, S., Yoshida, K., Natsume, S., Takagi, H., Kanzaki, H., Matsumura, H., Yoshida, K., Mitsuoka, C., Tamiru, M., et al. (2012). Genome sequencing reveals agronomically important loci in rice using MutMap. Nat. Biotechnol. *30*, 174-178.

Atwell, S., Huang, Y.S., Vilhjálmsson, B.J., Willems, G., Horton, M., Li, Y., Meng, D., Platt, A., Tarone, A.M., Hu, T.T., et al. (2010). Genome-wide association study of 107 phenotypes in *Arabidopsis thaliana* inbred lines. Nature *465*, 627-631.

Bandillo, N., Raghavan, C., Muyco, P.A., Sevilla, M.A.L., Lobina, I.T., Dilla-Ermita, C.J., Tung, C., McCouch, S., Thomson, M., Mauleon, R., et al. (2013). Multi-parent advanced generation inter-cross (MAGIC) populations in rice: progress and potential for genetics research and breeding. Rice *6*, 11.

Buckler, E.S., Holland, J.B., Bradbury, P.J., Acharya, C.B., Brown, P.J., Browne, C., Ersoz, E., Flint-Garcia, S., Garcia, A., Glaubitz, J.C., et al. (2009). The genetic architecture of maize flowering time. Science *325*, 714-718.

Bukowski, R., Guo, X., Lu, Y., Zou, C., He, B., Rong, Z., Wang, B., Xu, D., Yang, B., Xie, C., et al. (2018). Construction of the third-generation *Zea mays* haplotype map. GigaScience *7*, 1-12.

Chia, J., Song, C., Bradbury, P.J., Costich, D., de Leon, N., Doebley, J., Elshire, R.J., Gaut, B., Geller, L., Glaubitz, J.C., et al. (2012). Maize HapMap2 identifies extant variation from a genome in flux. Nat. Genet. *44*, 803-807.

Dell'Acqua, M., Gatti, D.M., Pea, G., Cattonaro, F., Coppens, F., Magris, G., Hlaing, A.L., Aung, H.H., Nelissen, H., Baute, J., et al. (2015). Genetic properties of the MAGIC maize population: a new platform for high definition QTL mapping in *Zea mays*. Genome Biol. *16*, 167.

Du, X., Huang, G., He, S., Yang, Z., Sun, G., Ma, X., Li, N., Zhang, X., Sun, J., Liu, M., et al. (2018). Resequencing of 243 diploid cotton accessions based on an updated A genome identifies the genetic basis of key agronomic traits. Nat. Genet. *50*, 796-802.

Fan, C., Xing, Y., Mao, H., Lu, T., Han, B., Xu, C., Li, X., and Zhang, Q. (2006). *GS3*, a major QTL for grain length and weight and minor QTL for grain width and thickness in rice, encodes a putative transmembrane protein. Theor. Appl. Genet. *112*, 1164-1171.

Fang, L., Wang, Q., Hu, Y., Jia, Y., Chen, J., Liu, B., Zhang, Z., Guan, X., Chen, S., Zhou, B., et al. (2017). Genomic analyses in cotton identify signatures of selection and loci associated with fiber quality and yield traits. Nat. Genet. *49*, 1089-1098.

Forsberg, S.K.G., Bloom, J.S., Sadhu, M.J., Kruglyak, L., and Carlborg, Ö. (2017). Accounting for genetic interactions improves modeling of individual quantitative trait phenotypes in yeast. Nat. Genet. *49*, 497-503.

Goodwin, S., McPherson, J.D., and McCombie, W.R. (2016). Coming of age: ten years of next-generation sequencing technologies. Nat. Rev. Genet. *17*, 333-351.

Gore, M.A., Chia, J., Elshire, R.J., Sun, Q., Ersoz, E.S., Hurwitz, B.L., Peiffer, J.A., McMullen, M.D., Grills, G.S., Ross-Ibarra, J., et al. (2009). A first-generation haplotype map of maize. Science *326*, 1115-1117.

Huang, B.E., Verbyla, K.L., Verbyla, A.P., Raghavan, C., Singh, V.K., Gaur, P., Leung, H., Varshney, R.K., and Cavanagh, C.R. (2015). MAGIC populations in crops: current status and future prospects. Theor. Appl. Genet. *128*, 999-1017.

Huang, X., Feng, Q., Qian, Q., Zhao, Q., Wang, L., Wang, A., Guan, J., Fan, D., Weng, Q., Huang, T., et al. (2009). High-throughput genotyping by whole-genome resequencing. Genome Res. *19*, 1068-1076.

Huang, X., Wei, X., Sang, T., Zhao, Q., Feng, Q., Zhao, Y., Li, C., Zhu, C., Lu, T., Zhang, Z., et al. (2010). Genome-wide association studies of 14 agronomic traits in rice landraces. Nat. Genet. *42*, 961-967.

Huang, X., Zhao, Y., Wei, X., Li, C., Wang, A., Zhao, Q., Li, W., Guo, Y., Deng, L., Zhu, C., et al. (2011). Genome-wide association study of flowering time and grain yield traits in a worldwide collection of rice germplasm. Nat. Genet. *44*, 32-39.

International Barley Genome Sequencing Consortium (2011). Genome sequence and analysis of the tuber crop potato. Nature *475*, 189-195.

International Barley Genome Sequencing Consortium (2012). A physical, genetic and functional sequence assembly of the barley genome. Nature *491*, 711-716.

Jiao, Y., Peluso, P., Shi, J., Liang, T., Stitzer, M.C., Wang, B., Campbell, M.S., Stein, J.C., Wei, X., Chin, C., et al. (2017). Improved maize reference genome with single-molecule technologies. Nature *546*, 524-527.

Jin, J., Hua, L., Zhu, Z., Tan, L., Zhao, X., Zhang, W., Liu, F., Fu, Y., Cai, H., Sun, X., et al. (2016). *GAD1* encodes a secreted peptide that regulates grain number, grain length, and awn development in rice domestication. Plant Cell *28*, 2453-2463.

Kump, K.L., Bradbury, P.J., Wisser, R.J., Buckler, E.S., Belcher, A.R., Oropeza-Rosas, M.A., Zwonitzer, J.C., Kresovich, S., McMullen, M.D., Ware, D., et al. (2011). Genome-wide association study of quantitative resistance to southern leaf blight in the maize nested association mapping population. Nat. Genet. *43*, 163-168.

Lai, J., Li, R., Xu, X., Jin, W., Xu, M., Zhao, H., Xiang, Z., Song, W., Ying, K., Zhang, M., et al. (2010). Genome-wide patterns of genetic variation among elite maize inbred lines. Nat. Genet. *42*, 1027-1030.

Li, H., Peng, Z., Yang, X., Wang, W., Fu, J., Wang, J., Han, Y., Chai, Y., Guo, T., Yang, N., et al. (2013). Genome-wide association study dissects the genetic architecture of oil biosynthesis in maize kernels. Nat. Genet. *45*, 43-50.

Ling, H., Zhao, S., Liu, D., Wang, J., Sun, H., Zhang, C., Fan, H., Li, D., Dong, L., Tao, Y., et al. (2013). Draft genome of the wheat A-genome progenitor *Triticum urartu*. Nature *496*, 87-90.

Liu, C., Ou, S., Mao, B., Tang, J., Wang, W., Wang, H., Cao, S., Schläppi, M.R., Zhao, B., Xiao, G., et al. (2018). Early selection of bZIP73 facilitated adaptation of japonica rice to cold climates. Nat. Commun. *9*, 3302.

Liu, J., Huang, J., Guo, H., Lan, L., Wang, H., Xu, Y., Yang, X., Li, W., Tong, H., Xiao, Y., et al. (2017). The conserved and unique genetic architecture of kernel size and weight in maize and rice. Plant Physiol. *175*, 774-785.

Ma, Z., He, S., Wang, X., Sun, J., Zhang, Y., Zhang, G., Wu, L., Li, Z., Liu, Z., Sun, G., et al. (2018). Resequencing a core collection of upland cotton identifies genomic variation and loci influencing fiber quality and yield. Nat. Genet. *50*, 803-813.

McCouch, S.R., Wright, M.H., Tung, C., Maron, L.G., McNally, K.L., Fitzgerald, M., Singh, N., DeClerck, G., Agosto-Perez, F., Korniliev, P., et al. (2016). Open access resources for genome-wide association mapping in rice. Nat. Commun. *7*, 10532.

Pan, Q., Xu, Y., Li, K., Peng, Y., Zhan, W., Li, W., Li, L., and Yan, J. (2017). The genetic basis of plant architecture in 10 maize recombinant inbred line populations. Plant Physiol. *175*, 858-873.

Schneeberger, K., Ossowski, S., Lanz, C., Juul, T., Petersen, A.H., Nielsen, K.L., Jørgensen, J., Weigel, D., and Andersen, S.U. (2009). SHOREmap: simultaneous mapping and mutation identification by deep sequencing. Nat. Methods *6*, 550-551.

Shomura, A., Izawa, T., Ebana, K., Ebitani, T., Kanegae, H., Konishi, S., and Yano, M. (2008). Deletion in a gene associated with grain size increased yields during rice domestication. Nat. Genet. *40*, 1023-1028.

Si, L., Chen, J., Huang, X., Gong, H., Luo, J., Hou, Q., Zhou, T., Lu, T., Zhu, J., Shangguan, Y., et al. (2016). *OsSPL13* controls grain size in cultivated rice. Nat. Genet. *48*, 447-456.

Sun, S., Zhou, Y., Chen, J., Shi, J., Zhao, H., Zhao, H., Song, W., Zhang, M., Cui, Y., Dong, X., et al. (2018). Extensive intraspecific gene order and gene structural variations between Mo17 and other maize genomes. Nat. Genet. *50*, 1289-1295.

Tian, F., Bradbury, P.J., Brown, P.J., Hung, H., Sun, Q., Flint-Garcia, S., Rocheford, T.R., McMullen, M.D., Holland, J.B., and Buckler, E.S. (2011). Genome-wide association study of leaf architecture in the maize nested association mapping population. Nat. Genet. *43*, 159-162.

Tomato Genome Consortium (2012). The tomato genome sequence provides insights into fleshy fruit evolution. Nature *485*, 635-641.

Wang, K., Wang, Z., Li, F., Ye, W., Wang, J., Song, G., Yue, Z., Cong, L., Shang, H., Zhu, S., et al. (2012). The draft genome of a diploid cotton *Gossypium raimondii*. Nat. Genet. *44*, 1098-1103.

Wang, W., Mauleon, R., Hu, Z., Chebotarov, D., Tai, S., Wu, Z., Li, M., Zheng, T., Fuentes, R.R., Zhang, F., et al. (2018). Genomic variation in 3010 diverse accessions of Asian cultivated rice. Nature *557*, 43-49.

Wang, X., Wang, H., Liu, S., Ferjani, A., Li, J., Yan, J., Yang, X., and Qin, F. (2016). Genetic variation in *ZmVPP1* contributes to drought tolerance in maize seedlings. Nat. Genet. *48*, 1233-1241.

Wellcome Trust Case Control Consortium (2007). Genome-wide association study of 14 000 cases of seven common diseases and 3000 shared controls. Nature *447*, 661-678.

Witek, K., Jupe, F., Witek, A.I., Baker, D., Clark, M.D., and Jones, J.D.G. (2016). Accelerated cloning of a potato late blight-resistance gene using RenSeq and SMRT sequencing. Nat.

Biotechnol. *34*, 656-660.

Xiao, Y., Liu, H., Wu, L., Warburton, M., and Yan, J. (2017). Genome-wide association studies in maize: praise and stargaze. Mol. Plant *10*, 359-374.

Xiao, Y., Tong, H., Yang, X., Xu, S., Pan, Q., Qiao, F., Raihan, M.S., Luo, Y., Liu, H., Zhang, X., et al. (2016). Genome-wide dissection of the maize ear genetic architecture using multiple populations. New Phytol. *210*, 1095-1106.

Xie, W., Feng, Q., Yu, H., Huang, X., Zhao, Q., Xing, Y., Yu, S., Han, B., and Zhang, Q. (2010). Parent-independent genotyping for constructing an ultrahigh-density linkage map based on population sequencing. Proc. Natl. Acad. Sci. USA *107*, 10578-10583.

Xu, X., Liu, X., Ge, S., Jensen, J.D., Hu, F., Li, X., Dong, Y., Gutenkunst, R.N., Fang, L., Huang, L., et al. (2011). Resequencing 50 accessions of cultivated and wild rice yields markers for identifying agronomically important genes. Nat. Biotechnol. *30*, 105-111.

Yang, N., Lu, Y., Yang, X., Huang, J., Zhou, Y., Ali, F., Wen, W., Liu, J., Li, J., and Yan, J. (2014). Genome wide association studies using a new nonparametric model reveal the genetic architecture of 17 agronomic traits in an enlarged maize association panel. PLoS Genet. *10*, e1004573.

Yang, W., Guo, Z., Huang, C., Duan, L., Chen, G., Jiang, N., Fang, W., Feng, H., Xie, W., Lian, X., et al. (2014). Combining high-throughput phenotyping and genome-wide association studies to reveal natural genetic variation in rice. Nat. Commun. *5*, 5087.

Yang, X., Gao, S., Xu, S., Zhang, Z., Prasanna, B.M., Li, L., Li, J., and Yan, J. (2011). Characterization of a global germplasm collection and its potential utilization for analysis of complex quantitative traits in maize. Mol. Breed. *28*, 511-526.

Yano, K., Yamamoto, E., Aya, K., Takeuchi, H., Lo, P., Hu, L., Yamasaki, M., Yoshida, S., Kitano, H., Hirano, K., et al. (2016). Genome-wide association study using whole-genome sequencing rapidly identifies new genes influencing agronomic traits in rice. Nat. Genet. *48*, 927-934.

Zhang, C., Wang, P., Tang, D., Yang, Z., Lu, F., Qi, J., Tawari, N.R., Shang, Y., Li, C., and Huang, S. (2019). The genetic basis of inbreeding depression in potato. Nat. Genet. *51*, 374-378.

Zhang, H., Wang, X., Pan, Q., Li, P., Liu, Y., Lu, X., Zhong, W., Li, M., Han, L., Li, J., et al.

(2019). QTG-Seq accelerates QTL fine mapping through QTL partitioning and whole-genome sequencing of bulked segregant samples. Mol. Plant *12*, 426-437.

Zhao, Q., Huang, X., Lin, Z., and Han, B. (2010). SEG-Map: a novel software for genotype calling and genetic map construction from next-generation sequencing. Rice *3*, 98-102.

Zhou, X., and Huang, X. (2019). Genome-wide association studies in rice: how to solve the low power problems? Mol. Plant *12*, 10-12.

Zhou, Z., Jiang, Y., Wang, Z., Gou, Z., Lyu, J., Li, W., Yu, Y., Shu, L., Zhao, Y., Ma, Y., et al. (2015). Resequencing 302 wild and cultivated accessions identifies genes related to domestication and improvement in soybean. Nat. Biotechnol. *33*, 408-414.

Zhu, G., Wang, S., Huang, Z., Zhang, S., Liao, Q., Zhang, C., Lin, T., Qin, M., Peng, M., Yang, C., et al. (2018). Rewiring of the fruit metabolome in tomato breeding. Cell *172*, 249-261.

Zuo, J., and Li, J. (2014). Molecular dissection of complex agronomic traits of rice: a team effort by Chinese scientists in recent years. Natl. Sci. Rev. *1*, 253-276.

基因组编辑

高彩霞

1 国内外研究进展

基因组编辑技术是实现精准育种的重要手段。通过现有的基因组编辑技术，可以在微观层面上对目标基因进行精准的编辑工作，相关工具具有简单、快速、高效、经济等优点。基因组编辑技术不仅推动了植物生物学基础研究的发展，而且在作物育种、创造新品种等应用领域展现出巨大的应用潜力。

1.1 基因组编辑技术的发展历程

基因组编辑技术是继转基因技术之后在生物遗传操作领域的又一革命性技术。基因组编辑技术是指通过序列特异核酸酶对基因组特定位点进行靶向修饰的技术。主要原理是通过特异切割 DNA 靶位点，产生 DNA 双链断裂，诱导 DNA 的损伤修复机制，实现对基因组的定向编辑。锌指核酸酶（zinc finger nuclease，ZFN）和类转录激活因子效应物核酸酶（transcription activator-like effector nuclease，TALEN）最先被应用于基因组定点编辑，但是这两项技术过程复杂且成本高，不利于推广应用。CRISPR/Cas 系统是继 ZFN 和 TALEN 技术之后出现的新的基因组编辑技术，该技术通过一段向导 RNA 和配套的核酸酶 Cas9 对特定的基因组序列进行定点编辑（Cong et al., 2013；Jinek et al., 2012），具有简单高效的优点。

2013 年 8 月，三个独立的课题组在 *Nature Biotechnology* 上分别报道了利用 CRISPR/Cas9 系统成功地在拟南芥、烟草、小麦、水稻中实现了基因组编辑（Li et al., 2013；Nekrasov et al., 2013；Shan et al., 2013），开启了 CRISPR/Cas 基因组编辑系统在植物中的研究。随后，CRISPR/Cas9 系统在许多重要的农作物和经济作物中实现了高效的基因组编辑（Hilscher et al., 2016）。此外，为了更有效地利用 CRISPR/Cas 对植物基因组编辑的研究，各国科学家开展了许多系统优化及新应用的开发。华南农业大学、美国宾夕法尼亚大学、中国农业大学、中国水稻研究所等分别开发了基于CRISPR/Cas9系统的多基因敲除系统（Ma et al.，2015；Wang et al.，2015；Xie et al.，2015；Xing et al.，2014），可以同时编辑多个基因，为同时改良多个农艺性状提供有效工具（Shen et al.，2017）。通过失活Cas9蛋白（deactivated Cas9，dCas9）融合具有转录调控功能的转录激活因子或转录抑制因子（Lowder et al.，2018；Piatek et al.，2015），如乙酰化酶、甲基化酶等功能元件，可以作为调控基因表达水平的控制元件，形成研究基因表达调控的有效工具（Hilton et al.，2015；Kearns et al.，2015；Sander and Joung，2014）。此外，许多新的CRISPR/Cas系统被开发，提供了新的基因组编辑工具并扩大了基因组编辑范围。例如，CRISPR/Cpf1系统是新一代基因组编辑工具，可以对CRISPR/Cas9系统进行有效补充（Hu et al.，2017；Xu et al.，2017；Zetsche et al.，2015；Zong et al.，2017）。

CRISPR/Cas系统目前主要被用于对靶位点的敲除。然而对植物内源基因进行更为精确地修饰仍然具有极大的挑战性，如基因定点替换及定点插入等，这严重限制了基因组编辑技术在植物基因组学研究和农作物分子设计育种中的应用。

1.2 无外源DNA的植物基因组编辑技术应用

大多数植物基因组编辑是通过农杆菌或基因枪的方法将CRISPR/Cas9 DNA表达框转入并整合到植物基因组中，进而发挥功能对目的基因进行编辑。尽管通过后代分离可获得无转基因成分的植物材料，但依然存在外源DNA污染的风险。韩国首尔大学（Woo et al.，2015）利用原生质体转化方法，将Cas9蛋白和gRNA在体外组装成的核糖核蛋白复合体（RNP）转入原生质体细胞，成功对目的基因进行了编辑，获得了无外源DNA的基因组编辑生菜植株；同样，利用基因枪法将RNP转入小麦和玉米细胞，中国科学院遗传与发育生物学研究所、杜邦先锋公司分别成功建立了全程无外源DNA的基因组编辑体系（Liang et al.，2017；Svitashev et al.，2016）。瑞典农业大学利用原生质体转化方法，成功将RNP转入马铃薯细胞，获得突变体材料（Andersson et al.，2018）。这一利用RNP实现基因组编辑的方法有助于最大限度地减少监管，建立起精准、生物安全的新一代育种技术体系，加快作物基因组编辑育种产业化进程。

1.3 单碱基编辑系统

核苷酸点突变是作物许多重要农艺性状发生变异的遗传基础。很多时候科研人员希望通过引入单碱基突变造成氨基酸的变化以实现基因功能的改变，而不是单纯地敲除基因功能。由于目前同源重组介导的基因组编辑效率较低，特异性单碱基突变编辑系统的研究就十分必要。单碱基编辑技术（base editor）是基于 CRISPR 系统的新型靶基因定点修饰技术，在不产生 DNA 双链断裂的情况下，利用胞嘧啶脱氨酶或人工进化的腺嘌呤脱氨酶等对靶位点进行精准的单碱基编辑，从而实现单碱基的替换。

美国哈佛大学 David Liu 实验室以 CRISPR/Cas9 系统为基础，融合大鼠胞嘧啶脱氨酶（rAPOBEC1）和尿嘧啶糖基化酶抑制剂（UGI），构成了高效的胞嘧啶单碱基编辑系统 CBE（cytosine base editor），在人类细胞系和小鼠细胞系中，成功将特定位置的碱基胞嘧啶（C）转化为胸腺嘧啶（T），实现了定点胞嘧啶单碱基编辑（Komor et al., 2016）。Nishida 等（2016）通过将 nCas9-D10A 与七鳃鳗（sea lamprey）的胞苷脱氨酶（activation induced cytidine deaminase, AID）融合表达，也成功实现了 C·G 碱基对向 T·A 碱基对的转换。2017 年，David Liu 实验室通过利用 Cas9 变体（nCas9-D10A）融合大肠杆菌野生型腺嘌呤脱氨酶（ecTadA）和人工定向进化的腺嘌呤脱氨酶（ecTadA*）二聚体，构建了腺嘌呤单碱基编辑系统 ABE（adenine base editor），成功实现了靶位点 A·T 碱基对转变成 G·C 碱基对（Gaudelli et al., 2017）。

单碱基编辑系统很快被成功的应用到植物中。多个实验室先后报道单碱基编辑系统可以在水稻、小麦、玉米、番茄等植物中实现 C·T 或 A·G 的替换（Chen et al., 2017; Hua et al., 2018; Li et al., 2017, 2018a; Lu and Zhu, 2017; Ren et al., 2017; Shimatani et al., 2017; Yan et al., 2018; Zong et al., 2017）。中国科学院遗传与发育生物学研究所还通过利用 Cas9 变体（nCas9-D10A）融合人类胞嘧啶脱氨酶 APOBEC3A（A3A）和 UGI，构成新的单碱基编辑系统 A3A-PBE，成功在小麦、水稻及马铃薯中实现更加高效、范围更广的 C·T 单碱基编辑（Zong et al., 2018）。单碱基编辑系统在植物中的成功建立和应用，为高效和大规模创制单碱基突变体提供了一个可靠方案，为作物遗传改良和新品种培育提供了重要技术支撑。

1.4 基因组编辑技术在作物育种中的应用

基因组编辑系统可以大大加快作物育种进程。与转基因育种不同，通过基因组编辑技术得到的品种没有引入外源基因，与常规诱变获得的产品类似。可以预见，基因组编辑技术在作物育种上具有十分广阔的应用前景。

2014 年，美国 Calyxt 公司利用 TALEN 在大豆中对 2 个脂肪酸脱氢酶基因 (*FAD2-1A* 和 *FAD2-1B*) 的 4 个等位基因进行编辑，成功培育了高油酸大豆品系 (Haun et al.，2014)。2015 年，中国科学院遗传与发育生物学研究所和微生物研究所合作，利用 TALEN 技术在六倍体小麦中对 3 个 *MLO* 基因拷贝同时进行了编辑，获得了对白粉菌具有广谱抗性的小麦品种 (Wang et al.，2014)。中国科学院遗传与发育生物学研究所还利用 TALEN 技术，在水稻中成功对米香控制基因 *BADH2* 进行了编辑，使普通稻米具有香味品质 (Shan et al.，2015)。

2016 年，华南农业大学在水稻中通过 CRISPR/Cas9 系统敲除温敏型雄性不育基因 (*TGMS*)，开发了 11 个新的雄性不育品系，仅在一年内就可以应用于两个水稻亚种的杂交育种，显著加快了不育系的培育过程 (Zhou et al.，2016)。湖南杂交水稻研究中心利用基因组编辑技术，研制出高产、优质、适于重度镉污染稻田种植的低镉水稻新品种'两优低镉 1 号'(Tang et al.，2017)。

美国冷泉港实验室在番茄中通过 CRISPR/Cas9 在调控花器官及控制分支形成的启动子区域产生了许多自然界原先不存在的新的遗传变异，使数量性状能以微调方式进行控制 (Rodriguez-Leal et al.，2017)。中国科学院遗传与发育生物学研究所在生菜中通过 CRISPR/Cas9 突变维生素 C 合成途径中关键基因上游的 uORF，使生菜叶片中维生素 C 含量提高约 150% (Zhang et al.，2018)。这些策略为创造新的农作物种质资源、加快作物改良提供了新的思路。

2018 年 10 月，中国科学院遗传与发育生物学研究所 (Li et al.，2018b) 及巴西圣保罗大学 (Zsogon et al.，2018) 分别选用野生醋栗番茄为基础材料，运用基因组编辑技术精准靶向多个产量和品质性状控制基因，将产量和品质性状精准导入野生番茄，加速了野生植物的人工驯化。该研究首次通过基因组编辑实现野生植物的快速驯化，为精准设计和创造全新作物提供了新的策略。

为了在未来的生物种业中抢夺先机，美国、阿根廷、巴西等国家已经明确宣布基因组编辑的作物不在转基因立法管辖范围之内。自2012年起，美国农业部已对ZFN敲除的低植酸玉米、TALEN创制高油酸大豆、耐低温储存提高加工品质的马铃薯、抗白粉病小麦，以及CRISPR/Cas9敲除的抗褐化双孢菇、高支链糯玉米、抗白粉病小麦、高油酸亚麻荠、晚花狗尾草等20余基因组编辑的农作物相继下达了转基因监管豁免权（Wolt, et al., 2016；Waltz, 2018）。2019年2月26日，Calyxt公司宣布TALEN敲除的高油酸大豆正式上市（http://www.calyxt.com/first-commercial-sale-of-calyxt-high-oleic-soybean-oil-on-the-u-s-market/），基因组编辑农产品正式走上了人们的餐桌，基因组编辑育种应用的产业化步伐越来越近了。

2 针对未来作物分子设计需求的研究内容与突破口

围绕实现未来作物设计需求，基因组编辑的主要研究内容和突破口在于开发具有自主知识产权的新型基因组编辑工具，研发高效精准安全的基因组编辑技术、基因时空量精准调控的新型衍生技术、精准基因定点替换和定点插入等基因组靶向修饰技术、基因组编辑工具高效导入植物细胞的方法及高通量基因组编辑技术等各种技术体系，实现对作物基因组的精准编辑及对基因表达时空量三维精准调控，为未来作物分子设计奠定技术基础，进而利用建立的基因组编辑技术，建立未来作物分子设计育种的技术框架与技术平台，创制未来作物。

2.1 开发具有自主知识产权的新型基因组编辑工具

从多种微生物中大规模筛选具有靶向功能的新型核酸免疫系统，挖掘新型Cas酶和核酸内切酶，开发能够替代现有技术，甚至是取代现有技术的颠覆式的创新的基因组编辑工具。结合系统生物学、计算生物学和结构生物学的人工设计途径，探寻建立新的基因组编辑工具，加强对现有CRISPR系统的分子进化和性能改造，获得具有自主知识产权的基因组编辑工具。

2.2 研发高效、精准、安全的基因组编辑技术

优化、提升现有的编辑技术体系，针对高保真、低脱靶为目标突破现有工具的局限，在精准性、安全性、稳定性、适用性等方面不断升级和发展完善，建立识别靶点更广泛、更高效、精准、安全的基因组编辑技术体系，提升植物基因组编辑工具安全性。同时在各种农作物中，建立DNA-free的基因定点敲除、单碱基编辑和基因替换等新型编辑技术体系，搭建生物安全的新型编辑农产品应用转化技术，为基因组编辑农产品应用铺平道路。

2.3 建立精准的基因组靶向修饰技术

一方面，继续挖掘新的单碱基编辑工具，优化现有CBE和ABE编辑器，在编辑效率、编辑窗口扩大或缩小等不同方向进行优化，建立农作物中广泛应用的单碱基编辑系统，实现基因精准定点替换。另一方面，深入研究相关的DNA损伤修复机理，通过对基因修饰酶的改造、优化，从同源重组途径入手多途径攻关，攻克基因定点替换、插入等植物基因组编辑技术难题，综合建立精准的基因组靶向修饰技术。

2.4 开发基因组编辑工具高效导入植物细胞的方法

基因组编辑工具的高效导入是实现植物基因组编辑的关键。在不同作物中，依托农杆菌、基因枪等传统遗传转化效率差异较大，而且需要经历周期长和效率低的组织培养的过程，不利于基因组编辑工具高效地进入植物细胞发挥作用。因此，亟待挖掘（如纳米材料递送法等）各种新材料递送体系、开发不依赖于基因型的编辑系统导入新方法，建立不依赖于组织培养和再生的植物基因组编辑的新方法，简化基因组编辑流程，突破不同基因型的限制，开发基因组编辑工具高效导入植物细胞的方法。

2.5 建立基因的时空量三维精准调控技术

充分拓展已有或新建的基因组编辑工具性能，开发各种新型转录调控因子模块、表观遗传调控模块、DNA或RNA标记模块、同源重组模块、染色

体结构调控模块，通过合理组合及与现有或新研发工具的融合，建立起各种衍生技术，结合常规编辑技术一起整合形成通过精准的条件控制表达调控技术，用以进行环境智能时间空间特意转录响应回路创制、基因组 3D 结构重编程等应用，满足不同层次的编辑需求，实现作物基因的时空量三维精准调控技术体系的建立。

2.6 建立高通量基因组编辑技术平台

利用优化的编辑技术及研发出的新型基因组编辑工具，建立相应的植物高通量基因替换、敲除、激活、抑制和单碱基编辑等不同技术体系；对不同的作物建立相应的全基因组层面定点敲除、激活、抑制等突变体高通量技术平台，以及全基因组饱和点突变技术平台，加强大规模实现控制作物重要农艺性状的关键基因功能及调控网络机制的解析及遗传性状的改良；重点推进对全基因组层面的基因时空量三维精准调控网络平台的建设。

2.7 建立未来作物分子设计育种的技术框架与技术平台

结合生物信息学、功能基因组学和现代育种学等多学科交叉，利用大数据分析、基因组技术和精准基因组编辑技术等与常规育种技术整合，形成未来作物分子设计技术新体系，建立不同作物适用性基因组编辑育种技术框架与技术平台。

3 阶段性目标

3.1 2035 年目标

本阶段的目标重点是抢占源头创新，主要实现以下 5 个目标。

第一，针对基因组编辑技术中存在的难题，以及我国缺乏基因组编辑的基础核心专利，挖掘一批具有靶向功能的新型核酸免疫系统并揭示其工作机制，开发具自主知识产权的新一代基因组编辑技术，形成可取代现有技术的突破性成果，获得源头创新的一批基因组编辑新工具。加强对现有 CRISPR 系统的分子进化和性能改造，获得一批更高效、精准、作用范围更广的具有自主知识产权的升级版基因组编辑工具。

第二，优化提升现有编辑技术，建立多种高效、精准、安全的基因组编辑系统，获得一批更高效、精准、安全的基因组编辑实用性工具。在水稻、玉米、小麦、大豆、油菜等多种重要粮食作物、经济作物中建立 DNA-free 的基因定点敲除、单碱基编辑和基因替换等多个新型编辑技术。

第三，获得可实现 A、C、T、G 4 个碱基的自由改换的多种新型单碱基编辑器，建立多种精准基因定点替换的植物单碱基编辑技术体系。解析 DNA 损伤修复机理，建立同源重组修复途径介导的多种基因靶向修饰技术，实现植物基因组的精准读写编辑。

第四，在水稻、玉米、小麦、大豆、油菜等重要粮食作物和经济作物中建立多种不依赖基因型、不依赖组织培养和再生的新方法，获得多种能在不同植物中高效导入基因组编辑工具的方法。

第五，建立多种新型基因组编辑衍生技术，拓展基因组编辑应用范围。在水稻、玉米、小麦、大豆、油菜等重要粮食作物和经济作物中构建多种全基因组、高通量编辑技术平台，获得多种重要农作物全基因组编辑种质资源库。

3.2 2050 年目标

本阶段的目标集中在未来作物分子设计育种的技术框架与技术平台的建立，主要实现以下 3 个目标。

第一，利用具有自主知识产权的基因组编辑新工具，全方位建设建立源头创新的基因时空量三维精准调控技术，实现真正意义上的人工定义的环境智能时间空间特意转录响应回路创制、基因组 3D 结构重编程等精准调控应用，达到未来作物的创制技术标准要求。

第二，实现基因组编辑与合成生物学等多学科整合，搭建未来作物分子设计育种的技术框架与产出平台，建立水稻、玉米、小麦、大豆、油菜等多种作物的重要农艺性状调控网络关键基因时空量三维精准调控种质资源库。综合基因组技术、基因组编辑技术、大数据处理等技术并结合常规育种方法，建设形成未来作物分子设计育种技术框架与技术平台。

第三，整合多个领域的研究成果，在相关基础分子机制的深入解析的基础上，设计未来极端环境，快速获得应对极端环境的突变类型及变异，绘制未来作物蓝图；利用未来作物分子设计育种的技术框架与技术平台，创制获得未来作物。

致谢：本章在撰写过程中得到了陈坤玲博士的协助，特此致谢！

参考文献

Andersson, M., Turesson, H., Olsson, N., Falt, A.S., Olsson, P., Gonzalez, M.N., Samuelsson, M., and Hofvander, P. (2018). Genome editing in potato via CRISPR-Cas9 ribonucleoprotein delivery. Physiol. Plant. *164*, 378-384.

Chen, Y., Wang, Z., Ni, H., Xu, Y., Chen, Q., and Jiang, L. (2017). CRISPR/Cas9-mediated base-editing system efficiently generates gain-of-function mutations in *Arabidopsis*. Sci. China Life Sci. *5*, 520-523.

Cong, L., Ran, F.A., Cox, D., Lin, S., Barretto, R., Habib, N., Hsu, P.D., Wu, X., Jiang, W., Marraffini, L.A., et al. (2013). Multiplex genome engineering using CRISPR/Cas systems. Science *339*, 819-823.

Gaudelli, N.M., Komor, A.C., Rees, H.A., Packer, M.S., Badran, A.H., Bryson, D.I., and Liu, D.R. (2017). Programmable base editing of A·T to G·C in genomic DNA without DNA cleavage. Nature *551*, 464-471.

Haun, W., Coffman, A., Clasen, B.M., Demorest, Z.L., Lowy, A., Ray, E., Retterath, A., Stoddard, T., Juillerat, A., Cedrone, F., et al. (2014). Improved soybean oil quality by targeted mutagenesis of the fatty acid desaturase 2 gene family. Plant Biotechnol. J. *12*, 934-940.

Hilscher, J., Burstmayr, H., and Stoger, E. (2016). Targeted modification of plant genomes for precision crop breeding. Biotechnol. J. *12*, 1600173.

Hilton, I.B., D'Ippolito, A.M., Vockley, C.M., Thakore, P.I., Crawford, G.E., Reddy, T.E., and Gersbach, C.A. (2015). Epigenome editing by a CRISPR-Cas9-based acetyltransferase activates genes from promoters and enhancers. Nat. Biotechnol. *33*, 510-517.

Hu, X., Wang, C., Liu, Q., Fu, Y., and Wang, K. (2017). Targeted mutagenesis in rice using CRISPR-Cpf1 system. J. Genet. Genomics *44*, 71-73.

Hua, K., Tao, X., Yuan, F., Wang, D., and Zhu, J.K. (2018). Precise A·T to G·C base editing in the rice genome. Mol. Plant *11*, 627-630.

Jinek, M., Chylinski, K., Fonfara, I., Hauer, M., Doudna, J.A., and Charpentier, E. (2012). A programmable dual-RNA-guided DNA endonuclease in adaptive bacterial immunity. Science *337*, 816-821.

Kearns, N.A., Pham, H., Tabak, B., Genga, R.M., Silverstein, N.J., Garber, M., and Maehr, R. (2015). Functional annotation of native enhancers with a Cas9-histone demethylase fusion. Nat. Methods *12*, 401-403.

Komor, A.C., Kim, Y.B., Packer, M.S., Zuris, J.A., and Liu, D.R. (2016). Programmable editing of a target base in genomic DNA without double-stranded DNA cleavage. Nature *533*, 420-424.

Li, C., Zong, Y., Wang, Y., Jin, S., Zhang, D., Song, Q., Zhang, R., and Gao, C. (2018a). Expanded base editing in rice and wheat using a Cas9-adenosine deaminase fusion. Genome Biol. *19*, 59.

Li, J., Sun, Y., Du, J., Zhao, Y., and Xia, L. (2017). Generation of targeted point mutations in rice by a modified CRISPR/Cas9 system. Mol. Plant *10*, 526-529.

Li, J.F., Norville, J.E., Aach, J., McCormack, M., Zhang, D., Bush, J., Church, G.M., and Sheen, J. (2013). Multiplex and homologous recombination-mediated genome editing in *Arabidopsis* and *Nicotiana benthamiana* using guide RNA and Cas9. Nat. Biotechnol. *31*, 688-691.

Li, T., Yang, X., Yu, Y., Si, X., Zhai, X., Zhang, H., Dong, W., Gao, C., and Xu, C. (2018b). Domestication of wild tomato is accelerated by genome editing. Nat. Biotechnol. *36*, 1160-1163.

Liang, Z., Chen, K., Li, T., Zhang, Y., Wang, Y., Zhao, Q., Liu, J., Zhang, H., Liu, C., Ran, Y., et al. (2017). Efficient DNA-free genome editing of bread wheat using CRISPR/Cas9 ribonucleoprotein complexes. Nat. Commun. *8*, 14261.

Lowder, L.G., Zhou, J., Zhang, Y., Malzahn, A., Zhong, Z., Hsieh, T.F., Voytas, D.F., Zhang, Y., and Qi, Y. (2018). Robust transcriptional activation in plants using multiplexed CRISPR-Act2.0 and mTALE-Act systems. Mol. Plant *11*, 245-256.

Lu, Y., and Zhu, J.K. (2017). Precise editing of a target base in the rice genome using a modified CRISPR/Cas9 system. Mol. Plant *10*, 523-525.

Ma, X., Zhang, Q., Zhu, Q., Liu, W., Chen, Y., Qiu, R., Wang, B., Yang, Z., Li, H., Lin, Y., et al. (2015). A robust CRISPR/Cas9 system for convenient, high-efficiency multiplex genome editing in monocot and dicot plants. Mol. Plant *8*, 1274-1284.

Nekrasov, V., Staskawicz, B., Weigel, D., Jones, J.D., and Kamoun, S. (2013). Targeted mutagenesis in the model plant *Nicotiana benthamiana* using Cas9 RNA-guided endonuclease. Nat. Biotechnol. *31*, 691-693.

Nishida, K., Arazoe, T., Yachie, N., Banno, S., Kakimoto, M., Tabata, M., Mochizuki, M., Miyabe, A., Araki, M., Hara, K.Y., et al. (2016). Targeted nucleotide editing using hybrid prokaryotic and vertebrate adaptive immune systems. Science *353*, aaf8729.

Piatek, A., Ali, Z., Baazim, H., Li, L., Abulfaraj, A., Al-Shareef, S., Aouida, M., and Mahfouz, M.M. (2015). RNA-guided transcriptional regulation in planta via synthetic dCas9-based transcription factors. Plant Biotechnol. J. *13*, 578-589.

Ren, B., Yan, F., Kuang, Y., Li, N., Zhang, D., Lin, H., and Zhou, H. (2017). A CRISPR/Cas9 toolkit for efficient targeted base editing to induce genetic variations in rice. Sci. China Life Sci. *60*, 516-519.

Rodriguez-Leal, D., Lemmon, Z.H., Man, J., Bartlett, M.E., and Lippman, Z.B. (2017). Engineering quantitative trait variation for crop improvement by genome editing. Cell *171*, 470.

Sander, J.D., and Joung, J.K. (2014). CRISPR-Cas systems for editing, regulating and targeting genomes. Nat. Biotechnol. *32*, 347-355.

Shan, Q., Wang, Y., Li, J., Zhang, Y., Chen, K., Liang, Z., Zhang, K., Liu, J., Xi, J.J., Qiu, J.L., et al. (2013). Targeted genome modification of crop plants using a CRISPR-Cas system. Nat. Biotechnol. *31*, 686-688.

Shan, Q., Zhang, Y., Chen, K., Zhang, K., and Gao, C. (2015). Creation of fragrant rice by targeted knockout of the *OsBADH2* gene using TALEN technology. Plant Biotechnol. J. *13*, 791-800.

Shen, L., Hua, Y., Fu, Y., Li, J., Liu, Q., Jiao, X., Xin, G., Wang, J., Wang, X., Yan, C., et al. (2017). Rapid generation of genetic diversity by multiplex CRISPR/Cas9 genome editing in rice. Sci. China Life Sci. *60*, 506-515.

Shimatani, Z., Kashojiya, S., Takayama, M., Terada, R., Arazoe, T., Ishii, H., Teramura, H.,

Yamamoto, T., Komatsu, H., Miura, K., et al. (2017). Targeted base editing in rice and tomato using a CRISPR-Cas9 cytidine deaminase fusion. Nat. Biotechnol. *35*, 441-443.

Svitashev, S., Schwartz, C., Lenderts, B., Young, J.K., and Mark Cigan, A. (2016). Genome editing in maize directed by CRISPR-Cas9 ribonucleoprotein complexes. Nat. Commun. *7*, 13274.

Tang, L., Mao, B., Li, Y., Lv, Q., Zhang, L., Chen, C., He, H., Wang, W., Zeng, X., Shao, Y., et al. (2017). Knockout of OsNramp5 using the CRISPR/Cas9 system produces low Cd-accumulating indica rice without compromising yield. Sci. Rep. *7*, 14438.

Waltz, E. (2018). With a free pass, CRISPR-edited plants reach market in record time. Nat. Biotechnol. *36*, 6-7.

Wang, C., Shen, L., Fu, Y., Yan, C., and Wang, K. (2015). A simple CRISPR/Cas9 system for multiplex genome editing in rice. J. Genet. Genomics *42*, 703-706.

Wang, Y., Cheng, X., Shan, Q., Zhang, Y., Liu, J., Gao, C., and Qiu, J.L. (2014). Simultaneous editing of three homoeoalleles in hexaploid bread wheat confers heritable resistance to powdery mildew. Nat. Biotechnol. *32*, 947-951.

Wolt, J.D., Wang, K., and Yang, B. (2016). The regulatory status of genome-edited crops. Plant Biotechnol. J. *14*, 510-518.

Woo, J.W., Kim, J., Kwon, S.I., Corvalan, C., Cho, S.W., Kim, H., Kim, S.G., Kim, S.T., Choe, S., and Kim, J.S. (2015). DNA-free genome editing in plants with preassembled CRISPR-Cas9 ribonucleoproteins. Nat. Biotechnol. *33*, 1162-1164.

Xie, K., Minkenberg, B., and Yang, Y. (2015). Boosting CRISPR/Cas9 multiplex editing capability with the endogenous tRNA-processing system. Proc. Natl. Acad. Sci. USA *112*, 3570-3575.

Xing, H.L., Dong, L., Wang, Z.P., Zhang, H.Y., Han, C.Y., Liu, B., Wang, X.C., and Chen, Q.J. (2014). A CRISPR/Cas9 toolkit for multiplex genome editing in plants. BMC Plant Biol. *14*, 327.

Xu, R., Qin, R., Li, H., Li, D., Li, L., Wei, P., and Yang, J. (2017). Generation of targeted mutant rice using a CRISPR-Cpf1 system. Plant Biotechnol. J. *15*, 713-717.

Yan, F., Kuang, Y., Ren, B., Wang, J., Zhang, D., Lin, H., Yang, B., Zhou, X., and Zhou, H. (2018). Highly efficient A·T to G·C base editing by Cas9n-guided tRNA adenosine deaminase in rice. Mol. Plant *11*, 631-634.

Zetsche, B., Gootenberg, J.S., Abudayyeh, O.O., Slaymaker, I.M., Makarova, K.S., Essletzbichler, P., Volz, S.E., Joung, J., van der Oost, J., Regev, A., et al. (2015). Cpf1 is a single RNA-guided endonuclease of a class 2 CRISPR-Cas system. Cell *163*, 759-771.

Zhang, H., Si, X., Ji, X., Fan, R., Liu, J., Chen, K., Wang, D., and Gao, C. (2018). Genome editing of upstream open reading frames enables translational control in plants. Nat. Biotechnol. *36*, 894-898.

Zhou, H., He, M., Li, J., Chen, L., Huang, Z., Zheng, S., Zhu, L., Ni, E., Jiang, D., Zhao, B., et al. (2016). Development of commercial thermo-sensitive genic male sterile rice accelerates hybrid rice breeding using the CRISPR/Cas9-mediated TMS5 editing system. Sci. Rep. *6*, 37395.

Zong, Y., Song, Q., Li, C., Jin, S., Zhang, D., Wang, Y., Qiu, J.L., and Gao, C. (2018). Efficient C-to-T base editing in plants using a fusion of nCas9 and human APOBEC3A. Nat. Biotechnol. *36*, 950-953.

Zong, Y., Wang, Y., Li, C., Zhang, R., Chen, K., Ran, Y., Qiu, J.L., Wang, D., and Gao, C. (2017). Precise base editing in rice, wheat and maize with a Cas9- cytidine deaminase fusion. Nat. Biotechnol. *35*, 438-440.

Zsogon, A., Cermak, T., Naves, E.R., Notini, M.M., Edel, K.H., Weinl, S., Freschi, L., Voytas, D.F., Kudla, J., and Peres, L.E.P. (2018). *De novo* domestication of wild tomato using genome editing. Nat. Biotechnol. *36*, 1211-1216.

驯化与多倍体育种

黄三文　韩方普　王汉中

1 作物的驯化

1.1 国内外研究进展

人类文明的诞生和发展离不开对主要农作物的驯化和改良，长期的选择使这些作物脱胎换骨，跟它们的野生祖先差异巨大，同时成就了人类社会今日的繁荣。了解作物驯化的过程和遗传基础，不仅可以加深我们对育种历史的理解，还能够为我们正在进行的育种和改良提供指导，从而提高作物的产量和品质，丰富食物的种类，满足人类社会对粮食和果蔬不断增长的物质需求。

1.1.1 作物的驯化历史

大约在一万年前，人类就开始对水稻、小麦、玉米等粮食作物的野生种进行栽培和驯化。大约在4000年前，人类便完成了这些主要作物的驯化，使野生种变成了栽培种（Doebley et al., 2006）。在这一过程中，作物的形态、生长习性、生理状态都发生了极其显著的变化。概括来讲，驯化后的作物与其野生种相比，具有分枝减少、顶端优势增强、长势和开花期及成熟期趋于一致、产品器官变大、不落粒、休眠期变短和抗病抗逆能力变弱等特征，这些特征被称为"驯化综合征"（Doebley et al., 2006; Olsen and Wendel, 2013）。

作物的驯化最早发生在其野生祖先的起源地。粳稻最早是从中国南方珠江流域的普通野生稻（*Oryza rufipogon*）驯化而来，籼稻则是由粳稻与东南亚及南亚的野生亲本种杂交后产生的（Huang et al., 2012）。小麦的祖先种一粒小麦（*Triticum monococcum*）和野生二粒小麦（*Triticum araraticum*）都起源于中亚的"新月沃土"区域，大约在一万年前完成了驯化（Heun et al.,

1997），随后它们被传播到了欧洲一带。玉米起源于南美洲的墨西哥一带，其野生祖先是大刍草（*Zea mays* ssp. *parviglumis*），约在 6300 年之前玉米便完成了驯化（Matsuoka et al.，2002；Piperno and Flannery，2001）。甘蓝型油菜（*Brassica napus*）的原始祖先起源于欧洲南部地中海一带的温暖地区，约在 5000 年前由白菜和甘蓝二倍体杂交自然加倍而来，随后被人类利用和驯化，形成了适应不同种植区域和生态地理环境的冬性、半冬性和春性油菜，并经历了双低、抗病、高产、高油和品质等性状的驯化和改良（Chalhoub et al.，2014；Sun et al.，2017）。马铃薯的野生种起源于南美洲，约在 8000 年前完成了野生种到地方栽培种的驯化，栽培种的直接驯化祖先是 *Solanum candolleanum*，最主要的一个驯化地是位于秘鲁和玻利维亚边界的的的喀喀湖（Titicaca）区域。驯化完成的四倍体马铃薯，在 16 世纪被传播到了欧洲和北美洲（Spooner et al.，2005）。番茄起源于南美洲的安第斯山脉，其直接驯化祖先是醋栗番茄（*Solanum pimpinellifolium*）。番茄在南美洲便完成驯化，果实的大小由 1～2 g 变为十几克的樱桃番茄，最后驯化为大果栽培番茄。番茄大约在 16 世纪传到了欧洲，随后传播到世界各地，在这个过程中由于变异的产生和不同的选择，番茄的大小、形状、颜色都产生了丰富分化，例如，因为生产用途的不同产生了鲜食和加工番茄两大类型（Lin et al.，2014）。黄瓜源自喜马拉雅山脉南麓，本是印度境内一种味苦、可入药的植物，其完成驯化之后在不同的地理环境下形成了欧亚类群、东亚类群和西双版纳类群。在我国，黄瓜经"丝绸之路"由西域传入中国，在不同的地域环境中产生了华南和华北两大类型（Qi et al.，2013）。

1.1.2 作物驯化的遗传学和基因组学基础

近代遗传学、分子生物学、基因组学的发展极大地提高了性状遗传解析的速度，许多控制重要驯化性状的基因得到了解析。例如，植物学性状基因包括控制水稻脱粒的位点 *sh4*（Li et al.，2006）、控制水稻匍匐/直立的关键基因 *PROG1*（Tan et al.，2008）、控制玉米分枝性和涉及顶端优势的基因 *teosinte branched 1*（Doebley et al.，1997）、控制油菜黄籽性状的基因（Hong et al.，2017）等；产品器官性状基因包括，控制番茄果实大小的位点 *fw2.2*（Frary et al.，2000）和 *fw3.2*（Chakrabarti et al.，2013）、控制油菜千粒重的位点（Li et al.，2018；Liu et al.，2015）和每角粒数的位点（Li et al.，2015）以及单株角果数的位点（Raboanatahiry et al.，2018）等；重要品质性状基因包括，控制黄瓜、西瓜和甜瓜苦味合成的基因簇（Shang et al.，2014；Zhou et al.，2016）、土豆和番茄中控制茄碱含量的基因和基因簇（Itkin et al.，2013；Zhu et al.，2018）。

作物驯化满足了人类对食物和农业生产需求，但同时也对作物本身的群体和其基因组产生了重要影响。作物的野生群体具有丰富的遗传变异，但因为人类在早期的驯化中往往只关注特定的性状，在长期的定向选择和淘汰后，作物群体内部性状趋于单一化，群体遗传变异急剧减少，导致全基因组范围的遗传多样性降低，即所谓的"瓶颈效应"（Tang et al.，2010）。不同的作物在驯化过程中经历了不同程度的"瓶颈效应"，如早期驯化的玉米与野生种大刍草相比，大约1200个基因受到驯化选择，而现代栽培种玉米的遗传多样性只占野生群体的57%（Wright，2005）。据估计，栽培番茄的遗传多样性仅占野生群体遗传多样性的50%（Miller and Tanksley，1990）。根据基因组同源区域的群体遗传多样性评估，甘蓝型油菜只有其二倍体亲本白菜和甘蓝的一半，其遗传多样性在物种形成和随后的驯化过程中变得狭窄（Schmutzer et al.，2015）。在驯化中，对控制某一特定目标性状基因区域的选择，使该区域和其相邻区间的遗传多样性显著降低，这一现象被称为"选择性清除"（selective sweep）。"瓶颈效应"和"选择性清除"是导致栽培品种遗传多样性显著降低的两个重要原因。遗传多样性的丧失导致了现代栽培品种遗传基础狭窄，限制了作物的进一步改良和品种的创新。

1.1.3 现代育种改良中存在的问题

近几十年的商业育种中，由于过度追求产量、抗病性、耐储运、外观等商品品质，导致许多作物的风味和营养品质显著下降。研究发现，现代番茄品种和早期的地方品种、农家种相比较，十几种主要的风味物质的成分显著下降，且这种表型的分化是源于对基因组的定向选择所导致的（Tieman et al.，2017）。在育种中，为了进一步提高品种的优良性状，野生资源被广泛地利用。例如，将野生水稻与栽培水稻杂交，向栽培种中导入许多具有优良性状的数量性状位点（QTL）（McCouch et al.，2007）；将野生大刍草与玉米优良自交系进行杂交并且反复回交得到在开花期、行粒数和粒重等性状上具有良好品质的可用于育种的近等基因系（Liu et al.，2016）；为了提高番茄抗病性，广泛利用了野生资源的抗病基因（如番茄花叶病毒抗病基因 *Tm-2*、根结线虫抗病基因 *Mi-1*）。油菜育种过程中也曾和二倍体白菜、甘蓝和黑芥等近缘种杂交，改良油菜的生育期、开花时间、菌核病和根肿病抗性等（Chatterjee et al.，2016；Wei et al.，2016；Zhan et al.，2017）。然而，基于基因组学分析发现，在目标抗性基因导入的同时，也渐渗了大片野生材料的其他"不良性状"基因，不仅导致了连锁累赘，同时也限制了此区域的进一步改良（Lin et al.，2014）。

1.2 研究内容和突破口

粮安天下，食物是人类最基本、最重要的需求。尽管在过去的 50 年中世界农业取得了巨大的进步，然而世界人口的不断增长、全球气候的不稳定性及人类不断增长的物质需求，都对农业的发展提出了更高的要求。作物的驯化和育种历史的解析，不仅让我们积累了过去成功经验，而且发现了过去育种实践的不足和缺陷，这为我们今后和未来的作物再驯化（re-domestication）提供了重要的指引。今天的育种应该是多样性的、进一步提升生产潜力的精准改良，突破点包括以下几个方面。

加强资源的收集和利用，丰富遗传变异。作物长期的驯化，导致群体内部同质化严重，多样性降低，要提高作物的生产潜力，就必须扩大作物的基因库。种质资源是育种改良的重要物质前提，也是世界各国都极为重视的研究领域，还被一些国家定为战略物资，受到国家的严格管控。材料的收集和利用是打破现有品种遗传瓶颈，实现品种更新的关键环节。例如，太谷核不育小麦是我国科技工作者发现的一个雄蕊完全败育的种质材料，我国育种家将育性基因和矮秆基因进行聚合培育了矮败小麦，利用这个材料培育了一系列小麦优良品种，累计增产几十亿千克（翟虎渠和刘秉华，2009）。以'中油 821'和'秦油 2 号'为骨干亲本，先后育种获得中双系列、阳光系列和秦油系列等油菜优异衍生系和品种达 30 多个，其中多个品种的推广成果获得国家发明奖和进步奖。野生资源也是育种的宝贵资源库，可溶性固形物对番茄品种有着重要的影响，其含量高低直接影响番茄的产量和品质。在番茄加工过程中，可溶性固形物含量每增加1%，最终番茄酱的产量就会增加25%。Fridman 等（2004）将含有野生番茄基因 *brix-9* 的一个渐渗系应用到栽培种番茄中，野生型基因编码的酶分子对蔗糖具有明显的亲和性，因此培育出的番茄具有较高含量的可溶性固形物。野生番茄基因 *brix-9* 已经被成功应用到优良加工番茄的育种中（Zamir，2008）。

　　加速性状的遗传解析，为全基因组分子设计育种奠定基础。高通量测序技术的快速发展，产生了大量群体基因组数据，高密度分子图谱极大地提高了作图和定位的精度，组学方法和技术可以快速定位和挖掘重要农艺性状的基因。近年来，借助基因组学发展起来的全基因组关联分析（GWAS）已经成为高通量性状遗传解析的有效工具，在作物和蔬菜的研究中已经体现其优势。通过对 1500 份水稻材料的全基因组重测序和关联分析，发现了几十个与谷粒性状、花期相关的性状的主效基因位点（Huang and Han，2014）。玉米籽粒油分含量是非常重要的农艺性状，通过对 368 份玉米材料的关联分析，鉴定到 74 个籽粒含油量及脂肪酸合成相关的位点，为培育含油量高的优质玉米提供了遗传基础（Li et al.，2013a）。通过对甘蓝型油菜群体转录组测序和关联分析，先后鉴定与产量、含油量、品种、开花时间等性状相关遗传位点 100 多个，为针对性的开发标记和开展全基因组选择育种提供了条件（Harper et al.，2012；Li et al.，2014；Smooker et al.，2011）深入解析重要农艺性状的遗传基础，可为全基因组分子设计育种提供大量的分子标记。

　　马铃薯是重要的快茎类粮食作物，四倍体马铃薯最早被带到欧洲和北美洲，随后的马铃薯生产和育种都是在四倍体水平进行，导致马铃薯产业面临两个结构性障碍。第一个障碍是四倍体的遗传非常复杂，导致马铃薯育种周期长，品种更新慢；第二个障碍是马铃薯以薯块进行繁殖，存在繁殖系数低、储运成本高、易携带病虫害等缺陷。基于对四倍体基因组的分析发现，马铃薯基因组中携带大量的有害突变，4 个等位基因并非都发挥功能，这为二倍体设计育种的可行性提供了重要理论指导。用二倍体替代四倍体，并用杂交种子替代薯块是马铃薯再驯化，是实现马铃薯产业"绿色革命"的有效途径。实现马铃薯的二倍体育种，首先需要解决的是自交不亲和与自交衰退问题。最近，基因组编辑技术为解决二倍体马铃薯自交不亲和的问题提供了重要方案（Ye et al.，2018），下一步将利用基因组学深入解析自交衰退的遗传机理和控制位点，清除有害突变，培育自交系，实现二倍体杂交马铃薯计划。

重视生物技术，用于精准育种。杂交是组合优良性状最为常用的方法，通过杂交和反复回交可以实现目标基因和性状的导入，但这种方式很难实现目标基因的精准导入，而且往往产生连锁累赘。例如，在玉米和小麦中，由于着丝粒附近的重复区间甲基化程度高，重组事件严重被抑制，这些区间中控制优良性状的基因很难通过重组交换来打破附近的连锁累赘（Bevan et al.，2017）。迅速发展的基因组编辑技术CRISPR/Cas9是一个高效并且操作简单的技术，该技术可以定点敲除不良基因和获得非转基因材料，且已经成功应用于水稻、玉米、小麦、油菜和部分园艺作物的遗传改良（Belhaj et al.，2015；Yang et al.，2018）。尽管在酵母中通过CRISPR/Cas9可以精确地靶向重组位点进行基因组编辑，但是在高等植物中实现通过同源重组介导的基因组编辑还有一定难度，利用重组时期特异表达的诱导型启动子启动 *Cas9* 基因或许可以解决这一问题（Bevan et al.，2017）。实现植物基因组的自由编辑，定点敲除不良基因，精确敲入有利基因，或者对基因调控位点进行定点改造，可以打破常规育种的局限性。

采用多种技术，扩展基因组遗传改良空间。作物的驯化由于"选择性清除"和"连锁效应"导致栽培材料中大量的基因组空间被固定，很难导入新的基因和性状，如何释放被"固定"的基因组空间，以便引入更多的有利性状是育种的一个重要方向。生物科学和技术的发展为解决这一问题提供了更多的手段。通过基因组学分析可以确定这些'固定'的区域，进一步利用分子生物学解析区域内基因的功能，明确哪些基因是有利的和有害的，最后通过大规模群体重组或者基因组编辑技术保留育种需要的基因，淘汰控制不良性状的基因。同时，对染色体重组机理的研究也将为基因在染色体上的组合提供更多遗传调控的方案。

1.3 阶段性目标

1.3.1 2035年目标

至2035年，阶段性目标主要如下。

全球种质资源的利用　在过去的一个世纪里，对种质资源（特别是野生种质资源）的收集和利用方面，进入了一个较为成功的阶段，这对于食物来源多样化和增加优良作物的遗传多样性来说是至关重要。国际组织在植物的收集和保存中发挥着重要作用。许多国家和组织每年在种质的获取和保存上花费数百万美元，但是绝大多数种质对于现代品种，特别是在产量和营养质量等复杂性状方面还没有贡献。因此，种质资源的建立和维护必须与这些材料积极利用的能力相结合，在更广泛的遗传资源上进行作物改良。基因组测序的广泛应用清楚地证明，种质库中存储的遗传多样性能够以比以前想象的更高的效率得到利用。为了实现这一目标，应当对约 3000 种人类用作食物的农作物进行测序和重测序（Borlaug，1983），从而进行可持续性和粮食安全的需求研究，并且这些数据应保存在一个统一的平台上，由科学界共享。

多倍体育种计划 由于基因组的高度重复性和近亲衰退，多倍体作物的基础研究和育种极具挑战性。马铃薯是全世界五种主要主粮之一。为了加快马铃薯的遗传改良，启动了"优薯计划"（"二倍体马铃薯杂交育种计划"），旨在利用现代农业技术将二倍体马铃薯重塑成种子作物（Li et al.，2013b）。计划在 2035 年前完成"优薯计划"的技术革新，以及整个马铃薯产业链的全面转型，这也将为多倍体作物的育种树立一个典范。

从头驯化 除产量外，农作物的质量和抗逆性已成为重要的研究重点（Francis et al.，2017；Martin and Li，2017）。当农作物的产量增加时，农作物的质量和抗性通常会下降，在一个优良品系中利用这些有利基因是一个巨大的挑战。例如，通过对谷物品质和产量有重大贡献的主要基因金字塔化，开发出了高产、优质的超级水稻（Zeng et al.，2017），但是这种方法花费了十多年的时间，以及大量的劳力和资源。鉴于这些困难，迫切需要对农作物进行从头驯化。从头驯化不仅可以驯化野生物种，而且因为没有驯化瓶颈，所以可以排除传统驯化过程中遗传多样性降低的不利影响，这将为作物驯化中的分子育种开辟新途径。对重要农艺性状涉及的基因和途径的系统理解，将对指导从头驯化及改良新作物至关重要。

1.3.2 2050 年目标

至 2050 年，阶段性目标主要如下。

可控重组 DNA 序列和表观遗传特征，以及重组机器工作模型的鉴定对于确定基因组中的重组热点和机制,以至于人工操纵重组是必需的。在酵母细胞中,已经能够精确靶向有丝分裂的重组(Sadhu et al., 2016)。至 2050 年,育种家们将鉴定出作物减数分裂重组的热点,并系统理解重组机器的工作模式,并开发和改造重组机器用于有效地在目标区域控制作物的减数分裂重组。

育种中的合成生物学　随着基因组学、基因组编辑和合成技术的飞速发展,合成生物学的出现标志着生命科学新革命的到来。至 2050 年,合成生物学将在未来人类农业生产中产生巨大影响,尤其是在光合作用和生物固氮中的应用。在 C_3 植物中,C_4 光合作用的实现将为人类社会提供更多的食物能量。使用合成生物学方法广泛研究生物固氮系统,从而实现该系统在细胞器之间或物种之间转移的最终目的。

2 多倍体育种

2.1 国内外研究进展

多倍体是指含有两套或两套以上完整染色体组的生物体。多倍体在动植物中广泛存在,是物种发生的一种重要方式。所有的被子植物在进化过程中都经历过一次或数次多倍化过程(Otto,2007)。按照染色体组的来源,多倍体可分为同源多倍体和异源多倍体。多倍体具有个体较大、抗逆性强等优点,在生产上具有重要的应用价值。植物中存在大量的多倍体类型,如普通小麦为天然的异源六倍体,棉花中栽培的陆地棉和海岛棉为异源四倍体,葡萄品种'巨峰'为四倍体,香蕉为三倍体,无籽西瓜为三倍体。多倍体水果的培育丰富了市场供应,带来了更多的经济效益。

2.1.1 多倍体优势及其基因组学基础

多倍体形成后主要有两个优势,即杂种优势和基因冗余。多倍体的杂种优势主要表现在 3 个方面。首先,异源多倍体中同源染色体的配对阻止了基因间的重组,有效地维持了相同水平的杂合性(Comai,2005)。其次,多倍体在配子体阶段就具有杂种优势(Butruille and Boiteux,2000;Groose and Bingham,1991)。再次,多倍体之间的杂交更有利于增强物种的杂种优势,使物种具有更广的适应范围(Auger et al.,2005;Birchler et al.,2003)。基因组加倍造成的基因冗余允许基因产生新的功能，以及出现其他新类型的变化（Lynch and Conery，2000），这种效应可以在两个生命阶段表现。第一个阶段是配子体的单倍体阶段，尽管这个阶段的生物复杂性已经降低，但是依然需要许多活跃的基因（McCormick，2004；Yadegari and Drews，2004）；相比之下，多倍体生物中有害的突变可能会被野生型等位基因所掩盖。第二个是染色体加倍阶段，多倍体可以减少纯合子衰退带来的影响（Mable and Otto，2001；Stadler，1929）。基因冗余带来的另一个优势是，通过改变基因的拷贝使基因功能多样化。在二倍体中，基因功能的多样化出现在极少的部分加倍事件中。另外，多倍体中几乎所有的基因都有重复的拷贝，这为基因功能的多样化提供可能（Moore and Purugganan，2005；Prince and Pickett，2002）。

2.1.2 多倍体的形成

物种在多倍化过程中，全基因组加倍往往给生物体带来了强烈的冲击，通常不能稳定遗传，新形成的多倍体需要经历二倍体化过程才能达到稳定状态，包括加倍基因的丢失或高度修饰，加倍染色体的重排及数目上的减少（Hollister，2015）。二倍化过程破坏了全基因组复制造成的明显的染色体组变化，同时也保留了一系列的重复基因（Wolfe，2001）。

多倍体在二倍化过程中，基因表达模式发生了很大的变化，转座子的表观遗传调控会发生变化，有丝分裂和减数分裂也必须通过遗传和表观遗传的调控达到较稳定的状态（Comai，2005）。减数分裂过程具有强烈的适应性，直接影响配子的形成。同源染色体的正确分离对新形成的多倍体的稳定具有重要影响，很多新形成的多倍体都伴随着减数分裂的异常，不同类型的多倍体需要克服不同的异常染色体行为带来的挑战（Cifuentes et al.，2010；Comai，2005）。

异源多倍体来自不同物种的杂交和加倍。不同物种的染色体组间存在部分同源染色体，部分同源染色体在减数分裂过程中发生配对从而形成多价体。同源多倍体来自同一物种自身染色体组的加倍，多对同源染色体在减数分裂过程中容易发生配对从而形成多价体。多价体的形成导致同源染色体在减数分裂后期发生不均等分离，从而产生不均衡的配子。异源多倍体中发生的部分同源染色体配对会导致染色体重排、基因的丢失和非整倍体的形成，而同源多倍体在减数分裂过程中形成的多价体更容易导致非整倍体的产生，因此成功进化的同源多倍体及异源多倍体具有重建的类似于二倍体的染色体行为（Hollister，2015）。

2.1.3 多倍体稳定形成机制的研究途径

到目前为止，研究多倍体的方法主要有三种途径。第一，通过使用秋水仙素（减数分裂的抑制剂）对二倍体材料或杂交 F_1 代植株进行染色体组加倍，创制人工合成的多倍体（Henry et al.，2014；Yant et al.，2013）。这种方法可用于研究多倍体形成初期的适应机制，也可以通过人工合成多倍体与天然稳定的多倍体杂交获得的 F_2 代构建群体，来寻找和多倍体稳定遗传相关的位点。第二，对于异源多倍体，可以通过配子培养的方法获得相应的单倍体材料，这种材料减数分裂过程中的重组交换过程只能发生在部分同源染色体之间，因此可以用来寻找控制部分同源染色体配对的基因位点（Jenczewski et al.，2003）。第三，通过全基因组测序的方法，比较天然二倍体和多倍体群体基因组的特点，寻找控制多倍体稳定遗传的基因位点（Hollister et al.，2012；Yant et al.，2013）。

2.2 研究内容和突破口

2.2.1 染色体形成二倍体化的遗传解析

小麦是世界最重要的粮食作物之一。普通小麦为异源六倍体，含有 A、B、D 三个染色体组，由野生二粒小麦（*Triticum turgidum*，genomes AABB）和山羊草（*Aegilops tauschii*，genomes DD）经杂交和自然加倍长期进化而来。普通小麦含有 3 套既高度同源又有明显分化的基因组，因此每一条染色体既有同源染色体也有部分同源染色体，为研究染色体行为提供了重要材料。小麦的部分同源染色体间也会发生配对和重组。1958 年，Riley 发现小麦 5B 染色体上存在的 *ph1* 基因位点能够抑制小麦部分同源染色体间的配对，在缺失 5B 染色体的条件下，小麦部分同源染色体间的重组频率会增加。*Ph1* 基因位点是调控细胞周期蛋白依赖性激酶（cyclin-dependent protein kinase，CDK）

的基因簇（Griffiths et al., 2006）。小麦染色体的二倍体化受多基因调控,其中,*Ph2*基因位点调控小麦同源染色体的配对（Sutton et al., 2003）。

芸薹属多倍体物种中的不同亚基因组间存在染色体对等和不对等交换（Osborn et al., 2003；Parkin, 2011；Piquemal et al., 2005；Udall et al., 2005）。经细胞学、分子生物学及基因组学分析和研究揭示，异源四倍体的甘蓝型油菜（AACC）的亚基因组之间存在染色体交换、易位和重组现象（Chalhoub et al., 2014；Grandont et al., 2014；Liu et al., 2014；Osborn et al., 2003；Parkin et al., 2011, 2014；Piquemal et al., 2005；Udall et al., 2005；Wang et al., 2011）。跟踪调查人工合成油菜的多个世代，研究人员发现A和C亚基因组之间具有频繁交换与重组（Chevre et al., 2007；Heneen et al., 2012；Ksiazczyk et al., 2011；Mason et al., 2014a, 2014b；Suay et al., 2014；Szadkowski et al., 2010）。总之，染色体重组导致的基因组结构变异在多倍体物种多样性、适应性等方面发挥重要作用（Chester et al., 2012；Edwards et al., 2013；Schiessl et al., 2014；Wang et al., 2012；Zou et al., 2011），为优异性状作物的筛选和驯化提供了广泛的材料基础。然而，控制和影响多倍体中不同亚基因组染色体之间的配对、重组和稳定遗传的调控机制仍不明确，急需深入研究和探讨。

2.2.2 利用人工合成多倍体解析多倍体二倍化的机理

细胞内两套或多套基因组在初期经过剧烈"基因组振荡"（genomic shock）后，多倍化的基因组进入长期的二倍化过程（Conant et al., 2014；Doyle et al., 2008；Freeling, 2009；Tian et al., 2005）。该二倍化过程的途径和机制主要包括，①DNA层次：基因组和基因序列大量丢失，大量不规则染色体重组交换，以及基因组结构变异发生（Schmutzer et al., 2015；Schnable et al., 2011）；②表达层次：多倍化后不同亚基因组之间的重复基因之间发生变化，或者说是表达的分化，可表现为亚功能化（含可变剪接分化）（Buggs et al., 2011；Combes et al., 2013；Grover et al., 2012；Yoo et al., 2014）或新功能化（Blanc and Wolfe, 2004；Buggs et al., 2011；Moore et al., 2005）；③大量的小RNA、DNA甲基化等表观遗传变化在多倍化二倍化过程中发生，用于调节和平衡不同亚基因组融合带来的基因表达变化（Chen,

2007；Cheng et al., 2016；Guan et al., 2014）。

多倍体二倍化机制的明确将加速多倍体的研究和育种, 人工合成多倍体为相关研究提供了重要的材料基础。Guo 和 Han (2014) 利用小麦人工异源四倍体 TL05-TMU06 (*Aegilop longissimi* ×*Triticum urartu*, SlSlA)、TMU38-TQ27 (*Triticum urartu*×*Aegilop tauschii*, AADD)、TQ27-TMU38 (*Aegilop tauschii*×*Triticum urartu*, DDAA)、TB01-TQ27 (*Aegilop bicornis*×*Aegilop tauschii*, SbSbDD) 等材料, 观察杂种后代从合成低代到高代过程中基因组的变化, 研究不同基因组 rDNA 序列的变化, 模拟自然界小麦的进化过程, 全面分析 rDNA 序列变化的特点及规律, 发现核糖体 rDNA 转录形成 rRNA, 参与核糖体的生物学功能, 其中 45S rDNA 被称为核仁组织区 (nucleolar organizing region, NOR)。在人工合成的四倍体小麦 SS(BB)AA 及 AADD 中, A 基因组的 NOR 位点从 S4 代开始丢失, 到 S7 代序列完全丢失, 这种变化趋势与自然界中四倍体小麦的进化结果一致。在人工合成六倍体小麦向普通小麦的进化过程中, D 组 NOR 位点的拷贝数显著降低。研究异源多倍体进化过程中 NOR 位点丢失的机制, 有助于明确多倍体的二倍化机理。

通过对小麦 AA、SS、DD 三个祖先同源四倍体, 小麦近缘种二倍体长穗偃麦草的同源四倍体的观察发现, 新合成的同源四倍体能够正常结实, 结实率稍低于二倍体材料, 说明通过染色体组加倍获得的同源四倍体材料在合成低代就能稳定遗传。研究人员可以通过研究人工合成同源四倍体的减数分裂过程来阐释同源多倍体迅速二倍化的机理, 或者通过全基因组测序及转录组测序分析寻找同源四倍体 DNA 序列及表达上的差异来阐释多倍体二倍化的机制。

科研人员在人工合成多倍体甘蓝型油菜中检测到了较高的染色体重组和基因组结构变异发生频率 (Schmutzer et al., 2015；Xiong et al., 2011)。Song 等 (1995) 利用 89 个核探针检测人工合成异源四倍体油菜和亲本多态性, 发现 F_2 至 F_5 所有世代基因组均发生变化, 包括亲本片段消失和出现亲本不存在的新片段。Gaeta 等 (2007) 利用 368 个 RFLP 和 65 个 SSR 标记, 发现第一代遗传组成变化较少, 但在随后的 F_1 至 F_5 代变化增多。科研人员对 53 个油菜品系的高通量重测序分析, 在人工合成的甘蓝型油菜中检测到了大规模的结构变异和大片段的丢失，其中很多为一至多条染色体或几百万碱基的缺失（Schmutzer et al.，2015；Xiong et al.，2011）。这些结果表明多倍体甘蓝型油菜在形成之初非常不稳定，大量基因组发生变异，进而导致性状发生变异。目前，科研人员可以借助基因组重测序技术追踪多世代人工合成甘蓝型油菜材料中的基因组序列丢失和结构变异的发生情况，研究发生的频率、规模和染色体位置，比较不同世代发生频率高低，从而解析多倍体二倍化过程中的基因组 DNA 序列层面的变化和相关进化机制。

基因组加倍后多拷贝基因的表达分化是多倍体二倍化的另一重要机制。探针芯片检测的研究结果表明油菜约 25% 的转录本出现相对于亲本基因组的非叠加表达（nonadditive expression，1+1>2 或 1+1<1，假定每个亲本多拷贝重复基因的表达量为 1 或中亲值为 1）（Albertin et al.，2007）。对根、茎和叶的转录组测序（RNA-seq）的研究结果表明甘蓝型油菜中 A 和 C 同源基因的 25% 左右的发生了表达分化，差异表达的重复基因主要集中在光合作用相关和糖类代谢相关的途径。值得一提的是，Zhou 等（2011）对 82 对甘蓝型油菜 A 和 C 重复基因进行可变剪切分化研究，发现 30% 的重复基因对发生了可变剪切分化，该结果提示多倍化发生后，基因可能丢失或获得新的可变剪切事件，从而形成重复基因间可变剪切的分化。然而，目前对此过程的研究还主要停留在少数案例和个体水平，缺乏多层次全基因组水平和基于大规模群体的系统调查。在当前基因组时代和高通量测序技术飞速发展和日新月异的形势下，开展对甘蓝型油菜多倍化后二倍化过程中的 DNA 和 RNA 水平的变化，对于解析多倍体二倍化的过程和机制具有重要意义。

2.2.3 异源多倍体的进化规律的研究有助于远缘杂交

小麦远缘杂交是利用小麦近缘种资源丰富小麦遗传多样性并进行品种改良的重要手段。通过小麦染色体工程的手段对小麦性状进行改良，将小麦近缘种与小麦杂交并进行染色体加倍是小麦染色体工程的第一步。小黑麦和小偃麦材料虽然不能直接应用于生产，但它们都是小麦育种过程中重要的遗传资源，既拥有野生资源的特殊农艺性状，又能融入小麦染色体背景。研究异源多倍体的进化规律及外源染色体的加入对小麦基因组带来的影响，有助于更好的发挥小麦近缘种在小麦育种中的功能。明确远源杂交过程中染色体组的变化能够为小麦远源杂交的育种应用打下良好的基础。

通过种间或远缘属间杂交后代的基因组片段渐渗可将近缘种的优异农艺性状导入到作物中，从而实现作物或性状的遗传改良。芸薹属物种种间杂交可在不同的二倍体之间，以及四倍体和二倍体亲本之间进行。不同二倍体物种之间杂交产生的杂种通常是不育的（仅有少量没有发生染色体减数的配子产生），进一步通过秋水仙素处理可以创造出人工合成多倍体。而四倍体和二倍体亲本种之间杂交产生的个体通常包含一个二倍体染色体组和一个单倍体染色体组，如四倍体甘蓝型油菜和二倍体白菜杂交将产生 AAC 杂种，即包含 10 对 A 染色体和 9 条 C 染色体（Leflon et al.，2006），而这些杂种植株通常是可育的，并可在 A 基因组上检测到大量同源重组的发生（Leflon et al.，2010）。具有不同基因组的芸薹属四倍体和二倍体物种之间杂交转育性状也是可行的，即首先产生 ABC 杂种，再通过不断回交和选择来实现优异性状的转育；也可以将 ABC 杂种通过秋水仙素加倍处理产生新的异缘六倍体，再与二倍体或四倍体物种进行杂交和遗传改良（Chen et al.，2010）。异缘六倍体是一个能够实现不同芸薹属物种性状转移的桥梁，因为它与其他物种杂交的后代基本是可育的。芸薹属不同四倍体物种之间的杂交可产生包含一个二倍体染色体组和另外两个单倍体染色体组的杂种（如 AABC、BBAC 和 CCAB 等类型），这些杂种在减数分裂中将产生比两个单倍体（如 AC，BC and AB）更多的多价体联会，增加不同亚基因组间同源重组的机会，从而更高效地进行作物遗传改良（Mason et al.，2010；Nagpal et al.，1996）。

目前，国内多个课题组都正在通过甘蓝型油菜同白菜、甘蓝、黑芥、萝卜、诸葛菜等近缘种杂交，将多倍体育种过程中丢失的但在这些近缘种中存在和保留的优异性状和遗传位点导入到油菜中，用于改良油菜的生态适应性、抗病性、花期长短、生育期长度、黄籽等性状，在全基因组测序、遗传变异和全基因组关联分析等研究的基础上，借助已解析的性状关联位点和区域，设计对应的分子标记组，在种间杂交过程中精确导入特定基因组片段，从而实现甘蓝型油菜关键性状的遗传改良。

2.3 阶段性目标

2.3.1 2035年目标

至2035年，多倍体育种的发展目标包括：建立材料创制的体系和规模、重新合成新的同源多倍体和异源多倍体，如创制玉米、水稻等新的同源四倍体，创制麦类新的八倍体类型（如八倍体小黑麦和小偃麦）等；筛选果蔬及林木新的四倍体新材料。

2.3.2 2050年目标

至2050年，多倍体育种的发展目标包括：将全新的作物多倍体应用于生产，真正做到从产量、品质到适应性有新的突破。

致谢：本章在撰写过程中得到了杨学勇和祝光涛的协助，特此致谢！

参考文献

翟虎渠, 刘秉华 (2009). 矮败小麦创制与应用. 中国农业科学 *42*, 4127-4131.

Albertin, W., Alix, K., Balliau, T., Brabant, P., Davanture, M., Malosse, C., Valot, B., and Thiellement, H. (2007). Differential regulation of gene products in newly synthesized *Brassica napus* allotetraploids is not related to protein function nor subcellular localization. BMC Genomics *8*, 56.

Auger, D.L., Gray, A.D., Ream, T.S., Kato, A., Coe, E.H.Jr., and Birchler, J.A. (2005). Nonadditive gene expression in diploid and triploid hybrids of maize. Genetics *169*, 389-397.

Belhaj, K., Chaparro-Garcia, A., Kamoun, S., Patron, N.J., and Nekrasov, V. (2015). Editing plant genomes with CRISPR/Cas9. Curr. Opin. Biotechnol. *32*, 76-84.

Bevan, M.W., Uauy, C., Wulff, B.B.H., Zhou, J., Krasileva, K., and Clark, M.D. (2017). Genomic innovation for crop improvement. Nature *543*, 346-354.

Birchler, J.A., Auger, D.L., and Riddle, N.C. (2003). In search of the molecular basis of heterosis. Plant Cell *15*, 2236-2239.

Blanc, G., and Wolfe, K.H. (2004). Functional divergence of duplicated genes formed by polyploidy during *Arabidopsis* evolution. Plant Cell *16*, 1679-1691.

Borlaug, N.E. (1983). Contributions of conventional plant-breeding to food-production. Science *219*, 689-693.

Buggs, R.J., Zhang, L., Miles, N., Tate, J.A., Gao, L., Wei, W., Schnable, P.S., Barbazuk, W.B., Soltis, P.S., and Soltis, D.E. (2011). Transcriptomic shock generates evolutionary novelty in a newly formed, natural allopolyploid plant. Curr. Biol. *21*, 551-556.

Butruille, D.V., and Boiteux, L.S. (2000). Selection-mutation balance in polysomic tetraploids: impact of double reduction and gametophytic selection on the frequency and subchromosomal localization of deleterious mutations. Proc. Natl. Acad. Sci. USA *97*, 6608-6613.

Cai, G., Yang, Q., Yang, Q., Zhao, Z., Chen, H., Wu, J., Fan, C., and Zhou, Y. (2012). Identification of candidate genes of QTLs for seed weight in *Brassica napus* through comparative mapping among *Arabidopsis* and *Brassica* species. BMC Genet. *13*, 105.

Chakrabarti, M., Zhang, N., Sauvage, C., Munos, S., Blanca, J., Canizares, J., Diez, M.J., Schneider, R., Mazourek, M., McClead, J., et al. (2013). A cytochrome P450 regulates a domestication trait in cultivated tomato. Proc. Natl. Acad. Sci. USA *110*, 17125-17130.

Chalhoub, B., Denoeud, F., Liu, S.Y., Parkin, I.A.P, Tang, H.B., Wang, X.Y., Chiquet, J., Belcram, H., Tong, C.B., Samans, B. (2014). Early allopolyploid evolution in the post-neolithic *Brassica napus* oilseed genome. Science *345*, 950-953.

Chatterjee, D., Banga, S., Gupta, M., Bharti, S., Salisbury P.A., and Banga S.S. (2016). Resynthesis of *Brassica napus* through hybridization between *B. juncea* and *B. carinata*. Theor. Appl. Genet. *129*, 977-990.

Chen, Z.J. (2007). Genetic and epigenetic mechanisms for gene expression and phenotypic variation in plant polyploids. Annu. Rev. Plant Biol. *58*, 377-406.

Chen, Z.J., and Ni, Z., (2010). Mechanisms of genomic rearrangements and gene expression changes in plant polyploids. Bioessays *28*, 240-252.

Cheng, F., Sun, C., Wu, J., Schnable, J., Woodhouse, M.R., Liang, J., Cai, C., Freeling, M., and Wang, X. (2016). Epigenetic regulation of subgenome dominance following whole genome triplication in *Brassica rapa*. New Phytol. *211*, 288-299.

Chester, M., Gallagher, J.P., Symonds, V.V., Cruz da Silva, A.V., Mavrodiev, E.V., Leitch, A.R., Soltis, P.S., and Soltis, D.E. (2012). Extensive chromosomal variation in a recently formed natural allopolyploid species, *Tragopogon miscellus* (Asteraceae). Proc. Natl. Acad. Sci. USA *109*, 1176-1181

Chevre, A.M., Adamczyk, K., Eber, F., Huteau, V., Coriton, O., Letanneur, J.C., Laredo, C., Jenczewski, E., and Monod, H. (2007). Modelling gene flow between oilseed rape and wild radish, I, evolution of chromosome structure. Theor. Appl. Genet. *114*, 209-221.

Cifuentes, M., Grandont, L., Moore, G., Chevre, A.M., and Jenczewski, E. (2010). Genetic regulation of meiosis in polyploid species: new insights into an old question. New Phytol. *186*, 29-36.

Comai, L. (2005). The advantages and disadvantages of being polyploid. Nat. Rev. Genet. *6*, 836-846.

Combes, M.C., Dereeper, A., Severac, D., Bertrand, B., and Lashermes, P. (2013). Contribution of subgenomes to the transcriptome and their intertwined regulation in the allopolyploid *Coffea arabica* grown at contrasted temperatures. New Phytol. *200*, 251-260.

Conant, G.C., Birchler, J.A., and Pires, J.C. (2014). Dosage, duplication, and diploidization: clarifying the interplay of multiple models for duplicate gene evolution over time. Curr. Opin. Plant Biol. *19*, 91-98.

Doebley, J., Stec, A., and Hubbard, L. (1997). The evolution of apical dominance in maize. Nature *386*, 485-488.

Doebley, J.F., Gaut, B.S., and Smith, B.D. (2006). The molecular genetics of crop domestication. Cell *127*, 1309-1321.

Dong, H., Tan, C., Li, Y., He, Y., Wei, S., Cui, Y., Chen, Y., Wei, D., Fu, Y., He, Y., Wan, H., Liu, Z., Xiong, Q., Lu, K., Li, J., and Qian, W. (2018). Genome-wide association study reveals both overlapping and independent genetic loci to control seed weight and silique length in *Brassica napus*. Front Plant Sci. *9*, 921.

Doyle, J.J., Flagel, L.E., Paterson, A.H., Rapp, R.A., Soltis, D.E., Soltis, P.S., and Wendel, J.F. (2008). Evolutionary genetics of genome merger and doubling in plants. Annu. Rev. Genet. *42*, 443-461.

Edwards, D., Batley, J., and Snowdon, J.R. (2013). Accessing complex crop genomes with next-generation sequencing. Theor. Appl. Genet. *126*, 1-11.

Francis, D., Finer, J.J., and Grotewold, E. (2017). Challenges and opportunities for improving food quality and nutrition through plant biotechnology. Curr. Opin. Biotech. *44*, 124-129.

Frary, A., Nesbitt, T.C., Frary, A., Grandillo, S., van der Knaap, E., Cong, B., Liu, J.P., Meller, J., Elber, R., Alpert, K.B., et al. (2000). *fw2.2*: a quantitative trait locus key to the evolution of tomato fruit size. Science *289*, 85-88.

Fredua-Agyeman, R., Coriton, O., Huteau, V., Parkin, I.A., Chèvre, A.M., and Rahman, H. (2014). Molecular cytogenetic identification of B genome chromosomes linked to blackleg disease resistance in *Brassica napus* × *B. carinata* interspecific hybrids. Theor. Appl. Genet. *127*, 1305-1318.

Freeling, M. (2009). Bias in plant gene content following different sorts of duplication: tandem, whole-genome, segmental, or by transposition. Annu. Rev. Plant Biol. *60*, 433-453.

Fridman, E., Carrari, F., Liu, Y.S., Fernie, A.R., and Zamir, D. (2004). Zooming in on a quantitative trait for tomato yield using interspecific introgressions. Science *305*, 1786-1789.

Fu, Y., Wei, D., Dong, H., He, Y., Cui, Y., Mei, J., Wan, H., Li, J., Snowdon, R., Friedt, W., Li, X., and Qian, W. (2015). Comparative quantitative trait loci for silique length and seed weight in *Brassica napus*. Sci. Rep. *5*, 14407.

Gaeta, R.T., Pires, J.C., Iniguez-Luy, F., Leon, E., and Osborn, T.C. (2007). Genomic changes in resynthesized *Brassica napus* and their effect on gene expression and phenotype. Plant Cell *19*, 3403-3417.

Girke, A., Schierholt, A., and Becker, H.C. (2012). Extending the rapeseed gene pool with resynthesized *Brassica napus* II: heterosis. Theor. Appl. Genet. *124*, 1017-1026.

Grandont, L., Cunado, N., Coriton, O., Huteau, V., Eber, F., Chevre, A.M., Grelon, M., Chelysheva, L., Jenczewskiet, E. (2014). Homoeologous chromosome sorting and progression of meiotic recombination in *Brassica napus*: ploidy does matter! Plant Cell *26*, 1448-1463.

Griffiths, S., Sharp, R., Foote, T.N., Bertin, I., Wanous, M., Reader, S., Colas, I., and Moore, G. (2006). Molecular characterization of *Ph1* as a major chromosome pairing locus in polyploid wheat. Nature *439*, 749-752.

Groose, R.W., and Bingham, E.T. (1991). Gametophytic heterosis for *in vitro* pollen traits in alfalfa. Crop Sci. *31*, 1510-1513.

Grover, C.E., Gallagher, J.P., Szadkowski, E.P., Yoo, M.J., Flagel, L.E., and Wendel, J.F. (2012). Homoeolog expression bias and expression level dominance in allopolyploids. New Phytol. *196*, 966-971.

Guan, X., Song, Q., and Chen, Z.J. (2014). Polyploidy and small RNA regulation of cotton fiber development. Trends Plant Sci. *19*, 516-528.

Guo, X., and Han, F.P. (2014). Asymmetric epigenetic modification and elimination of rDNA sequences by polyploidization in wheat. Plant Cell *26*, 4311-4327.

Harper, A.L., Trick, M., Higgins, J., Fraser, F., Clissold, L., Wells, R., Hattori, C., Werner, P., and Bancroft, I. (2012). Associative transcriptomics of traits in the polyploid crop species *Brassica napus*. Nat. Biotechnol. *30*, 798-802.

Heneen, W. K., Geleta, M., Brismar, K., Xiong, Z., Pires, J.C., Hasterok, R., Stoute, A.I., Scott, R.J., King, G.J., and Kurup, S. (2012). Seed colour loci, homoeology and linkage groups of the C genome chromosomes revealed in *Brassica rapa-B. oleracea* monosomic alien addition lines. Ann. Bot. *109*, 1227-1242.

Henry, I.M., Dilkes, B.P., Tyagi, A., Gao, J., Christensen, B., and Comai, L. (2014). The *BOY NAMED SUE* quantitative trait locus confers increased meiotic stability to an adapted natural allopolyploid of *Arabidopsis*. Plant Cell *26*, 181-194.

Heun, M., SchaferPregl, R., Klawan, D., Castagna, R., Accerbi, M., Borghi, B., and Salamini, F. (1997). Site of einkorn wheat domestication identified by DNA fingerprinting. Science *278*, 1312-1314.

Hollister, J.D. (2015). Polyploidy: adaptation to the genomic environment. New Phytol. *205*, 1034-1039.

Hollister, J.D., Arnold, B.J., Svedin, E., Xue, K.S., Dilkes, B.P., and Bomblies, K. (2012). Genetic adaptation associated with genome-doubling in autotetraploid *Arabidopsis arenosa*. PLoS Genet. *8*, e1003093.

Hong, M., Hu, K., Tian, T., Li, X., Chen, L., Zhang, Y., Yi, B., Wen, J., Ma, C., Shen, J., et al. (2017). Transcriptomic analysis of seed coats in yellow-seeded *Brassica napus* reveals novel genes that influence proanthocyanidin biosynthesis. Front. Plant Sci. *8*, 1674.

Huang, X., Kurata, N., Wei, X., Wang, Z.X., Wang, A., Zhao, Q., Zhao, Y., Liu, K., Lu, H., Li, W., et al. (2012). A map of rice genome variation reveals the origin of cultivated rice. Nature *490*, 497-501.

Huang, X.H., and Han, B. (2014). Natural variations and genome-wide association studies in crop plants. Ann. Rev. Plant Biol. *65*, 531-551.

Itkin, M., Heinig, U., Tzfadia, O., Bhide, A.J., Shinde, B., Cardenas, P.D., Bocobza, S.E., Unger, T., Malitsky, S., Finkers, R., et al. (2013). Biosynthesis of antinutritional alkaloids in solanaceous crops is mediated by clustered genes. Science *341*, 175-179.

Jansky, S.H., Charkowski, A.O., Douches, D.S., Gusmini, G., Richael, C., Bethke, P.C., Spooner, D.M., Novy, R.G., De Jong, H., De Jong, W.S., et al. (2016). Reinventing potato as a diploid inbred line-based crop. Crop Sci. *56*, 1412-1422.

Jenczewski, E., Eber, F., Grimaud, A., Huet, S., Lucas, M.O., Monod, H., and Chevre, A.M. (2003). *PrBn*, a major gene controlling homeologous pairing in oilseed rape (*Brassica napus*) haploids. Genetics *164*, 645-653.

Jiang, J., Wang, Y., Xie, T., Rong, H., Li, A., Fang, Y., and Wang, Y. (2015). Metabolic characteristics in meal of black rpeseed and yellow-seeded progeny of *Brassica napus-Sinapis alba* hybrids. Molecules *20*, 21204-21213.

Jiang, L., Li, D., Jin, L., Ruan, Y., Shen, W.H., and Liu, C. (2018). Histone lysine methyltransferases BnaSDG8.A and BnaSDG8.C are involved in the floral transition in *Brassica napus*. Plant J. *95*, 672-685.

Jugulam, M., Ziauddin, A., So, K.K., Chen, S., and Hall, J.C. (2015). Transfer of dicamba tolerance from *Sinapis arvensis* to *Brassica napus* via embryo rescue and recurrent backcross breeding. PLoS One *10*, e0141418.

Karim, M.M., Siddika, A., Tonu, N.N., Hossain, D.M., Meah, M.B., Kawanabe, T., Fujimoto, R., and Okazaki, K. (2015). Production of high yield short duration *Brassica napus* by interspecific hybridization between *B. oleracea* and *B. rapa*. Breed Sci. *63*, 495-502.

Ksiazczyk, T., Kovarik, A., Frédérique Eber, Huteau, V., Khaitova, L., Tesarikova, Z., Coriton, O., and Chevre, A.M. (2011). Immediate unidirectional epigenetic reprogramming of NORs occurs independently of rDNA rearrangements in synthetic and natural forms of a polyploid species *Brassica napus*. Chromosoma *120*, 557-571.

Leflon, M., Eber, F., Letanneur, J.C., Chelysheva, L., Coriton, O., Huteau, V., Ryder, C.D., Barker, G., Jenczewski, E., and Chevre, A.M. (2006). Pairing and recombination at meiosis of *Brassica rapa* (aa) × *Brassica napus* (aacc) hybrids. Theor. Appl. Genet. *113*, 1467-1480.

Leflon, M., Grandont, L., Eber, F., Huteau, V., Coriton, O., Chelysheva, L., Jenczewski, E., and Chèvre, A.M. (2010). Crossovers get a boost in *Brassica* allotriploid and allotetraploid hybrids. Plant Cell *22*, 2253-2264.

Li, C., Hao, M., Wang, W., Wang, H., Chen, F., Chu, W., Zhang, B., Mei, D., Cheng, H., and Hu, Q. (2018). An efficient CRISPR/Cas9 platform for rapidly generating simultaneous mutagenesis of multiple gene homoeologs in allotetraploid oilseed rape. Front. Plant Sci. *9*, 442.

Li, C.B., Zhou, A.L., and Sang, T. (2006). Rice domestication by reducing shattering. Science *311*, 1936-1939.

Li, F., Chen, B., Xu, K., Wu, J., Song, W., Bancroft, I., Harper, A.L., Trick, M., Liu, S., Gao, G., et al. (2014). Genome-wide association study dissects the genetic architecture of seed weight and seed quality in rapeseed (*Brassica napus* L.). DNA Res. *21*, 355-367.

Li, H., Peng, Z.Y., Yang, X.H., Wang, W.D., Fu, J.J., Wang, J.H., Han, Y.J., Chai, Y.C., Guo, T.T., Yang, N., et al. (2013a). Genome-wide association study dissects the genetic architecture of oil biosynthesis in maize kernels. Nat. Genet. *45*, 43-50.

Li, M.T., Li, Z.Y, , Zhang, C.Y., Qian, W., and Meng, J.L. (2005). Reproduction and cytogenetic characterization of interspecific hybrids derived from crosses between *Brassica carinata* and *B. rapa*. Theor. Appl. Genet. *110*, 1284-1289.

Li, N., Song, D., Peng, W., Zhan, J., Shi, J., Wang, X., Liu, G., and Wang, H. (2018). Maternal control of seed weight in rapeseed (*Brassica napus* L.): the causal link between the size of pod (mother, source) and seed (offspring, sink). Plant Biotech. J. *17*, 736-749.

Li, S., Chen, L., Zhang, .L, Li, X., Liu, Y., Wu, Z., Dong, F., Wan, L., Liu, K., Hong, D., et al. (2015). *BnaC9.SMG7b* functions as a positive regulator of the number of seeds per silique in *Brassica napus* by regulating the formation of functional female gametophytes. Plant Physiol. *169*, 2744-2760.

Li, Y., Li, G., Li, C., Qu, D., and Huang, S. (2013b). Prospects of diploid hybrid breeding in potato. Chin. Potato *27*, 96-99 (in Chinese).

Lin, T., Zhu, G.T., Zhang, J.H., Xu, X.Y., Yu, Q.H., Zheng, Z., Zhang, Z.H., Lun, Y.Y., Li, S.,

Wang, X.X., et al. (2014). Genomic analyses provide insights into the history of tomato breeding. Nat. Genet. *46*, 1220-1226.

Liu, J., Hua, W., Hu, Z., Yang, H., Zhang, L., Li, R., Deng, L., Sun, X., Wang, X., and Wang, H. (2015). Natural variation in *ARF18* gene simultaneously affects seed weight and silique length in polyploid rapeseed. Proc. Natl. Acad. Sci. USA *112*, E5123-E5132.

Liu, Y.B., Tang, Z.X., Darmency, H., Stewart, C.N. Jr, Di, K., Wei, W., and Ma, K.P. (2012). The effects of seed size on hybrids formed between oilseed rape (*Brassica napus*) and wild brown mustard (*B. juncea*). PLoS One *7*, e39705.

Liu, Y.Q., Shi, Z.H., Zalucki, M.P., and Liu, S.S. (2014). Conservation biological control and IPM practices in *Brassica* vegetable crops in China. Biol. Cont. *68*, 37-46.

Liu, Z.B., Cook, J., Melia-Hancock, S., Guill, K., Bottoms, C., Garcia, A., Ott, O., Nelson, R., Recker, J., Balint-Kurti, P., et al. (2016). Expanding maize genetic resources with predomestication alleles: maize-teosinte introgression populations. Plant Genome-Us *9*, 1-11.

Lynch, M., and Conery, J.S. (2000). The evolutionary fate and consequences of duplicate genes. Science *290*, 1151-1155.

Mable, B.K., and Otto, S.P. (2001). Masking and purging mutations following EMS treatment in haploid, diploid and tetraploid yeast (*Saccharomyces cerevisiae*). Genet. Res. *77*, 9-26.

Malek, M.A., Ismail, M.R., Rafii, M.Y., and Rahman, M. (2012). Synthetic *Brassica napus* L.: development and studies on morphological characters, yield attributes, and yield. Sci. World J., 416901.

Martin, C., and Li, J. (2017). Medicine is not health care, food is health care: plant metabolic engineering, diet and human health. New Phytol. *216*, 699-719.

Mason, A.S., Huteau, V., Eber, F., Coriton, O., Yan, G., Nelson, M.N., Cowling, W.A., and Chevre, A.M. (2010). Genome structure affects the rate of autosyndesis and allosyndesis in aabc, bbac and ccab brassica interspecific hybrids. Chromosome Res. *18*, 655-666.

Mason, A.S., Batley, J., Bayer, P.E., Hayward, A., Cowling, W.A., and Nelson, M.N. (2014a). High-resolution molecular karyotyping uncovers pairing between ancestrally related *Brassica* chromosomes. New Phytol. *202*, 964-974.

Mason, A.S., Nelson, M.N., Takahira, J., Cowling, W.A., Alves, G.M., Chaudhuri, A., et al.

(2014b). The fate of chromosomes and alleles in an allohexaploid *Brassica* population. Genetics *197*, 273-283.

Matsuoka, Y., Vigouroux, Y., Goodman, M.M., Sanchez, G.J., Buckler, E., and Doebley, J. (2002). A single domestication for maize shown by multilocus microsatellite genotyping. Proc. Natl. Acad. Sci. USA *99*, 6080-6084.

McCormick, S. (2004). Control of male gametophyte development. Plant Cell *16*, S142-S153.

McCouch, S.R., Sweeney, M., Li, J.M., Jiang, H., Thomson, M., Septiningsih, E., Edwards, J., Moncada, P., Xiao, J.H., Garris, A., et al. (2007). Through the genetic bottleneck: *O. rufipogon* as a source of trait-enhancing alleles for *O. sativa*. Euphytica *154*, 317-339.

Miller, J.C., and Tanksley, S.D. (1990). RFLP analysis of phylogenetic relationships and genetic variation in the genus Lycopersicon. Theor. Appl. Genet. *80*, 437-448.

Moore, R.C., and Purugganan, M.D. (2005). The evolutionary dynamics of plant duplicate genes. Curr. Opin. Plant Biol. *8*, 122-128.

Moore, R.C., Grant, S.R., and Purugganan, M.D. (2005). Molecular population genetics of redundant floral-regulatory genes in *Arabidopsis thaliana*. Mol. Biol. Evol. *22*, 91-103.

Nagpal, R., Raina, S.N., Sodhi, Y.S., Mukhopadhyay, A., Arumugam, N., and Pental, A.K.P. (1996). Transfer of *Brassica tournefartii* (tt) genes to allotetraploid oilseed *Brassica* species (*B. juncea* aabb, *B. napus* aacc, *B. carinata* bbcc): homoeologous pairing is more pronounced in the three-genome hybrids (tacc, tbaa, tcaa, tcbb) as compared to allodiploids (ta, tb, tc). Theor. Appl. Genet. *92*, 566-571.

Olsen, K.M., and Wendel, J.F. (2013). A bountiful harvest: genomic insights into crop domestication phenotypes. Annu. Rev. Plant Biol. *64*, 47-70.

Osborn, T.C., Pires, J.C., Birchler, J.A., Auger, D.L., Chen, J.Z., Lee, H.S., Comai, L, Madlung, A, Doerge, R.W., Colot, V., et al. (2003). Understanding mechanisms of novel gene expression in polyploids. Trends Genet. *19*, 141-147.

Otto, S.P. (2007). The evolutionary consequences of polyploidy. Cell *131*, 452-462.

Parkin, I.A., Koh, C., Tang, H., Robinson, S.J., Kagale, S., Clarke, W.E., Town, C.D., Nixon, J., Krishnakumar, V., and Bidwell, S.L. (2014). Transcriptome and methylome profiling reveals relics of genome dominance in the mesopolyploid *Brassica oleracea*. Genome Biol. *15*, 1-18.

Parkin, I.A.P., Robinson, S.J., Sadowski, J., Kole, C. (2011). Exploring the paradoxes of the *Brassica* genome architecture. CAB Direct, 328-348.

Piperno, D.R., and Flannery, K.V. (2001). The earliest archaeological maize (*Zea mays* L.) from highland Mexico: new accelerator mass spectrometry dates and their implications. Proc. Natl. Acad. Sci. USA *98*, 2101-2103.

Piquemal, J., Cinquin, E., Couton, F., Rondeau, C., Seignoret, E., Doucet, I., Villeger, M.J., Vincourt, P., Blanchard, P. (2005). Construction of an oilseed rape (*Brassica napus* L.) genetic map with ssr markers. Theor. Appl. Genet. *111*, 1514-1523.

Prince, V.E., and Pickett, F.B. (2002). Splitting pairs: the diverging fates of duplicated genes. Nat. Rev. Genet. *3*, 827-837.

Qi, J.J., Liu, X., Shen, D., Miao, H., Xie, B.Y., Li, X.X., Zeng, P., Wang, S.H., Shang, Y., Gu, X.F., et al. (2013). A genomic variation map provides insights into the genetic basis of cucumber domestication and diversity. Nat. Genet. *45*, 1510-1515.

Qian, W., Chen, X., Fu, D., Zou, J., and Meng, J. (2005). Intersubgenomic heterosis in seed yield potential observed in a new type of *Brassica napus* introgressed with partial *Brassica rapa* genome. Theor. Appl. Genet. *110*, 1187-1194.

Qian, W., Sass, O., Meng, J., Li, M., Frauen, M., and Jung, C. (2007). Heterotic patterns in rapeseed (*Brassica napus* L.): I, crosses between spring and Chinese semi-winter lines. Theor. Appl. Genet. *11*, 27-34.

Qu, C., Fu, F., Lu, K., Zhang, K., Wang, R., Xu, X., Wang, M., Lu, J., Wan, H., Zhanglin, T., et al. (2013). Differential accumulation of phenolic compounds and expression of related genes in black- and yellow-seeded *Brassica napus*. J. Exp. Bot. *64*, 2885-2898.

Raboanatahiry, N., Chao, H., Dalin, H., Pu, S., Yan, W., Yu, L., Wang, B., and Li, M. (2018). QTL alignment for seed yield and yield related traits in *Brassica napus*. Front Plant Sci. *9*, 1127.

Riley, R. (1958). Genetic control of the cytologically diploid behaviour of hexaploid wheat. Nature *182*, 713-715.

Sadhu, M.J., Bloom, J.S., Day, L., and Kruglyak, L. (2016). CRISPR-directed mitotic recombination enables genetic mapping without crosses. Science *352*, 1113-1116.

Schiessl, S., Samans, B., Hüttel, B., Reinhard, R., and Snowdon, R.J. (2014). Capturing sequence variation among flowering-time regulatory gene homologs in the allopolyploid crop species *Brassica napus*. Front. Plant Sci. *5*, 404.

Schmutzer, T., Samans, B., Dyrszka, E., Ulpinnis, C., Weise, S., Stengel, D., Colmsee, C., Lespinasse, D., Micic, Z., Abel, S., et al. (2015). Species-wide genome sequence and nucleotide polymorphisms from the model allopolyploid plant *Brassica napus*. Sci. Data *2*, 150072.

Schnable, J.C., Springer, N.M., and Freeling, M. (2011). Differentiation of the maize subgenomes by genome dominance and both ancient and ongoing gene loss. Proc. Natl. Acad. Sci. USA *108*, 4069-4074.

Shang, Y., Ma, Y., Zhou, Y., Zhang, H., Duan, L., Chen, H., Zeng, J., Zhou, Q., Wang, S., Gu, W., et al. (2014). Biosynthesis, regulation, and domestication of bitterness in cucumber. Science *346*, 1084-1088.

Smooker, A.M., Wells, R., Morgan, C., Beaudoin, F., Cho, K., Fraser, F., and Bancroft, I. (2011). The identification and mapping of candidate genes and QTL involved in the fatty acid desaturation pathway in *Brassica napus*. Theor. Appl. Genet. *122*, 1075-1090.

Somers, D.J., Rakow, G., Prabhu, V.K., and Friesen, K.R. (2001). Identification of a major gene and RAPD markers for yellow seed coat colour in *Brassica napus*. Genome *44*, 1077-1082.

Song, K., Lu, P., Tang, K., and Osborn, T.C. (1995). Rapid genome change in synthetic polyploids of *Brassica* and its implications for polyploid evolution. Proc. Natl. Acad. Sci. USA *92*, 7719-7723.

Spooner, D.M., McLean, K., Ramsay, G., Waugh, R., and Bryan, G.J. (2005). A single domestication for potato based on multilocus amplified fragment length polymorphism genotyping. Proc. Natl. Acad. Sci. USA *102*, 14694-14699.

Stadler, L.J. (1929). Chromosome number and the mutation rate in *Avena* and *Triticum*. Proc. Natl. Acad. Sci. USA *15*, 876-881.

Suay, L., Zhang, D., Eber, Frédérique, Jouy, Hélène, Lodé, Maryse, Huteau, V., Coriton, O., Szadkowski, E., Leflon, M., Martin, O.C., et al. (2014). Crossover rate between homologous chromosomes and interference are regulated by the addition of specific unpaired chromosomes in *Brassica*. New Phytol. *201*, 645-656.

Sun, F., Fan, G., Hu, Q., Zhou, Y., Guan, M., Tong, C., Li, J., Du, D., Qi, C., Jiang, L., et al. (2017). The high-quality genome of *Brassica napus* cultivar 'ZS11' reveals the introgression history in semi-winter morphotype. Plant J. *92*, 452-468.

Sutton, T., Whitford, R., Baumann, U., Dong, C.M., Able, J.A., and Langridge, P. (2003). The Ph2 pairing homoeologous locus of wheat (*Triticum aestivum*): identification of candidate meiotic genes using a comparative genetics approach. Plant J. *36*, 443-456.

Szadkowski, E., Eber, F., Huteau, V., Lodé, M., Huneau, C., Belcram, H., Coriton, O., Manzanares-Dauleux, M.J., Delourme, R., King, G.J., et al. (2010). The first meiosis of resynthesized *Brassica napus*, a genome blender. New Phytol. *186*, 102-112.

Tan, L., Li, X., Liu, F., Sun, X., Li, C., Zhu, Z., Fu, Y., Cai, H., Wang, X., Xie, D., et al. (2008). Control of a key transition from prostrate to erect growth in rice domestication. Nat. Genet. *40*, 1360-1364.

Tang, H., Sezen, U., and Paterson, A.H. (2010). Domestication and plant genomes. Curr. Opin. Plant Biol. *13*, 160-166.

Tian, C.G., Xiong, Y.Q., Liu, T.Y., Sun, S.H., Chen, L.B., and Chen, M.S. (2005). Evidence for an ancient whole-genome duplication event in rice and other cereals. Acta Genet. Sin. *32*, 519-527.

Tieman, D., Zhu, G.T., Resende, M.F.R., Lin, T., Taylor, M., Zhang, B., Ikeda, H., Liu, Z.Y., Fisher, J., Zemach, I., et al. (2017). A chemical genetic roadmap to improved tomato flavor. Science *355*, 391-394.

Udall, J.A., Quijada, P.A., Osborn, T.C. (2005). Detection of chromosomal rearrangements derived from homoeologous recombination in four mapping populations of *Brassica napus* L. Genetics *169*, 967-979.

Wang, H., Bennetzen, and Jeffrey, L. (2012). Centromere retention and loss during the descent of maize from a tetraploid ancestor. Proc. Natl. Acad. Sci. USA. *109*, 21004-21009.

Wang, J.L., Tang, M.Q., Chen, S., Zheng, X.F., Mo, H.X., Li, S.J., Wang, Z., Zhu, K.M., Ding, L.N., Liu, S.Y., et al. (2017). Down-regulation of BnDA1, whose gene locus is associated with the seeds weight, improves the seeds weight and organ size in *Brassica napus*. Plant Biotech. J. *15*, 1024-1033.

Wang, X., Wang, H., Wang, J., Sun, R., Wu, J., Liu, S., Bai, Y., Mun, J.H., Bancroft, I., Cheng, F., et al. (2011). The genome of the mesopolyploid crop species *Brassica rapa*. Nat. Genet. *43*, 1035-1039.

Wang, X.W., Li, J.M., Cheng, F., Liu, B., Wu, J., Aarts, M.G.M. (2014). Expression profiling reveals functionally redundant multiple-copy genes related to zinc, iron and cadmium responses in *Brassica rapa*. New Phytol. *203*, 182-194.

Wei, Z., Wang, M., Chang, S., Wu, C., Liu, P., Meng, J., and Zou, J. (2016). Introgressing subgenome components from *Brassica rapa* and *B. carinata* to *B. juncea* for broadening its genetic base and exploring intersubgenomic heterosis. Front Plant Sci. *7*, 1677.

Wen, J., Zhu, L., Qi, L., Ke, H., Yi, B., Shen, J., Tu, J., Ma, C., and Fu, T. (2012). Characterization of interploid hybrids from crosses between *Brassica juncea* and *B. oleracea* and the production of yellow-seeded *B. napus*. Theor. Appl. Genet. *125*, 19-32.

Wolfe, K.H. (2001). Yesterday's polyploids and the mystery of diploidization. Nat. Rev. Genet. *2*, 333-341.

Wright, S.I. (2005). The effects of artificial selection on the maize genome. Science *308*, 1310-1314.

Xiong, Z., Gaeta, R.T., and Pires, J.C. (2011). Homoeologous shuffling and chromosome compensation maintain genome balance in resynthesized allopolyploid *Brassica napus*. Proc. Natl. Acad. Sci. USA *108*, 7908-7913.

Yadegari, R., and Drews, G.N. (2004). Female gametophyte development. Plant Cell *16*, S133-S141.

Yang, H., Wu, J.J., Tang, T., Liu, K.D., and Dai, C. (2017). CRISPR/Cas9-mediated genome editing efficiently creates specific mutations at multiple loci using one sgRNA in *Brassica napus*. Sci Rep. *7*, 7489.

Yang, P., Shu, C., Chen, L., Xu, J., Wu, J., and Liu, K. (2012). Identification of a major QTL for silique length and seed weight in oilseed rape (*Brassica napus* L.). Theor. Appl. Genet. *125*, 285-296.

Yang, Y., Zhu, K., Li, H., Han, S., Meng, Q., Khan, S.U., Fan, C., Xie, K., and Zhou, Y. (2018). Precise editing of *CLAVATA* genes in *Brassica napus* L. regulates multilocular silique development. Plant Biotech. J. *16*, 1322-1335.

Yant, L., Hollister, J.D., Wright, K.M., Arnold, B.J., Higgins, J.D., Franklin, F.C.H., and Bomblies, K. (2013). Meiotic adaptation to genome duplication in *Arabidopsis arenosa*. Curr. Biol. *23*, 2151-2156.

Ye, M., Peng, Z., Tang, D., Yang, Z., Li, D., Xu, Y., Zhang, C., and Huang, S. (2018). Generation of self-compatible diploid potato by knockout of S-RNase. Nat. Plants *4*, 651-654.

Yoo, M.J., Liu, X., Pires, J.C., Soltis, P.S., and Soltis, D.E. (2014). Nonadditive gene expression in polyploids. Annu. Rev. Genet. *48*, 485-517.

Yu, C.Y. (2013). Molecular mechanism of manipulating seed coat coloration in oilseed *Brassica species*. J. Appl. Genet. *54*, 135-145.

Zamir, D. (2008). Plant breeders go back to nature. Nat. Genet. *40*, 269-270.

Zeng, D.L., Tian, Z.X., Rao, Y.C., Dong, G.J., Yang, Y.L., Huang, L.C., Leng, Y.J., Xu, J., Sun, C., Zhang, G.H., et al. (2017). Rational design of high-yield and superior-quality rice. Nat. Plants *3*, 17031.

Zhan, Z., Nwafor, C.C., Hou, Z., Gong, J., Zhu, B., Jiang, Y., Zhou, Y., Wu, J., Piao, Z., Tong, Y., et al. (2017). Cytological and morphological analysis of hybrids between *Brassicoraphanus*, and *Brassica napus* for introgression of clubroot resistant trait into *Brassica napus* L. PLoS One *12*, e0177470.

Zhang, J., Lu, Y., Yuan, Y., Zhang, X., Geng, J., Chen, Y., Cloutier, S., McVetty, P.B., and Li, G. (2009). Map-based cloning and characterization of a gene controlling hairiness and seed coat color traits in *Brassica rapa*. Plant Mol. Biol. *69*, 553-563.

Zhang, L., Li, S., Chen, L., and Yang, G. (2012). Identification and mapping of a major dominant quantitative trait locus controlling seeds per silique as a single Mendelian factor in *Brassica napus* L. Theor. Appl. Genet. *125*, 695-705.

Zhou, R., Moshgabadi, N., and Adams, K.L. (2011). Extensive changes to alternative splicing patterns following allopolyploidy in natural and resynthesized polyploids. Proc. Natl. Acad. Sci. USA *108*, 16122-16127.

Zhou, Y., Ma, Y., Zeng, J., Duan, L., Xue, X., Wang, H., Lin, T., Liu, Z., Zeng, K., Zhong, Y., et al. (2016). Convergence and divergence of bitterness biosynthesis and regulation in Cucurbitaceae. Nat. Plants *2*, 16183.

Zhu, G., Wang, S., Huang, Z., Zhang, S., Liao, Q., Zhang, C., Lin, T., Qin, M., Peng, M., Yang, C., et al. (2018). Rewiring of the fruit metabolome in tomato breeding. Cell *172*, 249-261.

Zou, H., Wu, Y., Liu, H., Lin, Z., Ye, X., Chen, X., and Yuan, Y.P. (2011). Development and identification of wheat-barley 2H chromosome translocation lines carrying the *Isa* gene.

Plant Breed. *131*, 69-74.

Zou, J., Zhu, J., Huang, S., Tian, E., Xiao, Y., Fu, D., Tu, J., Fu, T., and Meng, J. (2010). Broadening the avenue of intersubgenomic heterosis in oilseed *Brassica*. Theor. Appl. Genet. *120*, 283-290.

政策保障与环境支持

景海春 孙其信 吴孔明 万建民

1 作物育种创新相关研究的特征

1.1 作物育种基础研究和应用间隔周期长

技术成熟度（technology readiness level，TRL）是指将技术成熟过程划分为9个级别来评价技术发展状态，通常，1～3级为科技诞生和基础研究阶段，而7～9级则为科技实质生产应用的高峰期。作物育种创新是一个长期积累的过程，带有显著的生物学和地域分异属性，从基础研究到品种育成与大面积推广应用间隔期很长。已有研究表明，作物育种从资源挖掘、选育组配到育成品种获得生产力成果需要推广数十年。例如，玉米目前虽然已经成为世界第一大作物，但第一个杂交玉米品种育成就花费了20多年来进行基础与应用研究，之后又花费了20多年进行育成品种示范推广种植、多产品开发及产业链延伸。虽然现代生物技术，如分子标记辅助育种和转基因育种技术的应用加速了研发过程，但市场和公众的接受、监管部门的审批等会使研发滞后期延长5～10年。通过生物技术培育的玉米新品种，如转基因抗虫玉米和抗除草剂玉米等，创制过程依然花费20～30年，且这种过程迄今依然没有得到有效加速。

1.2 政府研究机构与跨国种业寡头在作物育种创新中发挥着不同的作用

由于作物育种在农业中的基础地位，政府研究机构与跨国种业寡头都发挥着重要作用，这显著区别于其他生命科学、工业科学研究领域。自20世纪50年代，杂种优势的发现与杂交育种技术的开发利用、知识产权概念的提出、政府监管和相关法律法规的确立以及专属利益保护与商业化，大大促进了政府研究机构增加作物育种的投入，同时也增加了私营种业企业关注作物育种研发。尤其是2015年以来，作物育种领域方兴未艾，超级大并购形成了跨国种业寡头，孟山都/拜耳、杜邦/陶氏及先正达/中国化工已经形成了更为明显的垄断模式，这三个超级巨头主导了主要作物的种业市场。另一方面，政府与私营机构在分工上有显著的不同，如政府公共研究机构更多关注食品安全、营养健康和环境保护等基础性科学研究，私营部门则越来越多参与到具有高市场回报的研发活动中，如生物技术、作物新品种、植物保护产品、农业机械等。

1.3 技术推广是作物育种创新成果应用的重要组成部分

作物育种创新的一个显著特征是需要开展不同规模的示范推广应用才能使成熟的品种得到广泛应用。通常，作物育种的重大基础理论与应用技术创新是由政府研发机构或植物科学基础研究机构完成，创新成果则是由政府或小型私营企业和科技人员来推广。作物新品种技术的推广是联系科技成果与应用的桥梁，也是农民掌握和应用创新成果的重要环节。

2 国内外作物育种创新的宏观政策与环境

科技创新已成为推动现代种业发展的主要力量。学科交叉融合和技术集成创新的特点，为作物育种提供了新理论、新方法和新技术，正在引领种业发展方式发生深刻变革。植物光合作用、生物固氮等研究的重大突破将会对作物育种产生深远、重要的影响；以基因组学等为核心的现代农业生物技术、大数据、智慧农业（互联网、移动互联网、云计算和物联网等信息技术）、合成生物技术等前沿和颠覆性技术等的交叉融合将带动种业产业格局重大调整和革命性突破。这些典型的科技创新发展新趋势同相关研究的政策保障与环境支撑密不可分。

2.1 农业结构及政策

任何政策的制定都有一定的社会条件和历史背景。当前，日本农业政策改革的基本底线是保障耕地面积和维护粮食安全，基本方向是培育新型农业经营主体，调整农业结构，政策手段主要包括培育骨干型农业生产主体、扩宽劳动力来源渠道、促进耕地的集约化和规模化管理和调整农业种植结构等

（王国华，2017）。对于新西兰和澳大利亚而言，两国自然资源丰富，新西兰农牧业发达，澳大利亚则是由原来的以畜牧业占绝对比重转变为农牧业大致平衡的农业结构，产业结构日趋合理。两国政府制定了完善的法律法规、农产品标准和严格的生产鉴定程序以保障农业发展。除《农业法》外，两国还分别制定了《出口控制法》和《营销法》等（李晓俐，2012）。美国的种植业和畜牧业占比相近。政府采取许多政策手段对农业发展进行宏观干预和调控，对农业的支持大致可分为直接投入、补贴、税收优惠、立法管理四大方面（https://wenku.baidu.com/view/ac3331f8c8d376eeaeaa3109.html）。同样，欧盟共同农业政策重要的一项内容也是农业补贴，尤其是价格补贴。早在1999年，欧盟成员国在《欧盟2000年议程》中就确定实施有限度的强制休耕规划，并奖励自愿休耕（王国进和刘姗姗，2012）。近年来，我国农业结构也有了明显的变化，种植业比例下降，畜牧业高速增长。党中央、国务院针对农业发展问题连续出台了多个关于"三农"的中央一号文件，并先后调整农业补贴政策，建立了以市场为导向的农产品价格形成机制。2017年党的十九大提出了实施乡村振兴战略。2018年中央一号文件《中共中央国务院关于实施乡村振兴战略的意见》围绕实施好乡村振兴战略，对三个阶段目标任务做出了重要部署。

在促进数字农业、智能化农业发展方面，世界各国推出了一系列政策和发展规划。早在2012年，欧盟发布了《信息技术与农业战略研究路线图》，旨在推动通信技术在欧盟的应用。2013年，英国政府和英国皇家化学学会先后发布《农业技术战略》和《可持续农业的土壤保护——科学引领战略》等战略指南，将农业大数据、精确检测生物信号传导列为农业技术投资重点，并于2015年成立世界首个农业食品大数据卓越中心Agrimetrics。法国也将开发快速、精准的农业机器人列入《农业创新2025计划》，用以推动法国农业创新发展。2016年，日本在《未来新型制造系统的发展方向》报告中提出以机器人带动农业等领域的结构变革。同年，欧盟委员会将园艺机器人列入"地平线2020"机器人计划（张晓，2013；赵春江等，2018）。2017年，澳大利亚科学院发布了《澳大利亚农业科学十年规划（2017～2026）》，提出基因组学的发展与利用、农业智能技术及大数据分析等六大研究领域在未来十年中最有可能大幅度地提高农业生产力。我国在2015年出台《中国制造 2025》，强调加快推动新一代技术信息和制造技术融合发展，着力发展智能装备。2016 年，中央一号文件首次提出"加快研发高端农机装备及关键核心零部件，提升主要农作物生产全程机械化水平，推进林业装备现代化"。

2.2 资源环境高效、环境友好的可持续发展

为使欧盟农业朝着更加绿色公平和更能适应市场挑战的方向发展，2013 年欧盟通过《2014～2020 年共同农业政策改革法案》，将直接支付与环境措施挂钩。2016 年法国发布《农业创新 2025 计划》，将发展可持续环境友好型农业列为优先重点方向。英国农业环境保护补贴有 100 多项，农场主可自愿申请（陈锴，2011）。澳大利亚和新西兰两国特别重视生态环境保护和自然资源管理，禁止任何对生态环境造成破坏的动植物入境。日本政府在资源环境双重约束下的产业结构升级中发挥了主导作用，其中高新技术园区和生态工业园的创建在推动日本产业结构向循环、低碳环保方向转变方面发挥了重要作用（王金波，2014）。美国农业部提出农业低投入可持续发展模式——生态渗入农业生产中的绿色生产方式，每年投入 2300 万美元用于补贴农民提高能源效率。美国金融机构还对发展绿色农业的企业提供抵押担保，实施相应的价格补贴和出口补贴等政策（王一鸣，2018）。我国的农业环境资源问题十分严峻，生态环境逐年恶化（许尔琪，2018），针对资源环境及可持续利用方面，政府颁布了一系列法律法规，包括《农业法》《森林法》《水土保持法》等（刘超，2017）。2015 年，我国颁布的《全国农业可持续发展规划（2015～2030 年）》文件中将优化发展布局、保护耕地资源、治理环境污染、节约高效用水和修复农业生态列为五大重点建设任务，以促进全国农业可持续发展。

2.3 转基因和基因编辑等生物技术

欧盟多数国家的政府和公众认为，尽管转基因作物能提高产量，在提高抗逆方面有更好表现，但并无科学证据确保其对人类健康和环境带来不良影响，因此欧盟自 20 世纪便着手建立转基因产品管理体系以监督和限制转基因作物的上市。美国是转基因食品主张派的代表国，对转基因食品基本持肯定态度，主张将转基因产品和传统农产品同等对待（于洲，2011）。

日本是典型的转基因产品进口国，该国出台了严格的转基因食品安全保障制度，包括安全性审查制度、区别性生产流通管理制度、上市审批制度和产品标识制度等（刘旭霞和欧阳邓亚，2009）。我国政府对转基因作物坚持立法先行、依法监管的原则，先后出台了一系列针对转基因的法规和规章等，对转基因农产品在研发、生产、加工及进出镜等方面做了严格规定（沈平等，2016）。

在基因编辑方面，欧盟国家将包括基因组编辑在内的基因诱变技术也认定为转基因技术。而日本专家认为，基因编辑原则上与良种选育或自然变异并无不同，因此销售这种食品无须进行特别的安全检查。基因编辑技术在新西兰属于严格管制范围。最近，澳大利亚出台的基因编辑监管规则指出，在不引入新的遗传物质的情况下，政府不会对植物、动物及人类细胞系中使用基因编辑技术进行管控（Mallapaty，2019）。根据《农业转基因生物安全管理条例》，运用基因编辑技术对作物进行改良在我国也归为转基因生物，但是具体的相关监管制度未见详细报道（付伟等，2016）。

2.4 企业孵化及人才培养

欧盟政府非常重视农业科学技术，建立了完整的教育、科研、试验与咨询和推广服务体系。在欧洲，农业相关的企业众多，各企业每年都投入大量资金和人力进行科技创新。法国于2015年整合农业领域12个国立科研机构与高等教育机构组建了法国农业、畜牧与林业研究院，旨在促进农业领域高等教育、技术教育与科研活动的紧密结合和协同发展，以实现国际一流水平的目标。为迎接科技进步对农业公司带来的挑战，各相关企业及科研单位的研发人才比例也不断增加。2019年，日本政府确定了旨在培育初创企业的综合战略，其中之一的措施是推进与大学的合作，鼓励大学从实际的公司经营角度出发开设公司注册、市场调查、融资方式等课程。政府将通过提供运营费等举措对于设置创业者培育课程的大学提供支援（https://tech.sina.com.cn/roll/2019-06-13/doc-ihvhiqay5347879.shtml）。新西兰与澳大利亚两国均拥有与农业发展相配套的社会化服务体系，其中农协等经济合作组织及农业研发公司起着非常重要的作用（柳明等，2015；Warren et al.，2014），还有合作研究中心把公共与私人领域最优秀人才联合起来，共同研究农业领域需要优先解决的科学问题（Core，2009）。此外，新西兰与澳大利亚两国还十分重视和强化农民的教育培训，不断提高农业劳动者素质。在我国，农业社会化服务体系、国家对农业的支持保护体系框架初步形成，包括科研体系、高校和企业的科研人员及基础农技推广人员在内的农业科研队伍不断壮大。近年来，我国还先后组建很多的创新研究团队，创新能力不断提升，另外，多种形式的新型农民专业合作经济组织大量出现，诸多企业也纷纷进军农业科技发展产业。2017年，中国化工集团完成了对先正达的收购，成为世界第七大农药公司。2018年，恒大集团与中国科学院签署全面合作协议，未来10年在人工智能、机器人和现代科技农业等重点领域投入10亿元以打造全球顶尖科学家的聚集地和科研基地。

3 作物育种创新的政策保障与环境支持建议

3.1 树立大农业、大食物观念，全面布局作物育种，优化我国农业结构

改革开放以来，我国居民食物消费结构发生显著变化，未来10年甚至更长时间我国居民对草食动物蛋白食品的消费需求规模仍将持续增加。在此背景下，我国以"大农业、大食物"发展理念推动农业供给侧结构性改革。建议在确定主粮作物育种的核心地位的基础上，进一步支持更为广泛的作物研究与育种，如饲草作物、特色作物等；在性状上，优先考虑品质改良以提高食品及饲料的安全和质量，增强作物适应多样化农业生产系统的能力等研究。

3.2 遵循作物育种创新规律，积极布局并持续支持

我国目前正处于发展方式转变的关键时期，要实现绿色、可持续发展，农业必须实现由传统农业向现代农业的转变。"种业是农业的芯片"，作物育种已进入主要依靠科技创新驱动的新阶段，要充分认识到种业科技创新发展是一个长期积累的过程。作物育种基础与应用研发投入严重不足、重大基础研究平台与技术支撑系统的匮乏、数据信息集成与资源共享利用效率低下等长期制约作物育种科技创新。基因组编辑和合成生物学被认为是新兴的前沿学科，将可能催生下一次生物技术革命，其应用前景非常广阔。建议成立学术研究联合团队，整合国内研究力量并长期持续支持，促进科研成果创新，

以实现从点到面的突破。

3.3 设立国家级作物育种创新与种业发展示范区，推动精准农业

建议国家出台相应的政策，在东北、华北、长江中下游、西南等区域，选择代表性市县和企业，打造"产学研、科工贸"于一体的种业创新示范区。通过试验示范，探索和解决政府研究机构和行业企业合作模式，作物育种创新与种业发展过程中所出现的各类政策、技术和工程问题，为不同区域发展现代种业提供样板。同时充分将新兴科技、智能技术、大数据分析等应用在农业生产中，驱动农业向数字化和智能化方向发展。

3.4 加强生物安全与伦理研究，制定合适的监管制度

针对维护国家生物安全的重大需求和生物安全威胁的战略挑战，规范作物育种科学研究伦理，完善转基因与基因编辑等生物技术育种的科普教育体系。同时积极开展安全性及监管制度的研究，制定政策法规加以正确引导，在国内为新兴技术的发展提供健康的环境，以推动其迅速、可持续发展。

致谢：本章在撰写过程中得到了董瑜、李志刚和郝怀庆的协助，特此致谢！

参考文献

陈锴 (2011). 欧洲现代农业发展现状及启示. 世界农业 7, 57-60.

付伟, 魏霜, 王晨光, 杜智欣, 朱鹏宇, 刘中勇, 吴希阳, 朱水芳 (2016). 基因编辑作物的发展及检测监管现状. 植物检疫 30, 1-8.

李晓俐 (2012). 新西兰和澳大利亚农业发展的特点及做法. 世界农业 10, 109-111.

刘超 (2017). 农业资源环境保护研究. 环境节约与环保 12, 88.

刘旭霞, 欧阳邓亚 (2009). 日本转基因食品安全法律制度对我国的启示. 法治研究 7, 42-46.

柳明, 李艳芬, 白林 (2015). 澳大利亚政府扶持中小企业政策分析及借鉴. 黑河学院学报 5, 59-63.

沈平, 章秋艳, 张丽, 卢新, 梁晋刚, 王颢潜, 刘培磊 (2016). 我国农业转基因生物安全管理法规回望和政策动态分析. 农业科技管理 35, 1-8.

王国华 (2017). 日本农业发展的现状及政策走向分析. 长春大学学报 27, 27-31.

王国进, 刘珊珊 (2012). 欧盟农业补贴政策对我国的启发. 管理观察 20, 3-4.

王金波 (2014). 资源环境约束下日本产业升级的低碳路径选择——以日本(生态)工业园的发展历程为例. 亚太经济 1, 64-69.

王一鸣 (2018). 美国日本如何扶持农业科技创新？全球商业经典 3, 74-75.

许尔琪 (2018). 中国农业资源环境分区. 中国工程科学 20, 57-62.

于洲 (2011). 各国转基因食品管理模式及政策法规. 北京：军事医学科学出版社.

张晓 (2013). 欧盟农业信息技术研发新战略及对中国的启示. 世界农业 9, 14-18.

赵春江, 李瑾, 冯献, 郭美荣 (2018). "互联网+"现代农业国内外应用现状与发展趋势. 中国工程科学 20, 50-56.

Core, P. (2009). A retrospective on rural R & D in Australia. Australian Government Department of Agriculture, Fisheries and Forestry, Canberra.

Mallapaty S. (2019). Australian gene-editing rules adopt 'middle ground'. Nature doi: 10.1038/d41586-019-01282-8.

Warren H., Colin B., Frank V., Jeff C. (2014). Recommendations arising from an analysis of changes to the Australian agricultural research, development and extension system. Food Policy 44, 129-141.

彩 图

彩图 1 多尺度高通量表型监测平台
(Chen et al., 2014; Deery et al., 2014; Pegoraro and Misteli, 2017)

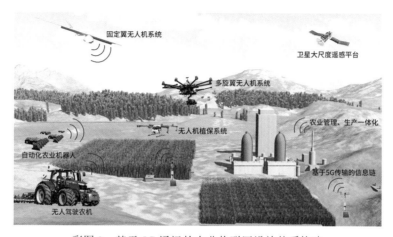

彩图 2 基于 5G 通讯的农业物联网设施体系构建